Forensic GIS

Editors
Gregory A. Elmes
Department of Geology and Geography
West Virginia University
Morgantown, WV, USA

George Roedl
Department of Geology and Geography
West Virginia University
Morgantown, WV, USA

Jamison Conley
Department of Geology and Geography
West Virginia University
Morgantown, WV, USA

ISBN 978-94-017-8756-7 ISBN 978-94-017-8757-4 (eBook)
DOI 10.1007/978-94-017-8757-4
Springer Dordrecht Heidelberg New York London

Library of Congress Control Number: 2014943255

Printed on acid-free paper

Springer is part of Springer Science+Business Media (www.springer.com)

Editorial

This book, *Forensic GIS: The Role of Geospatial Technologies for Investigating Crime and Providing Evidence*, presents ways in which geospatial technologies, including geographic information systems (GIS), global positioning systems (GPS), and remote sensing, contribute to the acquisition and analysis of forensic information today and provides timely illustrations in the form of case studies.

Initially, our interest in the forensic potential of geospatial technologies stemmed from a Forensic Science Initiative grant from the National Institute of Justice to the Forensic Science program at West Virginia University (2003-RC-CX-K001). Among many others, a task was initiated to investigate the characteristics of spatially enabled forensic science, with the goal of supporting the use of geographic information systems and science in a number of forensic-related research and teaching activities. During the course of the grant, I, Dr. Ge Lin, George Roedl, and a number of graduate students investigated the use of GIS and remote sensing in forensic contexts. Four important aspects of the spatial perspective of forensic science were examined: geographic profiling, time geography, high density scanning, and radio frequency identification. Among the outcomes were a special paper session on Forensic GIS at the 2009 Annual Meetings of the Association of American Geographers, a paper in the *Journal of Forensic Science* on space-time approaches to shoeprint matching; investigation of the coordination of measurements of indoor and outdoor spatial location (accommodating GPS); and experiments with time-space path analysis and point cloud analysis. These efforts included the development of a course in crime mapping and analysis. In 2010, Gregory A. Elmes and George Roedl were awarded a grant from the National Institute of Justice, Office of Justice Programs (2009-IJ-CX-0205) for "Increasing Student and Community Safety," which supported a research partnership with WVU and the City of Morgantown, WV Police Departments. The current work may be seen as a natural extension of the research interests generated by these NIJ-funded projects.

Strictly defined, forensic science is the use of scientific principles, methods, and techniques to establish facts or provide evidence used in a court of law. Here, we have adopted a broader working definition of forensics and forensic GIS, which includes the use of geospatial principles and techniques to establish facts or sequences of events, regardless of whether they are used in court. Thus, gathering and interpreting scientific data for regulation, intelligence, and national security purposes falls under this broader definition of forensic science. A further distinction may be made between criminal and civil forensics. Criminal law deals with offenses against the state—the prosecution of a person accused of breaking a law. Such offenses include crimes against persons and property. Civil law covers everything else, such as violations of contracts and lawsuits between two or more parties. The collection of data and presentation of evidence may be held to quite different standards, the process of data collection and imaging may be quite different, and the consequences of the case may have very different impacts. Examples of both will be found herein.

A relatively small literature on forensics exists in the discipline of geography, a larger one in environmental science, geology, soil science, and archaeology. In February 2009, the National Academy of Sciences published a report "Strengthening Forensic Science in the United States: A Path Forward" (National Academies Press, 2009). The committee had a mandate to identify the needs of the forensic sciences community. It was revealing that, in a search, the terms "spatial" and "geospatial" did not appear in this text of 352 pages and this observation provided further impetus toward the genesis of this book. It is evident that further research and a bridging of scientific communities are necessary to establish the best practice of the use of geospatial science and technology within forensic science. To that end, this book is directed to an audience comprising of law enforcement professionals, academics in a wide variety of fields, and students of criminology and forensic science.

Our initial call for submissions was made on October 15, 2012, with a chapter proposal submission deadline of January 31, 2013. After reviewing the submitted proposals, 14 manuscripts were invited to be developed into final chapters by June 2013. Each chapter was reviewed by the editor and two external reviewers in a double-blind process. Nine of the original 14 manuscripts were revised and resubmitted by August 2013. The chapters in Part I by Roedl, Elmes, and Conley were not included in the double-blind process and must therefore be considered as being monographic in nature. Our recognition and thanks are due to Dr. Michael Leitner for the guidance provided by his 2013 publication in this Springer series (Volume 8). Sincere acknowledgement is also due to the external reviewers of the manuscripts for their important contribution of time and effort.

In Part I, *Gregory A. Elmes*, *George Roedl*, and *Jamison Conley* draw attention to the theme and content of the book through a review of the various roles geospatial technologies provide in investigating crime, providing evidence, and developing policy within the legal system and how these roles have changed with advances in the technology itself and the challenges involved in using the technology for investigation and providing legal evidence.

The first chapter "Concepts, Principles, and Definitions" considers the ways in which geospatial information and technology (GIT) has significantly increased in prominence within the criminology and forensic fields in the last decade. Geospatial technology includes the tools and techniques applied to geographic or spatial data; additionally, the chapter recognizes the extensive nature and roles of GIT across many subject domains. In Chap. 2, "Geospatial Technologies in the Courtroom", *George Roedl*, *Gregory A. Elmes*, and *Jamison Conley* continue to develop the theme of the book by examining key rules such as federal rules for the admissibility of evidence, Frye, Daubert, and other court decisions that have influenced the potential admissibility of spatial data and technologies in a modern courtroom. Chapter 3 develops the theme of Spatial Tracking Applications, and Chap. 4 details the main tools of forensic GIT in Spatial Technology Applications.

Part II focuses on a selection of case studies illustrating the breadth of contemporary applications of GIT in criminal justice ranging from collecting evidence for presentation in court to an open software Web-enabled application bringing crime mapping and analysis to a larger audience than is possible with commercial packages.

In Chap. 5, *Ronald E. Wilson* and *Ann D. Fulmer* apply spatial and temporal categorizations of the "near repeat" concept to measure the extent of foreclosures in order to identify concentrations of mortgage fraud and predatory lending. They demonstrate that near repeat spatiotemporal analysis can be applied to help fraud investigators identify loans for scrutiny that show geographically systematic patterns of foreclosure. In a post-conviction setting, *Mark R. Leipnik* and *Xinyue Ye* examine geospatial strategies for the management of registered sex offenders in Chap. 6. Documenting current practice in the United States, the authors argue that, while four-fifths of US states use Web maps to provide notification to the public of the location and criminal history of registered sex offenders, the notifications vary considerably in form and content and are such notifications open to misuse. In Chap. 7, *Manuel Rodríguez Herrera* and *Daniel Salafranca Barreda* introduce the Science, Data, Intelligence, Knowledge (SDIK) project, a geo-information international security endeavor for making visible the "invisible" conditions of communities and neighborhoods. The SDIK project incorporates a set of technical-scientific and geospatial innovations to help understand newly emerging activities within communities and help uncover evidence of possible criminal activity.

Remaining at the neighborhood scale, *Jamison Conley* and *Rachel Stein* examine the relationships among the factors of neighborhood disorder and collective efficacy using measures of spatial correlation and spatial regression in Chap. 8. Their findings illustrate the potential of spatial analysis for informing policing strategies. They reason that the results of this type of analysis can lead to a better use of police resources to avert crime. Also recognizing the importance of place and neighborhood, in Chap. 9 *Matthew J. Hickman*, *Loren T. Atherley*, and *Geoffrey P. Alpert* describe the utility of geospatial analysis for monitoring, understanding, and responding to police use-of-force incidents. Their research stems from an investigation for the Seattle Police Department which had the aim of improving the quality of police monitoring and accountability. Through mapping spatiotemporal patterns of liquor law violation citations in the college town of La Crosse Wisconsin, *Gargi Chaudhuri*, *Steven Oxley*, and *Scott Wenzlaff* provide, in Chap. 10, a means to focus the deployment of intervention measures and increased vigilance to restrict alcohol consumption among underage youths and prevent associated crime and accidents.

Guiyun Zhou, *Jiayuan Lin*, and *Xiujun Ma* introduce in Chap. 11 a Web-based GIS for crime mapping, analysis, and decision support as an affordable option for small- and medium-sized police departments. The authors discuss the architecture, construction, and open software of the development of the prototype system. In Chap. 12 *William. C. Walker*, *Sunhui Sim*, and *Lisa Keys-Mathews* study the influence that a hurricane had on the space-time behavior of local patterns of crime. The authors explore the use of geographically weighted regression (GWR) for understanding aspects of the ecology of crime. The results reveal that more accurate prediction of crime types within cities is possible. Finally, Chap. 13, by *Irfan Ashraf*, *Urooj Saeed*, *Naeem Shahzad*, *Javed Gill*, *Shahid Parvez*, and *Akram Raja*, presents a detailed case study of the delineation of legal forest boundaries to identify and contest illegal forest encroachment. Forest boundaries and encroachments were mapped in coordination with the Punjab Forest Department, the Survey of Pakistan, the Punjab Revenue Department, and the World Wildlife Fund Pakistan. The study results were presented as evidence to the High Court in Lahore, Pakistan, as part of an effort to control illegal forest use.

The collection of insights and research presented here has advanced the literature on forensic GIS, albeit incrementally, and has raised the premise of the importance of continued research into spatially enabled forensics. The editors look forward to future developments in the integration of GIT and forensics. To that end we have included an annotated bibliography of court cases involving the legality of geospatial technology and its introduction in court. Selected cases involving GPS, remote sensing, and GIS have been included.

Contents

Part I
Fundamentals: Definitions, Concepts, Theories, and Principles

Chapter 1
Concepts, Principles, and Definitions

Gregory A. Elmes, George Roedl, and Jamison Conley

Abstract Forensics is the application of science to solve crime. Geographic Information Science, encompassing geospatial information and technology (GIT), has become established within the criminology and forensic fields in the last decade. Law enforcement agencies and forensic investigators embrace geospatial science and technologies for collecting, storing, manipulating, analyzing, and displaying spatial data, resulting in new information, procedures, and models for investigation, policy, and decision making. Applications, acceptability, relevance, and procedural legality of geospatial technologies vary substantially, leading to the assessment of their roles in law enforcement, rules of evidence, protection of privacy, and constitutional liberties. This chapter discusses the context and principles of geospatial technologies and the integration of geospatial tools, principles, and methods into a five-stage model of crime analysis and investigation.

Keywords Forensic science • Geographic Information Science • Geospatial technology • Geographic Information Systems (GIS) • Global Positioning System (GPS) • Remote sensing

1.1 Introduction

As geospatial science and technologies become ubiquitous in society, a wide range of disciplines and professions adopt them for collecting, storing, manipulating, analyzing, and displaying spatial data, resulting in the generation of new information and models for policy and decision making. Forensics is the application of science

G.A. Elmes (✉) • G. Roedl • J. Conley
Department of Geology and Geography, West Virginia University,
330 Brooks Hall, Morgantown, WV 26506, USA
e-mail: greg.elmes@mail.wvu.edu; groedl@gmail.com; jamison.conley@mail.wvu.edu

G.A. Elmes et al. (eds.), *Forensic GIS: The Role of Geospatial Technologies for Investigating Crime and Providing Evidence*, Geotechnologies and the Environment 11,
DOI 10.1007/978-94-017-8757-4_1, © Springer Science+Business Media Dordrecht 2014

to solve crime. It follows therefore that as geospatial science and its associated technologies emerge, they make a distinct and unique contribution to forensics (McKinley et al. 2008; Noond et al. 2002; Wolff and Asche 2009). Law enforcement agencies and forensic investigators have adopted geospatial technologies to profile serial offenders, track suspects, and guide crime reduction efforts, among other purposes. Legal experts utilize the analytical and visual capabilities of geospatial technologies to present, demonstrate, and explain complex information in the courtroom. Citizen groups have fought successfully against environmental discrimination and have engaged in class-action lawsuits, strengthened by the collection, analysis, and presentation of geospatial data.

Geospatial technologies have a long history of use in a broad range of applications such as environmental conservation, real estate, military and security, municipal planning, epidemiology, and agriculture. Recently, Listi et al. (2007) observed an increased popularity of what they described as the field of forensic spatial analysis, citing the evident increase in geospatial technology use presented at forensic-oriented conferences. Despite a relative lack of published research articles in criminology and forensic-related journals, geospatial technologies also serve as highly useful tools in criminal investigations. Geospatial technologies have unique capabilities which are ideally suited to collecting and analyzing spatial data. Traditional methods of investigation, such as pin maps, are largely unable to cope with volumes of multifaceted spatial information in any meaningful manner capable of assisting in identifying an offender or excluding possibilities. Digital spatial technologies result in a more efficient investigation, linking people, places, and objects in a way that assists in optimizing time and resources in pursuit of guilty parties. Conversely, traditional methods of investigation remain vitally important. The merging of spatial tools and methods into investigative practices to establish facts admissible in court is therefore a practice that should be recognized. The application of spatial tools to assist in established investigative practice by adding a spatial perspective is the focus of forensic GIS.

The widespread use of geospatial technologies has increasingly exposed courtrooms to the demonstrative powers of spatial technologies in civil and criminal cases. Section 1.5 of this chapter integrates a five-stage workflow for crime analysis and investigation with the comparable stages of GIS. In the United States of America, federal, state, district, and circuit courts, as well as the US Supreme Court, have ruled on the legality of the use of geospatial technologies under various circumstances. Higher courts have overruled the judgment of lower courts, which has introduced a degree of debate and contention into the legal uses of various geospatial technologies. This book is intended as a guide to understanding the various roles geospatial technologies provide in investigating crime, providing evidence, and developing policy within the legal system, broadly defined, and how these roles have changed with advances in the technology itself and the legal challenges involved in using the technology for investigation and providing evidence.

1.2 Geospatial Technologies

Recent innovations in information technology have had "dramatic and profound effects in the criminal justice system and will likely have both intended and unintended consequences" (Byrne 2008: 10). Many of these new technologies, incorporating advances in both hardware and software, have proven to be effective improvements over previous technologies for the purposes of investigating, prosecuting, convicting, and exonerating suspects, as well as pursuing civil suits (Jacobson 2004). Geospatial information and technology (GIT), or simply geospatial technology or geomatics outside the United States, has increased in prominence significantly within the criminology and forensic fields in the last decade. While geospatial technologies have been commonly associated with hundreds of applications over the past 50 years, they have only recently begun to be employed in investigative applications as a common practice, as their acceptance in the legal system increases and decreased costs make their use effective.

To identify geospatial technology competencies within the general workforce, Cyndi Gaudet assembled a focus group consisting of geospatial industry stakeholders. This group defined geospatial technology as "an information technology field of practice that acquires, manages, interprets, integrates, displays, analyzes, or otherwise uses data focusing on the geographic, temporal, and spatial context. It also includes development and life-cycle management of information technology tools to support the above" Gaudet et al. (2003: 24). Although such a definition is very general and all encompassing, it recognizes the extensive nature and roles of GIT across many subject domains. DiBiase et al. (2010) recognized that because of their breadth and diversity, geospatial technologies mean very different things to different people, ranging from a scientific discipline to a collection of tools and from a profession to an industry. However geospatial technologies are interpreted, they are identified as an opportunity for many fields and many disciplines. Goodchild (2008: 352) called geospatial technologies "powerful extensions of the senses, revealing things that would be impossible to obtain in any other way." At a minimum geospatial technologies include those tools or techniques which are commonly applied to geographic or spatial data. Frequently, Geographic Information Systems (GIS), Global Positioning System (GPS), and remote sensing are categorized as the three main geospatial technologies (see Bossler et al. 2010) which are well-established fields of study and in general public use across a wide range of applications (e.g., conservation, real estate, military applications, municipal planning, epidemiology, and agriculture).

1.2.1 Geographic Information Systems, Science, and Studies

Although the acronym "GIS" is typically reserved as an abbreviation for Geographic Information Systems, it is often applied somewhat ambiguously to distinctly different subfields of study. Goodchild (1992) suggested GIS was not only a "system" but also a "science" reflecting substantial differences in usage. Scholars (Forer and Unwin 1999; Longley et al. 2005) later associated the "S" with "studies," expressing the GIS acronym as either Geographic Information Systems, Geographic Information Science, or Geographic Information Studies. To avoid confusion and provide the precise context of the acronym, distinctions can be made. Geographic Information Science is often differentiated as GIScience, GISci, or GIs(c), and the field of Geographic Information Studies is typically denoted as either GIStudies or GIs.

Goodchild (2008) and Longley et al. (2005) elaborated the differences of meaning in the various GIS acronyms. A Geographic Information System (GIS) is defined as a computer system for capturing, managing, integrating, manipulating, analyzing, and displaying geographically referenced data. It can be seen as consisting of five components: hardware, software, data, methods, and people connected by a computer network (Longley et al. 2005). In a GIS, location becomes the common denominator between disparate datasets, enabling them to be correlated, merged together, and managed to explore relationships between data in order to identify patterns and trends in the form of maps, analytical reports, and charts. Geographic Information Science (GISci) emphasizes underlying principles and fundamental questions, as well as the research and development raised by the use of GIS through strong scientific and intellectual components (Goodchild 1992; Getis et al. 2000). Research into, and extension of the concepts of, scale, dimensionality, topology, and spatiotemporal representation drives GISci inquiry. In a synergetic relation, GISci affects the implementation of GIS while the functionality of GIS advances the theory and practice of GISci (Gold 2006). Mark (2003) provided a more thorough explanation of GISci, which he summarized as a multidisciplinary research field for redefining geographic theories, concepts, and uses. Geographic Information Studies (GIStudies) emphasize GIS in a societal context and examine issues such as privacy, ethics, legality, and return on investment (Forer and Unwin 1999; Chrisman 2005).

GIS, GISci, and GIStudies each has associated applications and challenges in crime investigation and the production of evidence. In Chap. 13, surveying, GIS, and remote sensing are combined to merge spatial datasets to create maps of legal forest boundaries in Pakistan. Using this information, researchers and law enforcement agencies are able to reproduce, update, and distribute standardized maps and atlases to identify and monitor areas of illegal encroachment on protected forest land. The maps identifying encroachment within the legal forest boundaries are then used to provide evidence to the court system. Chapter 11 focuses on GISci as the authors develop a prototype crime mapping and decision support system for law enforcement agencies in the Web environment. A focus on publicly available Web maps of registered sex offender residences in Chap. 6 explores aspects of GIStudies to examine societal issues of geospatial technology that are both beneficial to public safety and potentially injurious to offenders who may become targeted victims of harassment, arson, assault, and murder. The challenges are numerous when applying GIS, GISci, or GIStudies to crime investigation or providing evidence arising from concerns with associated hardware, software, data, methods, and people. Problems arise with the selection of appropriate data collection, storage, analysis, integration, and display. Underlying theories and principles have been questioned as well as the interpretation of results and societal impacts. Chapter 2 explores such challenges to illustrate what is required under the rules of evidence and admissibility for geospatial technology.

1.2.2 Global Positioning Systems

Among the family of geospatial technologies, satellite navigation systems are perhaps the most familiar, of which the Global Positioning System (GPS) is the most frequently used. The availability of GPS devices for outdoor activities, such as hiking and geocaching, as well as vehicle navigation and location-based services provided through cellular phones and mobile computing devices permeate contemporary society. Equally, no other geospatial technology has affected the geospatial industry and discipline as profoundly as GPS (Bossler et al. 2010). Made public in the United States in the 1980s, the GPS has advanced to provide real-time, centimeter-level positional accuracy across much of the globe through the use of networks of 30+ satellites that orbit the earth. Galileo, GLONASS, and BeiDou (BDS) are the European, Russian, and Chinese equivalent satellite navigation systems in various stages of development and accessibility.

The advantages of real-time positional data for investigating crime and providing evidence are abundant. Law enforcement agencies, judicial systems, and citizens have embraced the tracking ability of GPS; for example, law enforcement agencies have solved criminal investigations armed with a record of a suspect's movements captured by a GPS device (Byrne 2008). Courts commonly order the attachment of ankle bracelets fitted with a GPS to offenders to ensure compliance with restrictions (e.g., in-home confinement or restraining orders). Citizens equip vehicle and equipment with devices such as LoJack which allow for tracking and recovery of stolen equipment; however, as with many other technologies, GPS has an established track record of misuse (e.g., for stalking victims) and has been subject to numerous legal controversies and challenges (e.g., violations of the fourth amendment (United States v. Jones 2012).

1.2.3 Remote Sensing

The oldest geospatial technology, remote sensing is a mature discipline with well-understood concepts and principles and a range of versatile applications ranging from aerial photography to LiDAR (Warner et al. 2009). Jensen (2007) provided eight definitions of remote sensing with the common premise that remote sensing is the art and science of acquiring data and information about an object without making physical contact with that object, ensuring the object is not disturbed or altered through the act of measurement. There are two types of remote sensing: passive and active. Passive remote sensing detects and collects energy emitted from an object, such as reflected sunlight captured in photographs or sensors. Active remote sensing emits energy in order to measure the reflected energy from the object of observation. Some active remote sensing (e.g., X-ray) has the potential of affecting targeted objects, but longer wavelengths such as visible light or RADAR do not. Remote sensing sensors can be employed on ground platforms (e.g., vehicles or structures), in the air (e.g., attached to aircraft such as unmanned drones), and in space (e.g., satellites).

Just as GIS and GPS have benefited investigations and provided evidence, the long history of remote sensing has ensured its widespread use. As new technological advances and capabilities continue to emerge, the potential applications for investigations and evidence will continue to increase. However, there are, and will continue to be, controversies and challenges associated with the surveillance and data storage capabilities inherent to remote sensing and other geospatial technologies.

1.3 Spatial Data

Spatial data are distinctive in that they contain linked information about location and attributes (Gabrosek and Cressie 2002). Goodchild et al. (2000) noted that the term spatial is in fact shorthand for spatiotemporal, as it refers to data and phenomenon that have both spatial and temporal dimensions of variation. In having integrated components of location (places and times) and attributes, spatial data require specialized data storage, handling, and management. Although the terms geospatial and spatial are often used interchangeably, spatial is the more generalized term denoting locational context at any scale, while geospatial refers to data with a location component that can be referenced to a point on the earth (Longley et al. 2005). Spatial data is an extremely valuable and versatile resource, which is subject to widely accepted international standards (Boer et al. 2007). Rooted in the spatial science tradition of the discipline of geography (see Martin and James 1993), the spatial perspective emphasizes the role of location and scale as important variables in understanding patterns, processes, and causality.

Because of its unique composition, spatial data are handled differently from nonspatial data to address the role of location. For example, any location has the potential for the occurrence of a crime incident; however, certain spatial relationships significantly increase the chances of places having an incident, such as *within* blighted neighborhoods, *near* alcohol outlets, *outside* gated communities, *along* particular streets, *adjacent* to wooded lots, or combinations of such factors. As may be gathered from these examples, location is an important component in not only understanding crime but also in the investigation and prosecution of crime. The use of geographic principles and spatial perspectives assist in efforts to investigate crime and apprehend offenders and increasingly takes on a forensic role, especially in the form of geographic profiling (Brantingham and Brantingham 1981; Rossmo 2000).

Spatial data have several advantages over nonspatial data. Given known location and attributes, spatial data can be mapped and analyzed readily. Mapping allows for the visual recognition and empirical demonstration of patterns from complex datasets, such as the detection of clusters of crime incidents within a large database of all police events. Crime mapping and analysis takes advantage of these clusters or hot-spot areas for tactical and strategic planning to prevent and reduce crime (Boba 2005; Bruce 2001; Getis et al. 2000). A growing number of police departments routinely map and analyze crime data for different purposes (Mamalian and LaVigne 1999; Markovic et al. 2006; Nelson 1999; Wang 2005; Wartell 2003). Through georeferencing, spatial data also provides a commonality between disparate datasets based on their common geographic location or footprint. Consider the Internet for example – a user can search any set of locations and retrieve data on people, businesses, attractions, items for sale, weather, and news events for the set of locations. An investigator may become more informed in identifying vehicles, weapons, parolees, warrants, or calls for service to a particular location by querying disparate but georeferenced databases using one or more addresses. Similarly, the investigator is able to expand the query to include all records within a given distance of the address, without the need to know and query every specific individual address within the given distance.

Clearly modern spatial data are also digital. Whether collected by, or stored in computers, these data can be easily exchanged and shared through clearing houses like the National Spatial Data Infrastructure (NSDI). Vast amounts of spatial data have been produced by the federal government and state governments with many local government partners. In the United States, much of the governmentally produced spatial data is in the public domain and covers large geographic areas at ever-increasing spatial resolution, which allows for cross-jurisdictional analysis with little difficulty. The wide availability of spatial data generates a wealth of information for decision making, policy, and management, and the dynamic and interactive nature of spatial data provides an additional advantage. Given dynamic data, the user can update results and explore new relationships and trends. For example, a common application is to identify routes using an Internet map service. A user can specify whether the user wants the quickest route, the shortest route, the route with most or least use of interstate highways, or even a route which avoids tolls and construction activities. Regardless of the final choice, GIS presents the user with options, informs the user with results, and produces new information, while not altering the original underlying data in generating the results.

1.4 Forensic GIS

According to the American Heritage Dictionary of the English Language,[1] forensics is "the use of science and technology to investigate and establish facts in criminal or civil courts of law." More broadly interpreted, forensics is the application of any science or technology used to investigate and establish facts. As a result, the

[1] http://www.freedictionary.com

concepts of location, place, and scale are intrinsically embedded within forensic investigations, analyses, and through the presentation of evidence, as are spatiotemporal relationships.

Burrough and McDonnell (1986: 11) defined GIS as "a powerful tool capable of storing, retrieving, transforming and displaying spatial data for a particular purpose." Applying the Burrough and McDonnell definition with its emphasis on a particular purpose, the phrase "forensic GIS" can be said to establish that purpose. A distinction made for forensic GIS is necessary due to the overwhelming volume of literature pertaining to GIS in general and in other specific fields. The term and concept of forensic GIS is introduced here, not as a new or unique field so much as a way to provide the forensic and criminal justice-related communities with a distinction pertinent to their purposes and interests. As such, it is anticipated that new ideas and research in this area will not get lost among the volumes of GIS literature in general, but will stand out as a forensic GIS literature. The main utility of forensic GIS is to provide associative evidence, which assists in either proving or disproving links between people, places, and objects as they relate to the court of law. It should be stressed that a forensic GIS acts to assist traditional investigative techniques by adding a spatial perspective and should not be considered as a technological replacement for them.

Access to GIS technology and spatial data has accelerated over the last two decades and is clearly impacting the ways investigations are performed and evidence is presented in court. Technology has become a more reliable and efficient means of generating evidence than police hunches (Jacobson 2004). The National Institute of Justice (2009, 2010) identified high-priority needs for criminal justice technology within the scope of five functional areas, one of which was enabling informed decision making. The NIJ reports listed a range of needs for spatial analysis tools and technologies that included the need to analyze geographic linkages among people of interest to criminal justice agencies, extend the current capabilities of crime-related databases, provide compatibility with mobile/handheld devices, and identify and extract hidden relationships in large and complex datasets.

Although a definition of forensic GIS has yet to be formally established, Morrow-Jones et al. (2005: 19) set forth the concept of a "forensic GIS" as a way to "collect, explore, and analyze spatial data in order to detect irregularities that may violate law or fair practice." However, this definition is rather narrowly based on the terms "detect" and "irregularities." An alternative definition is the application of geographic and spatial tools, principles, and methodologies to investigate and establish facts within the boundaries of forensics. As such, under the basic definition of forensics, spatial science serves as a specified science, and geospatial technology is the technology used to investigate and establish facts that may be presented in criminal or civil courts of law.

The boundaries of forensic GIS are not clearly defined and may overlap with other forensic or spatial science subdisciplines. However, an examination of closely related fields reveals important characteristic differences. Forensic geography, geoforensics, forensic geoscience, forensic mapping, and environmental forensics are often closely associated with geospatial technologies and geographic theories and principles. Forensic GIS is set apart in that the use of geospatial technologies is not discipline specific and the geospatial analyst does not necessarily need to be an expert trained in spatial science. For example, anyone (from any discipline with any level of spatial expertise) can use an online Web mapping service to determine distances and time requirements for travel between two locations to determine the possibility, or improbability, of a suspect conforming to an established timeline of events.

Forensic geography is a subdiscipline of geography wherein a geographer or other expert does research and provides expert testimony appropriate to a court of law based on geographic theory and principles (Brodsky 2003; Schmitz et al. 2013). DeVorsey (1973; 1980) also referred to this as both forensic historical geography and historic geography when historical geographers conducted the research and testimony. Geoforensics and forensic geoscience are synonymous with each other and, although there are varying definitions, refer to the application of "geo" or earth sciences to forensic investigations (Morgan and Bull 2007; Pye and Croft 2004; Ruffell 2006; Ruffell and McKinley 2008). Schmitz et al. (2013) categorized forensic mapping as a field of forensic geography that maps criminal activity using location data from GPS devices and cell phone usage data. Environmental forensics focuses exclusively on environmental concerns and enforcement (Brilis et al. 2000, 2001; Grip et al. 2000; Ruffell and McKinley 2008). There are numerous other "forensic" fields, such as forensic archeology (Obledo 2009), forensic geology (Murray 2004), forensic geomorphology (Ruffell and McKinley 2005), forensic seismology (Ruffell and McKinley 2005), and forensic palynology (Mathewes 2006; Mildenhall et al. 2006). The time is ripe to develop the study of forensic GIS.

1.5 Stages of Crime Analysis and Investigation

Zhao et al. (2003) divided crime analysis and investigation into five workflow stages: (1) collecting crime data; (2) processing and storing crime data and documents; (3) searching, retrieving, and collecting additional information for crime analysis; (4) analyzing information to find clues; and (5) using information to prosecute (or defend) individuals. These workflow stages are organized to follow a seamless timeline (temporal continuum) beginning before a crime is committed and culminating in the obtainment of either a prosecution or acquittal before the cycle is repeated. Noticeably, the five workflow stages of crime analysis and investigation are similar to the six workflow stages used to define GIS (capture, manage, integrate, manipulate, analyze, and display spatial data). Calling it the "geographic approach," Dangermond (2007) referred to the integration of GIS as a new way of thinking. Conceptually, forensic GIS can be viewed as a new way of thinking which integrates the five workflow stages of crime analysis and investigation with the defining workflow stages of GIS along a temporal continuum that extends from a spatially enabled pre-crime capability to the courtroom presentation of spatial data.

1.5.1 Stage I: Collecting Crime Data

Crime scenes exist in time and space. Concepts of location, place, and relative location (next, near, overlapping) are intrinsic to investigation and evidence. Location, place, and scale are embedded within the data, as are spatiotemporal relationships. The challenge in establishing a spatially enabled forensic GIS crime database is first to collect and store the information and then to use and reuse it during the investigative process (Oatley et al. 2006; Rossmo 2006). Information about any particular crime should be collected with the understanding that it will be used in future analysis, including spatial analysis. Oatley et al. (2006) contended that while a single recorded crime may have little value, when a diversity of information from an event is collected and stored in a comprehensive crime database, it provides a powerful retrospective investigative tool. Rossmo (2000, 2006) similarly stated that information must be properly collected, analyzed, and shared if it is to be of any value to investigators. Because a forensic GIS identifies spatial relationships, it is important to have a precise and accurate location of a crime scene to facilitate confidence in the results of any future analysis.

1.5.2 Stage II: Processing and Storing Crime Data

As Nelson (1999) pointed out, law enforcement is about information management and location. Unlike traditional record management databases, a GIS database supports traditional tabular-based data, such as spreadsheets, and additional multimedia-based information, such as word processing documents, digital pictures, and video and audio recordings. However, analysis is currently restricted to the spatial and nonspatial component of the tabular data, excluding the images, video, and audio recordings. This limitation may eventually disappear with advances in technology. For example, integrating facial recognition software with GIS will provide further linkages between suspects and crime scenes based on collected and stored photographic evidence.

1.5.3 Stage III: Searching, Retrieving, and Collecting Additional Information for Crime Analysis

The investigation process involves asking specific questions, and the responses are essential in establishing links in criminal activity. A simple example is determining if a crime scene has distinctive characteristics that are similar to those at other crime scenes. Two crime scenes which share distinctive characteristics might indicate to investigators that the same offender was possibly involved in both incidents and provide a link between an offender and crime scenes; however, the ability of a forensic GIS to link criminal activities efficiently relies on an investigator formulating the relevant questions (Adderley and Musgrove 2001). Traditional GIS investigations have focused on suspect identification and pattern analysis in linking activities with offenses (Groff and La Vigne 1998; LaVigne and Groff 2001). Descriptive factors and locations of crime scenes serve as starting points to identify potential suspects from additional databases which can be linked to the GIS.

During this third phase, information collected is examined to determine what is available and what remains to be collected for further investigation. Any necessary additional information is then obtained. A common means of obtaining additional information is through the process of querying additional law enforcement databases, such as sex offender registries or parolee databanks, as well as government record databanks such as motor vehicle registration or through third-party databanks, such as utility companies and cell phone providers. A forensic GIS enables investigators to search for information not only based upon attribute linkages but also by spatial linkages. An example might be to identify registered sex offenders living nearest to a sexual assault scene (spatial link) or to identify a list of previous offenders on parole/probation who may not be properly registered (attribute link). The importance of dispersed systems to maximize the search is evident.

Discovering which crime incidents belong to a series is an essential and important step in serial crime investigation (Rossmo 2006). Link analysis, also known as comparative case analysis, is the process used to tie a series of crimes together based on three main methods: (1) physical evidence, (2) offender description, and (3) crime scene behavior (Rossmo 2000). A forensic GIS can be used to establish such links and display the locations where the potentially linked incidents occurred. Using a forensic GIS to link similar crimes or suspected offenders within a limited geographic area provides further associative evidence of a relationship. One of the benefits of linking cases is that it informs an investigator of a potential serial criminal, which permits the focus of the investigation to shift from multiple unrelated incidents to identifying a single suspect responsible for many incidents. As such, apprehension of one individual would have the potential of solving numerous linked cases. Once a suspect is apprehended, the link analysis informs the investigators to question the suspect about each of the linked cases and seek to prove guilt through a confession, physical evidence, or witnesses (Rossmo 2006). In contract to establishing links between multiple incidents and a single offender, using a forensic GIS in the link analysis process may also inform investigators of potential multiple offenders for similar incidents based on geographic location.

1.5.4 Stage IV: Analyzing Information

Whether investigating a major crime or a crime series, vast amounts of data are generated, and the sheer volume of these data invariably obscures possible underlying relationships and linkages (Adderley and Musgrove 2001). Nelson (1999) stated without analysis that data is useless. A GIS allows a database and a map to be linked for the purpose of data analysis and visualization. Markovic et al. (2006) described such GIS maps as "heuristic devices, or effective tools for stimulating investigatory processes, exploration, and reexamination" with the advantage of being able to empirically validate hypotheses generated through discussion and an interactive discovery process performed by patrol officers, detectives, and crime analysts. While collected information alone is useful within an investigation, further analysis can allow investigators to confirm or reject identified links. Within a forensic GIS, the analysis is centered on spatial information to inform the investigative process. Longley et al. (2005) suggest six generalized types of spatial analysis common to GIS: (1) queries, (2) measurements, (3) transformations, (4) descriptive summaries, (5) optimization, and (6) hypothesis testing. A detailed discussion on the various types of spatial analysis remains outside the scope of this chapter; suffice it to say there is a wide range of spatial analysis techniques available to investigators, and the choice of which types of analysis to use would depend upon each unique investigation case.

1.5.5 Stage V: Using Information

GIS technology allows the analysis of data to identify, apprehend, and prosecute suspects (Nelson 1999). Suggs et al. (2002) discussed the benefit of link analysis as a widely accepted tool for criminal and environmental investigations which assists in providing an understanding of complex relationships during trial presentations. Wilson et al. (1997) described computer-generated evidence from GIS as being highly reliable. One of the most common applications of GIS in criminal prosecution cases is in validating distance measurements. For example, when a drug sale occurs near a school, a GIS map clearly marking whether the location of the suspect's arrest falls within a GIS-measured drug-free buffer zone often suffices to enact enhanced penalties. In an investigation of high-crime areas, Leipnik and Albert (2003) discussed the application of GIS to prove significant relationships between the location of crime and certain establishments, such as liquor stores. In these cases, civil enforcement actions and license revocation were actions taken in lieu of proving criminal activity on the part of the establishments, although the GIS provided both graphic and geographic proof of localized crime when taken to court or liquor control board proceedings.

1.6 Summary

The management of the large amounts of raw data and derived information generated during criminal investigations call for new approaches using spatial information technology (Adderley and Musgrove 2001). In order to perform meaningful analysis, practitioners are finding an increasing need for the transfer of new knowledge and technologies from other disciplines (Haggerty 2004). Many of these new technologies have proven to be improvements over existing forensic technologies for the purposes of demonstrating compelling evidence in lawsuits (Jacobson 2004). Different types of geospatial technology have been used to investigate crime, prosecute and convict offenders, and exonerate suspects. Common examples of geospatial technologies include GIS, remote sensing, ground-penetrating radar, high-definition 3-D laser scanning, Light Detection and Ranging (LiDAR), thermal imagery, radar, sonar, magnetic resonance imaging, X-ray, GPS-related tracking, and radio-frequency identification. The applications, acceptability relevance, and procedural legality of each technology vary substantially, leading to a number of considerations still being addressed by the court system, ranging from the rules of evidence to the protection of privacy and constitutional liberties. Although considerable precedent exists for the use of geospatial technology, new issues and challenges are emerging as the technology evolves, generating new legal considerations.

The main utility of geospatial technology has been to provide associative evidence to assist in proving or disproving links between people, places, and objects as they relate to the court of law. Spatial data, inherent to geospatial technologies, are a valuable and versatile resource when used to investigate and establish facts in a court of law. Forensic GIS, defined here for the first time, is the application of geographic and spatial tools, principles, and methodologies to investigate and establish facts within the boundaries of forensics. The main utility of forensic GIS is to provide correlated evidence, which assists in either proving or disproving geographic, spatial, or temporal links between people, places, and objects as they relate to the court of law.

References

Adderley RW, Musgrove P (2001) Police crime recording and investigation systems – a user's view. [General review]. Policing Int J Police Strateg Manag 24(1):100–114

Boba R (2005) Crime analysis and crime mapping, 1st edn. Sage, Thousand Oaks

Boer A, Engers TV, Peters R, Winkels R (2007) Separating law from geography in GIS-based eGovernment services. Artif Intell Law 15(1):49–76. http://dx.doi.org/10.1007/s10506-007-9042-4

Bossler JD, McMaster RB, Rizos C, Campbell JB (2010) Manual of geospatial science and technology, 2nd edn. Taylor & Francis, Boca Raton

Brantingham P, Brantingham P (1981) Environmental criminology. Sage, Thousand Oaks

Brilis GM, Gerlach CL, van Waasbergen RJ (2000) Remote sensing tools assist in environmental forensics. J Environ Forensic 1:63–67. doi:10.1006/enfo.2000.0009

Brilis GM, van Waasbergen RJ, Stokely PM, Gerlach CL (2001) Remote sensing tools assist in environmental forensics: part II digital tools. Environ Forensic. doi:10.1006/enfo.2000.0033

Brodsky H (2003) Retail area overlap: a case in forensic geography. Prof Geogr 55(2):250, 259

Bruce CW (2001) A thousand words for a picture. Is the overvaluation of GIS disrupting a critical balance in crime analysis? [Crime Analysts' Round Table, Spring 2001, Massachusetts Association of Crime Analysts, Hyannisport Massachusetts]. http://www.macrimeanalysts.com/rt/f2cb1p.htm. Accesssed 15 Feb 2012

Burrough PA, McDonnell RA (1986) Principles of geographical information systems. Oxford University Press, Oxford

Byrne JM (2008) The best laid plans: an assessment of the varied consequences of new technologies for crime and social control (Newsletter). Fed Probat 72(3):10–21

Chrisman N (2005) Full circle: more than just social implications of GIS. Cartogr Int J Geogr Inform Geovis 40(4):23–35. doi:10.3138/8u64-k7m1-5xw3-2677

Dangermond J (2007) A framework for understanding, managing, and improving our world. GIS – the geographic approach. ArcNews 29(3):1–3

De Vorsey L (1973) Florida's seaward boundary: a problem in applied historical geography. Prof Geogr 25(3):214–220. doi:10.1111/j.0033-0124.1973.00214.x

De Vorsey LJ (1980) Dating the emergence of a Savannah River island: an hypothesis in forensic historical geography. Environ Rev 4(2):6–19

DiBiase D, Corbin T, Fox T, Francica J, Green K, Jackson J et al (2010) The new geospatial technology competency model: bringing workforce needs into focus. URISA J 22(2):55–72

Forer P, Unwin D (1999) Enabling progress in GIS and education. In: Longley P, Goodchild MF, Maguire DJ, Rhind DW (eds) Geographic information systems: management issues and applications. Wiley, New York, pp 747–756

Gabrosek J, Cressie N (2002) The effect on attribute prediction of location uncertainty in spatial data. Geogr Anal 34(3):262–285

Gaudet CH, Annulis HM, Carr JC (2003) Building the geospatial workforce. URISA J 15(1):21–30

Getis A, Drummy P, Gartin J, Gorr W, Harries K, Rogerson P et al (2000) Geographic information science and crime analysis. URISA J 12(2):7–14

Gold CM (2006) What is GIS and what is not? Trans GIS 10(4):505–519. doi:10.1111/j.1467-9671.2006.01009.x

Goodchild MF (1992) Geographic information science. Int J Geogr Inform Sci 6(1):31–45

Goodchild MF (2008) Geospatial technologies and homeland security research. In: Sui DZ (ed) Geospatial technologies and homeland security: frontiers and future challenges, vol 94, GeoJournal library. Springer, Dordrecht, pp 345–353

Goodchild MF, Anselin L, Appelbaum RP, Harthorn BH (2000) Toward spatially integrated social science. Int Reg Sci Rev 23(2):139–159. doi:10.1177/016001760002300201

Grip WM, Grip RW, Morrison RD (2000) Application of aerial photography and photogrammetry in environmental forensic investigations. Environ Forensic 1(3):121–129

Groff ER, Vigne NGL (1998) The use of geographic information systems (GIS) for state and local crime analysis. Paper presented at the conference of European statisticians, Ottawa, 5–7 Oct 1998

Haggerty KD (2004) Displaced expertise: three constraints on the policy relevance of criminological thought. Theor Criminol 8(2):211–231. doi:10.1177/1362480604042244

Jacobson M (2004) Reply to Kevin D. Haggerty. Theor Criminol 8(2):233–238. doi:10.1177/1362480604042246

Jensen J (2007) Remote sensing of the environment, 2nd edn. Pearson Prentice Hall, Upper Saddle River New Jersey

LaVigne NG, Groff ER (2001) The evolution of crime mapping in the United States: from the descriptive to the analytic. In: Hirschfield A, Bowers K (eds) Mapping and analysing crime data: lessons from research and practice. Taylor & Francis, New York, pp 203–222

Leipnik MR, Albert DP (2003) Overview of implementation issues. In: Leipnik MR, Albert DP (eds) GIS in law enforcement: implementation issues and case studies. Taylor & Francis, New York

Listi GA, Manhein MH, Leitner M (2007) Use of the global positioning system in the field recovery of scattered human remains. J Forensic Sci 52(1):11–15

Longley P, Goodchild MF, Maguire DJ, Rhind DW (2005) Geographical information systems and science, 2nd edn. Wiley, New York

Mamalian CA, LaVigne NG (1999) The use of computerized crime mapping by law enforcement: survey results. National Institute of Justice Journal Research Preview, January, Washington, DC

Mark DM (2003) Geographic information science: defining the field. In: Duckham M, Goodchild MF, Worboys M (eds) Foundations of geographic information science. Taylor & Francis, New York, pp 1–15

Markovic J, Bueermann J, Smith K (2006) Coming to terms with geographical information systems. Police Chief 73(6):60–73

Martin GJ, James PE (1993) All possible worlds: a history of geographical ideas, 3rd edn. Wiley, New York

Mathewes RW (2006) Forensic palynology in Canada: an overview with emphasis on archaeology and anthropology. Forensic Sci Int 163(3):198–203

McKinley J, Ruffell A, Harrison M, Meier W, Kemp H, Graham C, Barry L (2008) Spatial thinking in search methodology: a case study of the "No body murder enquiry", West of Ireland. In: Ritz K, Dawson L, Miller D (eds) Criminal and environmental soil forensics. Springer, New York, pp 285–302

Mildenhall DC, Wiltshire PEJ, Bryant VM (2006) Forensic palynology: why do it and how it works. Forensic Sci Int 163(3):163–172

Morgan RM, Bull PA (2007) Forensic geoscience and crime detection. Identification, interpretation and presentation in forensic geoscience. Minerva Medicoleg 127(2):73–90

Morrow-Jones HA, Anacker KB, Chen W, Horner MW, Malecki EJ (2005) Rapid property transactions and price changes: an exploratory study in forensic Geographic Information System (GIS). FOCUS Geogr 48(3):19–25

Murray RC (2004) Evidence from the earth: forensic geology and criminal investigation. Mountain Press, Missoula

National Institute of Justice (2009) High-priority criminal justice technology needs. In: Office of Justice Programs (ed). U.S. Department of Justice, Washington, DC, p 6. http://www.ncjrs.gov/pdffiles1/fs000237.pdf. Accessed 20 Jan 2009

National Institute of Justice (2010) High-priority criminal justice technology needs. In: Office of Justice Programs (ed) U.S. Department of Justice, Washington, DC, p 37

Nelson L (1999) GIS. A powerful weapon for law enforcement. ArcUser, Jan–Mar 1999

Noond J, Schofield D, March J (2002) Visualising the scene: computer graphics and evidence presentation. Sci Justice 42(2):89–95

Oatley G, Ewart B, Zeleznikow J (2006) Decision support systems for police: lessons from the application of data mining techniques to "soft" forensic evidence. Artif Intell Law 14(1):35–100. doi:10.1007/s10506-006-9023-z

Obledo MN (2009) Forensic archeology in criminal and civil cases. Forensic Mag 6(4):31–34

Pye K, Croft DJ (2004) Forensic geoscience: introduction and overview. Geol Soc Lond Spec Publ 232(1):1–5. doi:10.1144/gsl.sp.2004.232.01.01

Rossmo DK (2000) Geographic profiling. CRC Press, Boca Raton

Rossmo DK (2006) Geographic profiling in cold case investigations. In: Walton RH (ed) COLD CASE HOMICIDES: practical investigative techniques. CRC Press, Boca Raton, pp 537–560

Ruffell A (2006) Feature: forensic geoscience. Geol Today 22(2):68–70

Ruffell A, McKinley J (2005) Forensic geoscience: applications of geology, geomorphology and geophysics to criminal investigations. Earth-Sci Rev 69(3–4):235–247

Ruffell A, McKinley J (2008) Geoforensics. Wiley, Chichester

Schmitz P, Eloff C, Talmakkies R, Linnen C, Lourens R (2013) Forensic mapping in South Africa: four examples. Cartogr Geogr Inform Sci 40(3):238–247. doi:10.1080/15230406.2013.800273

Suggs JA, Beam EW, Biggs DE, Collins W Jr, Dusenbury MR, MacLeish PP et al (2002) Guidelines and resources for conducting an environmental crime investigation in the United States [Guest Editorial]. Environ Forensic 3(2):91–113

United States v. Jones (2012) 615 F.3d 544

Wang F (2005) Geographic information systems and crime analysis. Idea Group Publishing, Hershey, p 345

Warner TA, Nellis MD, Foody GM (2009) Remote sensing scale and data selection issues. In: Foody G, Warner T, Nellis MD (eds) The Sage handbook of remote sensing. Sage, New York, pp 1–17

Wartell J (2003) Crime mapping and data sharing. In: Leipnik MR, Albert DP (eds) GIS in law enforcement: implementation issues and case studies. Taylor & Francis, New York

Wilson AC, Jones SG, Smith MA, Liles R (1997) Tracking spills and releases: high-tech in the courtroom. Tulane Environ Law J 10:371–394

Wolff M, Asche H (2009) Towards geovisual analysis of crime scenes – a 3D crime mapping approach. In: Sester M, Bernard L, Paelke V (eds) Advances in GIScience. Springer, Berlin/Heidelberg

Zhao JL, Bi HH, Chen, HC (2003) Collaborative workflow management for interagency crime analysis. In: Intelligence and security informatics, proceedings. Lecture notes in computer science, vol 2665. Springer, Berlin, pp 266–280

Chapter 2
Geospatial Technologies in the Courtroom

George Roedl, Gregory A. Elmes, and Jamison Conley

Abstract The function of a court is to resolve disputes through a legal process. With few exceptions, the progression of a legal case will follow the strict guidelines of rules and codes developed from numerous court decisions to fairly and efficiently securing a just determination. All federal courts adhere to a flexible set of rules published in the Federal Rules of Evidence (FRE). The FRE provides rules and definitions governing general provisions, judicial notice, presumptions, relevance, privileges, witnesses, expert witnesses, hearsay, and authentication. However, there are as yet no special rules governing the use of geospatial technologies or spatial data. From a pragmatic legal perspective, spatial data differs immensely from the traditional form of evidence. However, the power of spatial information is extremely persuasive and compelling in litigation. While the acceptance of spatial data and methods has increased in litigation, there are also several issues that merit careful consideration when using spatial data. This chapter examines key rules and court decisions that impact the potential admissibility of spatial data and technologies in a modern courtroom.

Keywords Admissibility • Rules of evidence • Demonstrative evidence • Scientific evidence • Computer-generated evidence • Frye test • Daubert • Expert witness

2.1 Introduction

Crowsey (2002a) described spatial information as one of the most powerful comprehension and communication tools available to legal practitioners. The likelihood of a successful litigation is greatly enhanced through an effective means

G. Roedl (✉) • G.A. Elmes • J. Conley
Department of Geology and Geography, West Virginia University,
330 Brooks Hall, Morgantown, WV 26506, USA
e-mail: groedl@gmail.com; greg.elmes@mail.wvu.edu; jamison.conley@mail.wvu.edu

G.A. Elmes et al. (eds.), *Forensic GIS: The Role of Geospatial Technologies for Investigating Crime and Providing Evidence*, Geotechnologies and the Environment 11,
DOI 10.1007/978-94-017-8757-4_2,

of communication and visualization that aids in a greater understanding of the truth (Crowsey 2002b; Cohen 2008). Spatial technologies, and the associated spatial information derived from spatial technologies, provide the tools and methods for persuasive communications by those who adopt them as well as harm those who are hesitant to adopt them (Crowsey 2002a; Gonzalez 2009). When effectively and properly used, spatial information is capable of providing a persuasive understanding of the facts of a case (Crowsey 2002a, b; Cohen 2008). Fiedler (2003) described humans as visual learners who prefer visual evidence, citing an 87 % retention rate of visual information as opposed to only a 10 % retention rate of oral information alone. Pratt (2001) acknowledged information retention rate statistics as the difference between winning and losing a case. On the other hand, however, effective visual information communication may be the admissibility of spatial tools and data in legal proceedings requires an understanding of proper procedures. Since judiciaries from one country to the next can vary, a discussion of procedural processes and the admission of spatial data in courts is necessarily restricted. Cho (2005) provides an introduction to spatial law, legal systems, and legal theories of various jurisdictions around the world, whereas this chapter addresses the legal system in the United States exclusively.

The function of a court is to resolve disputes through a legal process. Based on the English common law court model, the US court system uses an adversarial process in which facts are presented to a judge and jury in a persuasive manner intended to support an argument effectively or to refute an opponent's argument. The role of the court in an adversarial system is to act as an impartial referee for the parties presenting facts in a dispute to ensure due process is followed. There are two basic types of courts: criminal court[1] and civil court.[2] In a criminal court, the adversaries presenting facts are the prosecutor and the defender. The prosecutor provides compelling evidence to support accusations of illegal activities by the defender. If the prosecutor successfully presents a persuasive argument, the defender can be punished through fines or imprisonment. Therefore, criminal court cases require a prosecutor to provide proof beyond a reasonable doubt to help ensure an innocent defender is not punished. By contrast, a civil court resolves disputes between parties in which one party prevails by providing the most compelling argument which supports their position more favorably than the opponent's argument. The judge serves as the fact finder to first determine if a party was wronged and then either assesses damages or issues a judicial order to start or stop an action petitioned to the court (Cohen 2008). For example, in a car accident between two citizens, the judge would listen to arguments on each side to first determine who was at fault and then make a decision as to what the remedy should be (e.g., repair costs).

Both criminal and civil courts can hold two types of trials. The first type of trial is a trial by jury. In a trial by jury, ordinary citizens are called upon to become fact finders. For criminal cases, a trial by jury is standard with few exceptions. Because

[1] See also http://www.uscourts.gov/uscourts/rules/criminal-procedure.pdf

[2] See also http://www.uscourts.gov/uscourts/rules/civil-procedure.pdf

selecting jurors is a lengthy process and not all trials (e.g., probate) necessitate a jury, bench trials are an alternative. In a bench trial, a judge serves as the finder of facts and administrator of justice. Every US state has a court system making criminal and civil trials available. In addition to state courts, there are federal district courts which preside over trials of specific matters. District courts preside over cases involving constitutional law, treaties, maritime law, interstate law and cases, and cases in which the US government is either the plaintiff or defendant. There are currently 89 districts and 94 district courts in the United States. There are also an additional 13 (11 district, 1 D.C., and 1 federal) judicial courts (also called circuit courts or courts of appeals) which preside over district court cases that have been appealed when issues regarding errors in the law are suspected (e.g., misinterpretation of a constitutional amendment or overstepping/lack of jurisdictional authority).[3] A US court of appeals typically serves as the final arbitrator for most federal cases and often sets legal precedent through its decisions (Hemmens et al. 2007). Unlike state and federal courts, an appeals court does not hold trials. Decisions are determined entirely through records presented at the lower court, although attorneys may be permitted to provide a brief oral argument. After an appeals court reaches a decision, either party may apply to the US Supreme Court to review the decision. Although a lengthy discussion of the judicial hierarchy may seem extraneous in a work on forensic GIS, it is important to establish the implications of decisions. It should now be clear that district courts only hear certain types of cases, while an appeals court is a much more powerful entity that sets legal precedent and policy after reviewing select cases, typically dealing with constitutional liberties.

2.2 Admissibility of Evidence

With few exceptions, the progression of a legal case will follow the strict guidelines of rules and codes developed from numerous court decisions (Cohen 2008). All federal courts adhere to the rules for admissibility adopted by the US Supreme Court and published in the Federal Rules of Evidence (FRE).[4] FRE Rule 1101 specifies which courts, judges, cases, and proceedings are required to adhere to the FRE, as well as the only three exceptions (a preliminary question of fact governing admissibility, grand-jury proceedings, and miscellaneous proceedings such as issuing an arrest warrant). The purpose of the FRE, as stated in Rule 102, is to "administer every proceeding fairly, eliminate unjustifiable expense and delay, and promote the development of evidence law, to the end of ascertaining the truth and securing a just determination." State court systems adhere to similar sets of published rules (e.g., Rules of Civil Procedures) based on the FRE (Onsrud 1992). The current edition of FRE contains 68 rules.

[3] See also http://www.uscourts.gov/uscourts/rules/appellate-procedure.pdf

[4] Federal Rules of Evidence. Amended 01 December 2012. Available at http://www.uscourts.gov/uscourts/rules/rules-evidence.pdf. Accessed 20 September 2013.

In theory, evidence which is relevant and not excluded by any of the FRE is admissible (Levi et al. 2013). Evidence may be admissible as either scientific evidence or demonstrative evidence. Scientific evidence is substantive evidence having probative value in and of itself, while demonstrative evidence is meant to clarify or illustrate testimony and has no probative value (Dischinger and Wallace 2005; Pratt 2001). Presented to prove or disprove a matter at issue in court, scientific (substantive) evidence comes in the form of testimonial evidence (e.g., oral testimony), documentary evidence (e.g., written testimony), or real evidence (e.g., a physical object) (Pratt 2001). Demonstrative evidence comes in the form of illustrative evidence (e.g., photographs) and actual evidence (e.g., confiscated items). Since scientific evidence has independent probative value, it becomes part of the formal record which may be examined by a deliberating jury. The jury is then able to review the facts revealed through the scientific evidence to aid in determination of guilt or innocence and liability (Bird 2001). In contrast, demonstrative evidence is rarely available during deliberations since it has no probative value.

Both scientific evidence and demonstrative evidence must be deemed relevant, authentic, and accurate and reliable (commonly referred to as foundation) in order to be admissible (Markowitz 2002; Dischinger and Wallace 2005). Depending upon the manner in which they are used, geospatial technologies can provide scientific evidence, demonstrative evidence, or both. When spatial data is merely used to illustrate and clarify testimony, it is admissible as demonstrative evidence (i.e., illustrative evidence). However, when used to prove the existence of a fact (e.g., a GPS tracking log) or as the basis of an expert opinion or conclusion, spatial data becomes scientific (Pratt 2001). Admitting spatial data as scientific evidence in a trial is more difficult than admission as demonstrative evidence and is largely dependent on having an expert witness testify about the facts (e.g., authenticity and accuracy) of the data (Dischinger and Wallace 2005). However, spatial data gathered from geospatial technologies may be granted greater admissible probative value when there is minimal human interaction with the data, such as raw satellite images (Krouse et al. 2000).

Spatial data generated from geospatial technologies, such as GIS, GPS, and remotely sensed data, are considered computer-generated data as well as digital data since computer technology is used to create the data which is stored digitally. Pratt (2001) referred to CGE as an abbreviation applicable to computer-generated evidence (also referred to as computer-generated exhibits), a broad term encompassing any use of computers in producing evidence for litigation. The Committee on Identifying the Needs of the Forensic Sciences Community (2009) referred to digital data as the digital evidence (e.g., photographs, call logs, and location records of a device) that could be gathered, processed, or interpreted from digital devices, such as desktop and laptop computers, cell phones, digital cameras, GPS devices, portable media players (e.g., IPods), etc.

Although Chap. 1 characterized spatial data as digital data, it is worthwhile to make a distinction between spatial data and the computer-generated end product derived from spatial data (generated by geospatial technologies) that may ultimately be admitted into court. For example, digital maps would be considered digital data representations of reality, whereas animated maps may be considered computer-generated data. Additional examples could include models which combine spatial data with mathematical formulas and 3-dimensional visualizations or virtual reality. The implications of this distinction may have a direct bearing on ease of admissibility under the FRE. Consider a simple digital photograph displaying erosion caused by a stream. The photograph represents the stream as it appeared when the image was taken for the given spatial location. Next consider a model which combines several spatial data sources (e.g., elevation, soil classifications, slope, average rainfall, land cover) with a mathematical model to predict the erosion of a stream. The accuracy of the source data and any transformations of the data; the choice; the quality and reliability of the mathematical model, e.g., the Water Erosion Prediction Project Model or the Universal Soil Loss Equation (Flanagan et al. 2001); the uncertainty of parameters input into the model; the error associated with the procedure; and the qualifications and experience of the modeler can drastically reduce the chances of the model being admitted into court as an accurate representation of erosion, since any of the pathways leading to speculative results could be challenged. Since the digital photograph experiences less human intervention, courts are more inclined to consider admitting it into evidence (Crowsey 2002a).

Before geospatial technologies can be used in court, they must first be allowed to be admitted into court as evidence. Although all rules are applicable to legal proceedings (Rule 1101 provides for applicability and exceptions), Spencer (2006) identified eight specific rules potentially applying to the question of admissibility of data derived from geospatial technologies. However, an additional four rules may also be applicable. The 12 rules are 401, 402, 403, 611, 702, 703, 801, 802, 803, 804, 901, and 1006.

2.2.1 Rules 401, 402, and 403

FRE Rule 401 is the test for relevance of evidence in court. Prior to the current version of the FRE, Rule 401 simply stated the definition of "relevant evidence" as "evidence having any tendency to make the existence of any fact that is of consequence to the determination of the action more probable or less probable than it would be without the evidence." The new Rule 401 has been amended to include two conditions: (1) "it has any tendency to make a fact more or less probable than it would be without the evidence" and (2) "the fact is of consequence in determining the action." In essence, the revised rule stresses that both conditions must be met in order to deem evidence as relevant.

Digital photographs are commonly admitted as evidence, provided they are relevant for proving or disproving a material fact. There is no requirement for the photographer to testify about the photograph provided a witness with knowledge of the photograph is able to testify that it fairly and accurately represents the depicted scene (Gonzalez 2009). Similarly, satellite images are likely to be deemed relevant and admissible due to their unique ability of providing evidence which satisfies Rule 401 (Crowsey 2002a). The admissibility of satellite images and aerial photographs as evidence is the same as ordinary photographs (Craig 2007). As for GIS evidence, whether in the form of maps, models, or simulations, relevance again poses a minimal barrier to admission provided consequential facts can be deemed more or less probable via the GIS evidence (Dischinger and Wallace 2005). It would understandably be a waste of time and resources to attempt to introduce irrelevant evidence into a case. Rule 401 therefore should not be considered an obstacle to the admissibility of evidence but rather a mechanism for ensuring an efficient judicial process without adding confusing or ambiguous evidence that provides no supportive assistance in a case.

FRE Rule 402 states that relevant evidence, which meets the criteria for Rule 401 (irrelevant evidence is not admissible), is admissible as evidence unless any of following provides otherwise: (1) the Constitution, (2) a federal statute (e.g., an Act of Congress), (3) the Federal Rules of Evidence, or (4) other rules prescribed by the Supreme Court. Under most circumstances, Rule 402 does not provide a barrier to the admissibility of spatial data. However, Rule 402 does make it clear that there may be circumstances in which relevant evidence cannot be admissible. One clear example would be instances in which national security could be compromised (e.g., a satellite image of a military base). Public safety, privacy, and warrantless searches are the main constitutional issues facing admissibility of spatial data (Markowitz 2002). According to the Fourth Amendment, "the right of the people to be secure in their persons, houses, papers, and effects, against unreasonable searches and seizures, shall not be violated" (US Constitution amend. IV). Three landmark cases decided by the US Supreme Court can be cited as precedent for contesting the admissibility of data derived from geospatial technologies on the basis of constitutional protection guaranteed by the Fourth Amendment (Markowitz 2002). In Dow Chemical Co. v. United States (1986), the Supreme Court ruled the use of commercial aerial photography used to search private property did not constitute a search if the private property was observable with the naked eye and therefore the photography was admissible as evidence (Craig 2007). In Kyllo v. United States (2001), the Supreme Court ruled the use of a thermal imaging device did require a warrant since the technology revealed details of the interior via technology that was not in the general public use (Craig 2007). However, the Supreme Court did not define "general use" technology, leaving lower courts to rule on a standard (Markowitz 2002). In United States v. Jones (2012) 132 S.Ct. 945, the court ruled that the installation of a Global Positioning System (GPS) tracking device on a vehicle and then using the device to monitor the vehicle's movements constitutes a search under the Fourth Amendment. As such action constitutes a "search," a warrant must be issued.

Aside from Rule 402 denying the admissibility of evidence, FRE Rule 403 cites additional reasons why evidence may not be admissible in court. Specifically, Rule 403 states that evidence in which the probative value is substantially outweighed by a danger of unfair prejudice, confusing the issues, misleading the jury, undue delay, wasting time, or needlessly presenting cumulative evidence may be excluded. Clearly, photographs and images have the potential to invoke emotion, leading to either favorable or unfavorable prejudice. Similarly, it is not hard to imagine how the use of images could be abused to cause confusion or deceit or how an excess of similar images could be needless. However, as Krouse et al. (2000) pointed out, the interpretation of Rule 403 is highly discretionary when it comes to weighing the probative value against the undesirable affects. Therefore, whether a court allows or denies the admissibility of evidence on the basis of Rule 403 is strictly at the discretion of individual judges and an uncertainty to be prepared for.

Just as photographs can be seen to cause prejudice, confusion, and misleading testimony, spatial data can also be presented in a manner leading to the same results. For example, the representation of a suspect's home on a map symbolized by a skull and crossbones or a mean-faced emoticon symbol conveys the implication of guilt. Other potential examples could be more subtle, but effective implants of prejudice, confusion, or misleading, such as the choice of color on a map or the use of a map originally intended for a very different purpose. When he wrote "How to Lie with Maps," Monmonier (1996) recognized the fact that maps signified authority and people trusted them unquestioningly, despite the ease of deceitful representation. The reality is that judges, juries, attorneys, plaintiffs, and defendants are part of the population Monmonier suggested would have no reason not to trust a map. However, as Wood has recently reemphasized, "the map is not reality" (Krygier and Wood 2011); therefore, recognition of unfair prejudice or misleading testimony provided through maps may not be evident within the judicial system without the aid of an expert.

Pratt (2001) discussed the potential for CGE to be misleading or prejudicial. CGE is commonly used during the presentation of evidence and in opening and closing arguments to enhance the quality and persuasiveness of the argument. As long as CGE accurately and fairly represents the testimony without being prejudicial or misleading, CGE is admissible to aid in understanding the truth. However, CGE technology has a great potential for promoting prejudice on the part of jurors who often consider CGE to be the truth instead of an aid for understanding the truth. As a heightened sense of reality becomes feasible through CGE technology, such as simulations, 3-D visualization, and virtual reality, the concern for undue prejudice is also increased through the emotions communicated, intentionally or otherwise, to the jurors. In order to minimize the concerns for prejudice and misleading testimony, which may disallow the admissibility of CGE, there are now standard practices in place. First, it should be evident to the court that the CGE is not real. For example, the representation of individuals should not resemble individuals in the courtroom. The more realistic the CGE, the more likely it is to unduly influence jurors. Second, jurors should only be allowed to view CGE, not participate in it. Third, inflammatory sound effects, such as screaming or weapons fire, should be omitted. And finally, the court should request limiting instructions on the jury's consideration of the CGE to make it absolutely clear the CGE is not reality, but rather an illustrative aid for understanding facts.

2.2.2 Rule 611

FRE Rule 611 deals with the mode and order of examining witnesses and presenting evidence. The court is given "reasonable" control over the mode and order of examining witnesses to make the procedures effective for determining the truth, avoid wasting time, and to protect the witnesses from harassment or undue embarrassment. This rule further limits witness cross-examinations and provides guidelines concerning leading questions during testimony. Spencer (2006) cited Rule 611 as a basis for the exclusion of evidence after it has been admitted into testimony due to the evidence being cumulative of other testimony. Although no explanation was provided, Flamm and Solomon (2004) suggested demonstrative evidence involving the interpretation or manipulation of the underlying data (e.g., animations, simulations, or models) would also fall under Rule 611. Among other terminology, demonstrative evidence introduced under Rule 611 is also known as demonstrative summaries, demonstrative charts, or pedagogical summaries. As such, the purpose is not to provide evidence or summarize voluminous data, but rather to summarize evidence already introduced in a manner intended to persuade a jury into accepting an argument. Demonstrative summaries may be opinions or inferences but are not admissible as evidence, and the court must make it clear to a jury that demonstrative summaries are aids only, not evidence (Levi et al. 2013).

Explaining that the purpose of Rule 611 is to prevent surprise testimony, Delaney and McMahon (2000) recommended showing CGE to opposing counsel before the jury is allowed to see it. The rationale is twofold. First, a witness may use demonstrative evidence to convey an interpreted opinion. Rule 611 is held to a less stringent standard of demonstrative evidence accuracy than Rule 1006 (Levi et al. 2013). However, the use of demonstrative evidence by an expert witness to convey an opinion could unduly influence a jury by being too persuasive or argumentative, but it would still meet Rule 403 standards since there is no probative value. Therefore, the jury should receive instructions to listen to the testimony and regard the demonstrative evidence as an interpretation that has not been subjected to the rigorous standards imposed by Rule 1006.

Second, a witness should be subject to cross-examination. The "surprise" admission of demonstrative evidence by a witness would not allow an opposing council sufficient time to formulate questions for cross-examination. By showing the CGE to the opposing counsel prior to presenting it, any objections or issues can be resolved and avoided. One final consideration more specific to animations and simulations is the fact that they may run uninterrupted (imagine an animation with a long run time) while a witness is on the stand (Delaney and McMahon 2000). Obviously, uninterrupted animations/simulation could have implications on the ability to effectively cross-examine a witness, waste time, or lead to embarrassment if opposing counsel objects to asserted truths. When introducing demonstrative evidence under Rule 611, it must be recognized that the court (i.e., judge) has complete discretion. While some courts could be very liberal in the amount and types of demonstrative evidence, other courts could be very strict.

2.2.3 *Rules 702 and 703*

Due to the complexity inherent in the nature of geospatial technologies and the evidence they produce, it is probable that an expert witness will be needed to explain and clarify spatial evidence (Markowitz 2002). The issues of opinions and expert testimony are governed by FRE Rules 701–706. Rule 702 allows a qualified expert to testify in the form of an opinion if four criteria are met: (1) the expert's scientific, technical, or other specialized knowledge will help the trier of fact to understand the evidence or to determine a fact in issue; (2) the testimony is based on sufficient facts or data; (3) the testimony is the product of reliable principles and methods; and (4) the expert has reliably applied the principles and methods to the facts of the case. Furthermore, Rule 702 defines an "expert" witness as a person having knowledge, skills, experience, training, or education qualifying the witness as an expert. Rule 703 is the rule pertaining to the basis of an opinion testimony given by an expert. Amended in the 2011 FRE, Rule 703 now states that "an expert may base an opinion on facts or data in the case that the expert has been made aware of or personally observed." Furthermore, Rule 703 clarifies that "if experts in the particular field would reasonably rely on those kinds of facts or data in forming an opinion on the subject, they need not be admissible for the opinion to be admitted. But if the facts or data would otherwise be inadmissible, the proponent of the opinion may disclose them to the jury only if their probative value in helping the jury evaluate the opinion substantially outweighs their prejudicial effect."

Rules 702 and 703 limit not only the testimony offered by experts but also evidence offered through an expert witness (Spencer 2006). In conjunction with Rule 402, judges serve as "gatekeepers" who apply Rules 702 and 703 to potential testimony to analyze the reliability and relevance of an expert witness (Krouse et al. 2000; Crowsey 2002a; NLECTC 2001). Although Rule 702 does not specify what constitutes "reliable principles and methods," Rule 703 addresses the issue by clarifying the principles and methods should be reasonably relied upon within the experts' particular field (see also Sects. 2.3.1 and 2.3.2 Frye Test and Daubert). While Rule 702 envisioned a relaxed test for admitting scientific, technical, or other specialized knowledge, the "gatekeeper" obligation of judges requires an assessment of the scientific validity of the reasoning or methodology underlying expert testimony to ensure such reasoning or methodology can be properly applied to the facts at issue (NLECTC 2001; Krouse et al. 2000; Crowsey 2002a; Pratt 2001). With respect to CGE, the implications of Rules 702 and 703 are that the qualifications and scientific background of an expert develop into a greater concern than the technology used to surmise facts or opinions (NLECTC 2001). In effect, new and complex methodologies require that an expert not just be a court-qualified expert under Rule 702 but rather a true and recognized expert within the expert's field (Cohen 2008; NLECTC 2001). Based on his extensive experience as an expert witness, Cohen (2008) added that it is the background and experience of the expert witness that are often a decisive factor in determining the outcome of a case.

2.2.4 Rules 802, 803, and 804

FRE Rules 801 through 807 are concerned with hearsay. Defined by Rule 801, hearsay is secondhand information, provided out of court, by someone other than the person testifying in court, which is offered as evidence to prove the truth of the matter asserted (Cohen 2008; Dischinger and Wallace 2005). Generally, hearsay is not admissible evidence (Rule 802) unless it can be categorized as an exception to the hearsay rule under Rule 803 or Rule 804 (Markowitz 2002; Dischinger and Wallace 2005). Rule 804 is primarily concerned with the admissibility of hearsay evidence when a declarant (the person who made the statement) is unavailable for testimony (e.g., physically or mentally ill). There are six exceptions that allow hearsay to be presented in court under Rule 804 if the conditions of declarant unavailability set forth by Rule 804 are met. Rule 803 allows for 23 exceptions (exception 24 has been transferred to Rule 807) allowing hearsay to become admissible.

Percipient witness (eyewitness) testimony, expert witness testimony, and CGE are each subject to the rule against hearsay. A percipient witness is generally only allowed to provide firsthand knowledge of events without opinions or conjecture (Cohen 2008). However, exceptions to the hearsay rule are available to expert witnesses who may rely on scientific articles and other similar pieces of information not personally known to the expert (Cohen 2008). In conjunction with Rule 703, testimony presented by an expert witness may include hearsay if the basis of the testimony is reasonably relied upon by other members of the expert witness's field of expertise (Crowsey 2002a). When offered in court for the truth of what they assert, CGE, which includes digital spatial data and the products generated from them (e.g., maps and charts), is almost always considered to be out-of-court statements and therefore hearsay (Onsrud 1992; Dischinger and Wallace 2005). Flamm and Solomon (2004) provided a distinction between electronically stored computer data created by humans and data automatically generated by computers, such as the Internet service provider (ISP) log-in or telephone call records. While the evidentiary concerns of automatic data generation by computers raise issues of authenticity and proper functionality, the human-generated electronic data records raise concerns about false and inaccurate out-of-court statements and therefore are considered hearsay. Unless a declarant designed the computer, wrote the software, and generated, manipulated, and stored the computer-generated data end product, an exception to the hearsay rule is needed to qualify the CGE as admissible for proving the truth of the matter asserted (Onsrud 1992).

Assuming CGE is deemed authentic and reliable (based on the discretion of the judge), it may be admissible under the business record or public record hearsay exception (Dischinger and Wallace 2005; Markowitz 2002). Rule 803(6) specifies the criteria necessary to qualify business records as an exception to the hearsay rule and admissible in court. Despite the hearsay nature of spatial data, Rule 803 (6) provides that it may be considered reliable and admissible when a qualified witness can testify it was gathered in a routine manner as part of ordinary everyday activities of a business, organization, occupation, or calling (Crowsey 2002a). Furthermore, the data need to be made at or near the time in question by a person with knowledge of the system and not indicate a lack of trustworthiness from either the source of information or the methods or circumstances of preparation. The business record exception is most often applicable to the successful admission of CGE (Onsrud 1992). Rule 803(8) governs the criteria necessary to qualify public records as an exception to the hearsay rule. Public records are simply the records or statements of a public office. Similar to business records, public records should be properly kept and not indicate a lack of trustworthiness. When a copy of an electronic document submitted for admission comes from files collected and maintained by a public agency and the copy is certified as a correct copy by an authorized person (e.g., custodian of the records), the copy is self-authenticating, and no extrinsic evidence of authenticity is needed (Onsrud 1992). In other words, if the electronic records can be shown to be those provided by a public agency, the records are considered admissible as CGE.

2.2.5 Rules 901 and 902

The previous discussion pertaining to the admissibility of business or public records stipulated that the data needed to be qualified via testimony that the data records were in fact the purported records. Any evidence submitted for admission must be identified sufficiently to satisfy the judge that the evidence is what it is claimed to be. The process of identifying an item of evidence as the purported item is known as authenticating evidence. According to Rule 901, "to satisfy the requirements of authenticating or identifying an item of evidence, the proponent must produce evidence sufficient to support a finding that the item is what the proponent claims it is."

A nonexclusive list of ten examples for authenticating evidence is provided by Rule 901(b). Among the ten examples, five are relevant to digital data and CGE: (1) testimony by a witness with knowledge, (2) comparison with an authenticated specimen by an expert witness or the trier of fact, (3) distinctive characteristics, (4) evidence about public records, and (5) evidence describing a process or system and showing that it produces an accurate result. Unless authenticity is challenged, prima facie evidence (e.g., a declaration of accuracy by a witness) is sufficient for admission under Rule 901 (Barakat and Miller 2004). In other words, the threshold of authenticity is relatively low. The requirement is that the evidence could be authentic, not that the evidence is proven to be truly authentic. However, in the event authenticity is challenged, a proponent must be prepared to conclusively prove it is what is claimed. Rule 902 stipulates certain forms of evidence are self-authenticating. Twelve items are listed with provisions attached. Depending on their creation and publication, most of the 12 items could be relevant to paper maps (Markowitz 2002). Although CGE may be self-authenticating, CGE is much more likely to be challenged unless printouts are certified as correct copies by a custodian of the records (Onsrud 1992).

Unless considered self-authenticating by Rule 902, the extrinsic authentication standard of Rule 901 applies to all items introduced as evidence. Although spatial data and CGE have unusual characteristics, authentication problems are similar to those encountered by any other computer records (Onsrud 1992; Flamm and Solomon 2004). One major challenge to authenticating evidence from digital data is establishing accuracy and reliability of results from the process or system producing the data. Proper authentication requires that reliability of computer processes and accuracy of results be subject to scrutiny (State of Connecticut v. Alfred Swinton 2004).

A particularly significant issue involved in establishing accuracy and reliability to support a finding that an item of evidence is what the proponent claims it is involves establishing a chain of custody (Pratt 2001). Whether evidence is physical or digital, establishing a chain of command is essential if tampering is alleged. While physical evidence should be accompanied by a required chain of custody form that documents custody, control, transfers, analyses, or disposition of each item, digital data may not have a chain of custody form. If the originality of evidence is questioned, each and every custodian may be required to testify to the integrity of the evidence during their possession if authentication cannot be provided otherwise by Rule 901(b) (Cohen 2008). Although digital data is highly susceptible to tampering (e.g., alteration of digital photos), another concern is errors generated during collection, storage/management, analyses, manipulation, or presentation. In order to establish a chain of custody for digital data, the proponent must show (1) the accuracy and reliability of the original source data to include all assumptions, formulas, and calculations used in defining and analyzing the data; (2) the accuracy of the source data entered into the computer; (3) the reliability and capability of the hardware and software used; (4) the process of software used for the computer graphics; (5) the methods used to produce the graphics in court; and (6) the accuracy and reliability of the final presentation (Crowsey 2002a; Krouse et al. 2000; Dischinger and Wallace 2005). Crowsey (2002a) recommended testimony by expert witnesses who could reliably authenticate the accuracy and reliability of digital data to establish a chain of custody.

Citing three previous landmark cases, the Connecticut Supreme Court (State of Connecticut v. Alfred Swinton 2004) established legal precedent for authenticating CGE (NDAA 2004). The court opined that there must be "testimony by a person with some degree of computer expertise, who has sufficient knowledge to be examined and cross-examined about the functioning of the computer." Stating that the salient issue was not only the reliability of the evidence, but also the reliability of the procedures involved, the court decided that the opposing counsel must have the opportunity to cross-examine a witness as to the methods used. The court noted that reliability issues may arise through or in (1) the underlying information, (2) the entering of information into a computer, (3) the computer hardware, (4) the computer software, (5) the execution of instructions which transforms information, (6) the output, (7) the security system used to control access to the computer, and (8) the user errors which may arise at any stage. Addressing reliability of computers, the court adopted six factors to be used as guidelines for authenticating CGE evidence: (1) the computer equipment is accepted in the field as standard and competent and was in good working order; (2) qualified computer operators were employed; (3) proper procedures were followed in connection with the input and output of information; (4) a reliable software program was utilized; (5) the equipment was programmed and operated correctly; and (6) the exhibit is properly identified as the output in question. Furthermore, the court opined there should be no distinction between various CGE. Animations and simulations should be held to the same stringent standards of reliability despite claims that animations are illustrative in nature while simulations provide substantive evidence.

2.2.6 *Rule 1006*

The two potential avenues for admitting CGE derived from spatial data are as scientific evidence under Rules 702 and 703 or as demonstrative evidence under Rule 1006 (Crowsey 2002a; Krouse et al. 2000). Demonstrative evidence used to clarify or illustrate complex information is allowed under Rule 1006. According to Rule 1006, summaries, charts, or calculations to prove the content of voluminous writings, recordings, or photographs that cannot be conveniently examined in court can be used, provided the originals or duplicates are available for examination. Demonstrative evidence is essential for successful litigation since juries expect it and it appeals to the five senses (Marks 2003). Although demonstrative evidence has no probative value and cannot be examined during jury deliberation, it is frequently and opportunistically used as a persuasive aid in keeping a jury interested and focused on what is being presented during trial (Gonzalez 2009).

Demonstrative evidence is subject to FRE tests for admissibility governed mainly by Rules 401–403 and 901 or 902. Accordingly, demonstrative evidence must be relevant (Rule 401), admissible (Rule 402), fair (Rule 403), and authenticated (Rule 901) or self-authenticating (Rule 902), regardless of whether or not it is CGE (Pratt 2001). Flamm and Solomon (2004) recommended pretrial disclosure of the intended use of demonstrative evidence to the courts and opposing council to hasten proceedings with regard to evidentiary standards of relevance, fairness, admissibility, and authentication. Crowsey (2003) constructed a pretrial disclosure checklist exclusive to geospatial technologies to avoid oversight as well to as identify potential weaknesses that would result in exclusion of exhibits. Conversely, Pratt (2001) discussed numerous potential reasons to object to the inclusion of demonstrative evidence in general, and CGE in specific, by an opposing council. Ultimately, the admissibility of demonstrative evidence is left to the sound discretion of the trial judge (Spencer 2006).

2.3 Additional Considerations

This section discusses additional evidentiary considerations that may apply to geospatial data. In practice, several legal tests control admission of evidence in the United States (Markowitz 2002). The FRE have already been discussed as the rules of admissibility adopted by the US Supreme Court which apply to all federal courts in order to ascertain the truth and secure a just determination from proceedings. Federal courts must also adhere to the US Constitution, which is the supreme law within the United States. Following the rules set forth within the US Constitution also requires following constitutional law, which are the interpretations defining the scope and application of the Constitution. An additional legal test of admissibility used by federal courts is known as the Daubert standard. Unlike the FRE, which are a defined set of rules governing numerous topics, the Daubert standard is derived from a court case decided by the US Supreme Court in which Rule 104 was reviewed to interpret Rule 702. Prior to the Daubert standard, the Frye test was used by federal courts. Many state courts today still apply the Frye test as a means of testing the admissibility of evidence.

Furthermore, Cheng and Yoon (2005) cited research suggesting that in practice, Daubert courts essentially still perform what is considered a Frye test. Since the Frye test was originally a federal test, it remains the basis of the Daubert standard, is still used by state courts, is prominent in the literature, and is applicable to spatial data. On these grounds, it must be recognized as an important consideration for further discussion. After reviewing the Frye test and Daubert standard, this section will discuss the pictorial testimony and silent witness theories which govern the admissibility of digital photographs and video.

At this point, it would be prudent to recognize the considerations discussed within this section are not the only additional considerations regarding the admissibility of evidence. There are many further considerations relevant to the admissibility of evidence. For example, the previous discussion of the FRE already explained chain of custody issues. While some additional considerations have already been discussed elsewhere, identifying and discussing all possible considerations are well beyond the scope of this chapter.

2.3.1 Frye Test

As science advanced, the legal system attempted to develop coherent tests that could be applied to scientific evidence (Committee on Identifying the Needs of the Forensic Sciences Community 2009). The first notable development occurred with a landmark decision in 1923 that ruled scientific evidence needed to be "generally accepted." Scientific evidence interpreted by a court as "generally accepted" by a meaningful segment of the associated scientific community meets the requirements of the Frye test (Cheng and Yoon 2005). This standard comes from the case Frye v. United States. The Frye test applies to procedures, principles, or techniques that may be presented in the proceedings of a court case and is still the de jure standard in several states.

The Frye test emerged from a murder appeal in which a lie detector was used. Expert witnesses elaborated on scientific experiments which revealed an increase in blood pressure when someone lied, concealed facts, or was guilty of a crime. The proffered theory was that the truth was a spontaneous event while a deception was an intentional event, which could be identified through an increase in blood pressure. Although the court conceded difficulty in establishing when a scientific principal advances from the experimental stage to the demonstrable stage, it was ruled that scientific recognition and acceptance was necessary for the admissibility of expert testimony (Frye v. United States 1923). Simply stated, the court must determine if an expert's testimony is based on recognized and accepted science. The acceptance of science within the associated community, also known as the scientific validity, emerged as a legal test referred to as the Frye test. Despite the modern significance, the Frye case went unnoticed for decades (Faigman et al. 2006). No other court cases cited Frye, and no law review articles discussed Frye until the 1970s and the release of the first version of the FRE.

It should be noted that the Frye test specifically addressed the underlying principles and methods used by experts to form an opinion. Courts and commentators found the general acceptance test to have significant limitations, particularly the vagueness of its conditions (see Faigman et al. 2006 for criticisms of Frye). The underlying methods were required to be scientifically valid; however, the Frye test did not require the opinion of the expert to be generally accepted by the scientific community (Committee on Identifying the Needs of the Forensic Sciences Community 2009). The issue of whether or not an expert opinion needed to be generally accepted within the expert's field was later decided in Berry v. CSX Transportation, Inc. (1998) when the court ruled that as long as the opinion was based on scientific validity, it was not necessary for the opinion to be generally accepted.

2.3.2 *Daubert*

After the FRE were first enacted in 1975, many people wondered whether or not the Frye test had become obsolete (NLECTC 2001). The first version of Rule 702 provided that the mere "assistance" to the trier of fact appeared as the touchstone of admissibility. Litigants, judges, and scholars were uncertain if Rule 702 embraced the Frye test or established a new standard (Committee on Identifying the Needs of Forensic Sciences Community 2009). The Supreme Court ruling in the case of Daubert v. Merrell Dow Pharmaceuticals, Inc. (1992) clarified the ambiguity between Rule 702 and the Frye test by adopting a new framework for evaluating the reliability of scientific evidence (Committee on Identifying the Needs of the Forensic Sciences Community 2009; NLECTC 2001; Cheng and Yoon 2005).

In Daubert, the Supreme Court held that Rule 702, not the Frye test, controlled the admission of expert testimony in federal courts. Noting Rule 702 made no mention of the Frye test, the Daubert court ruling provided judges with guidance in a newly established role of "gatekeepers" of scientific evidence (NLECTC 2001; Committee on Identifying the Needs of the Forensic Sciences Community 2009). The Daubert standard set forth that "any and all scientific testimony or evidence admitted is not only relevant, but reliable" (Daubert v. Merrell Dow Pharmaceuticals, Inc. 1992). This "gatekeeper" approach had judges analyzing the relevance and reliability of potential testimony (NLECTC 2001). While relevance was determined from Rule 104(b), reliability was to be based on four considerations interpreted in the context of each case: (1) falsifiability, (2) peer review, (3) error rates, and (4) acceptability in the relevant scientific community (Cheng and Yoon 2005; NLECTC 2001; Farber 2008). The falsifiability consideration allowed judges to consider whether a theory or technique could be, or had been, tested. The peer review consideration allowed judges to examine whether a theory or technique was subjected to peer review and publication. Error rates considered the known or potential rate of error of a particular

imposing adequate safeguards against manipulation and tampering, the courts impose a burden upon the offering party of the photographic evidence to prove authenticity above and beyond the requirements of the "pictorial testimony" theory.

Digital photographic evidence may be admitted as either demonstrative evidence, pursuant to the "pictorial testimony" theory, or as substantive evidence pursuant to the "silent witness" theory (Pratt 2001). Manipulations and alterations of digital photographic evidence represent a major concern, especially when there is no witness to testify about the authenticity. Barakat and Miller (2004) warn that digital photographic evidence will be subject to significant cross-examination, particularly in regard to chain of custody issues. Once the digital photographic evidence is collected, it should be documented carefully and all changes logged to support authenticity of the digital evidence. The proffering party should be prepared to prove the digital photographic evidence is unedited. In the case of image enhancements, each step must be documented and repeatable from a copy of the original.

2.4 Summary

In today's society, computer technology is used not only to commit crimes but also in litigation (Pratt 2001). From the most basic use (e.g., drafting a motion) to more advanced uses (e.g., producing digital photographs and maps) through the most sophisticated uses (e.g., creating animations, simulations, and virtual reality), it is rare to discover a case that does not involve the use of computers. While remaining cautious about the trustworthiness and reliability of CGE, judges at all levels of the judicial hierarchy have demonstrated an increased comfort level with CGE use (Flamm and Solomon 2004). However, it is clear from a pragmatic legal perspective that CGE differs immensely from traditional paper documents (Onsrud 1992). It is also clear, at least for now, there are no special rules governing the use and admissibility of CGE in court (Pratt 2001). The most recent revision to the FRE makes only two references to computer data: Rules 901(b)(9) and 1001(3). As this chapter has pointed out, the FRE have proven flexible enough to cover the use and admission of CGE in the same manner as they would use traditional paper exhibits. Referring specifically to geospatial technologies, Foote and Lynch (2000) acknowledge that if GIS could not be used to prove legal cases, it would lose much of its value to the forensics system. This chapter has also identified several issues which have complicated the admissibility of CGE (e.g., authenticity) and the FRE requirements (e.g., relevance and reliability) and tests (e.g., Daubert trilogy) necessary to satisfy a judge that any CGE evidence is what it is claimed to be. However, the combinations of rules and tests have thus far helped to establish a more precise test for scientific testimony that does not focus so much on technology as it does the qualification and background of an expert witness (NLECTC 2001).

The goal of this chapter has been to provide a source of generalized information that can be used as a guide for a more thorough examination of specific issues related to admissibility of geospatial technologies. Most of the information has been collected from scholarly articles from diverse journals and disciplines. Scholars and practitioners are only now beginning to publish books about the legal aspects of spatial data and technologies. For a further in-depth discussion on the admissibility of scientific evidence, see Faigman et al. (Faigman et al. 2006). Cho (2005, 2012) has provided comprehensive information specific to GIS legal issues, while Janssen and Crompvoets (2012) addressed geographic data. Additionally, Ito (2011) exclusively discussed the legal issues of satellite remote sensing.

References

Barakat B, Miller B (2004) Authentication of digital photographs under the "pictorial testimony" theory: a response to critics. Fla Bar J 78(7):38–43

Berry v. CSX Transportation, Inc (1998) 709 So. 2d 552 (Fla. 1st DCA 1998)

Bird S (2001) Scientific certainty: research versus forensic perspectives. J Forensic Sci 46(4):978–981

Cheng EK, Yoon AH (2005) Does Frye or Daubert matter? A study of scientific admissibility standards. Va Law Rev 91(2):471–513. doi:10.2307/3649430

Cho G (2005) Geographic information science: mastering the legal issues. Katholieke University, Leuven

Cho G (2012) Geographic data and legal liability issues. In: Janssen K (ed) Legal aspects of geographic data and spatial data infrastructures. Katholik University, Leuven, pp 153–166

Cohen KS (2008) Expert witnessing and scientific testimony: surviving in the courtroom. CRC Press, Boca Raton

Committee on Identifying the Needs of the Forensic Sciences Community (2009) Strengthening forensic science in the United States: a path forward. The National Academies Press, Washington, DC

Craig BJ (2007) Online satellite and aerial images: issues and analysis. [Minnesota Legal Studies Research Paper No. 08–11]. N. D. Law Rev 83:547–578

Crowsey R (2002a) A legal assistant's guide to legal applications of geospatial information. http://www.crowsey.com/pdf/spatialInformation.pdf. Accessed 30 Sept 2008

Crowsey R (2002b) Using spatial information. http://www.crowsey.com/pdf/litigatorsGuide.pdf. Accessed 30 Sept 2008

Crowsey R (2003) Geographic intelligence risk reduction checklist. http://www.crowsey.com/pdf/CheckList.pdf. Accessed 20 Jan 2009

Daubert v. Merrell Dow Pharmaceuticals, Inc (1992) 509 U.S. 579, 113 S.Ct. 2786

Delaney TQ, McMahon CM (2000) Jumping over the evidence hurdle at trial. Natl Law J. http://www.brinksgilson.com/news_events/index.php?action=view&publication_id=116. Accessed 28 Aug 2013

Dischinger SS, Wallace LA (2005) Geographic information systems: coming to a courtroom near you. Colo Lawyer 34(4):11–23

Dow Chemical Company v. United States (1986) 476 U.S. 227

Faigman DL, Saks MJ, Sanders J (eds) (2006) Admissibility of scientific evidence. In: Modern scientific evidence: the law and science of expert testimony, vol 1. West Pub Co, Eagan, pp 1–124

Farber DA (2008) Harnessing the power of information for the next generation of environmental law: II Use and abuse of information: modeling climate change and its impacts: law, policy, and science. Tex Law Rev 86:1655

Fiedler BS (2003) Are your eyes deceiving you?: the evidentiary crisis regarding the admissibility of computer generated evidence. N Y Law Sch Law Rev 48:295–321

Flamm S, Solomon SH (2004) Admissibility of digital exhibits in litigation. In: Samuel C, Solomon H (eds) Lynbrook. DOAR Litigation Consulting, New York

Flanagan DC, Ascough JC, Nearing MA, Laflen JM (2001) The Water Erosion Prediction Project (WEPP) model. In: Harmon RS, Doe WW III (eds) Landscape erosion and evolution modeling. Springer, Dordrecht, pp 145–199
Foote KE, Lynch M (2000) Legal issues relating to GIS. The geographer's craft. University of Colorado at Boulder, Boulder
Frye v. United States (1923) 293 F. 1013, 1014 (D.C.. Cir 1923)
General Electric Co. et al. v. Joiner et ux (1997) 522 U.S. 136, 118 S. Ct. 512, 139 L. Ed. 2d 508
Gonzalez EA (2009) Advanced trial handbook – demonstrative evidence. http://www.caught.net/prose/advtt/hbdemons.htm. Accessed 14 Jan 2009
Hemmens C, Cooper J, Hatch V (2007) Law enforcement case law. Crim Justice Rev 32(3):303–328. doi:10.1177/0734016807304917
Ito A (2011) Legal aspects of satellite remote sensing. Martinus Nijhoff Publishers, Boston
Janssen K, Crompvoets J (eds) (2012) Geographic data and the law – defining new challenges. Leuven University Press, Leuven
Krouse AJ, Ferry MM, Crowsey RC (2000) Satellite imagery: the space odyssey arrives in the courtroom. http://www.crowsey.com/pdf/spaceOdyssey.pdf. Accessed 30 Sept 2008
Krygier J, Wood D (2011) Making maps: a visual guide to map design for GIS. Guilford Press, New York
Kumho Tire Company, Ltd. v. Carmichael (1999) 526 U.S. 137, 119 S.Ct. 1167
Kyllo v. United States (2001) 533 U.S. 27
Levi DF, Nowinski PA, Killefer G (2013) Federal trial objections, revision 7. James Pub., Inc., Costa Mesa
Markowitz KJ (2002) Legal challenges and market rewards to the use and acceptance of remote sensing and digital information as evidence. Duke Environ Law Policy Forum 12(2):219–264
Marks SC (2003) The admissibility and use of demonstrative aids. The Brief 32:4
Melendez-Diaz v. Massachusetts (2009) 557 U.S. 305
Monmonier MS (1996) How to lie with maps. University Of Chicago Press, Chicago
NDAA (2004) Connecticut Supreme Court issues new standard for computer-generated evidence. Update Express, May
NLECTC (2001) Technology goes to court. TechBeat (Spring)
Onsrud HJ (1992) Evidence generated from GIS. GIS Law 1(3):1–9
Pratt FH (2001) The use of computer-generated exhibits in federal criminal cases. Defender Services Division Training Branch, Washington, DC
Spencer EL (2006) Use and misuse of technical data telling the scientific story to scientific virgins. In: American Law Institute – American Bar Association Continuing Legal Education ALI-ABA Course of Study, pp 1–9. http://www.grahamdunn.com/go/articles/use-and-misuse-of-technical-data-telling-the-scientific-story-to-scientific-virgins. Accessed 26 Feb 2009
State of Connecticut v. Alfred Swinton (2004) 268 Conn. 781; 847 A.2d 921
US Constitution. Amendment IV
United States v. Jones (2012) 132 S.Ct. 945
Wells D (2012) In brief: Law 101: legal guide for the forensic expert. NIJ J (269):24–25. http://www.nij.gov/nij/journals/269/inbrief.htm. Accessed 23 Apr 2012

Chapter 3
Spatial Tracking Applications

George Roedl, Gregory A. Elmes, and Jamison Conley

Abstract In this chapter, various technologies that permit law enforcement agencies to track people and objects through time and space are discussed. Innovative applications of spatial technologies have been adopted by law enforcement agencies to reduce crime, apprehend offenders, and keep officers safe. A brief review of each technology is given followed by examples in which the technologies have been successfully utilized. Specifically, this chapter examines the following: (1) Global Positioning Systems, (2) cellular phone tracking, (3) unmanned aerial vehicles, (4) automated license plate recognition technology, and (5) radio frequency identification technology.

Keywords Tracking technologies • Global Positioning System (GPS) • Unmanned aerial vehicles (UAV) • Drones • License plate readers • Radio frequency identification (RFID)

3.1 Introduction

Tracking technologies are used to locate people, vehicles, and other assets (Pease 2005), a process also known as geolocation. Recent technological advances enable a precise and real-time location to be determined. Tracking technologies are often used by consumers for navigation and to assist in tracking family members (e.g., children who wander off). The potential applications of tracking technologies are limitless (Edmundson 2005). The data collected from tracking technologies are capable of generating a precise and comprehensive record of an individual's movements that reflect detailed information pertaining to possible political, professional,

---

G. Roedl (✉) • G.A. Elmes • J. Conley
Department of Geology and Geography, West Virginia University,
330 Brooks Hall, Morgantown, WV 26506, USA
e-mail: groedl@gmail.com; greg.elmes@mail.wvu.edu; jamison.conley@mail.wvu.edu

G.A. Elmes et al. (eds.), *Forensic GIS: The Role of Geospatial Technologies for Investigating Crime and Providing Evidence*, Geotechnologies and the Environment 11,
DOI 10.1007/978-94-017-8757-4_3,

familial, religious, and sexual associations (Fulton 2013). Although locational information is valuable to third parties, such as marketers and businesses, it is invaluable to law enforcement agencies. As the capabilities and accessibility of location tracking technologies have increased, so has the ability of law enforcement agencies to utilize the technologies in routine investigations. The prevalence of GPS in vehicles, cell phones, and other handheld devices has given law enforcement authorities a powerful tool for tracking suspects (Stacy 2008a, b).

More recent innovations in spatial tracking have further extended the potential for law enforcement agencies to more easily apprehend offenders and deter crime while keeping officers safe. Unmanned aerial vehicles have provided increased situational awareness during dangerous operations. Automated license plate readers and radio frequency identification technology have emerged as vital tools for the recovery of stolen items. Each technology has its own advantages and disadvantages; however, the combination of technologies provides law enforcement agencies with a powerful set of tools in which the potential is only now beginning to be realized.

3.2 Global Positioning System

A Global Positioning System, or GPS, is available to the general public through handheld and vehicle navigation devices. Developed by the United States military, GPS devices entered the consumer market in the 1990s (Stacy 2008a, b). GPS devices are currently available in a variety of sizes and shapes, capable of centimeter accuracy (Edmundson 2005). Reliable GPS data tracks people and objects in both space and time (Ball 2008). Throughout the world, GPS and its counterparts are being used more than ever (Moenssens 2006). Not surprisingly, as the general public has realized the benefits of GPS, so too has law enforcement (Dowdell 2005). Prior to the advent of GPS tracking technology, law enforcement agencies needed to invest an enormous amount of time and resources in an effort to track and collect detailed information on location and movement patterns of individuals, often serving to deter against individual tracking practices (Fulton 2013). The increased availability of personal tracking and monitoring technology has greatly enhanced the ability of law enforcement agencies to accurately monitor individual actions (Ganz 2005; Werbach 2007). GPS-based surveillance systems provide law enforcement agencies with an inexpensive means of easily gathering intelligence and evidence electronically and surreptitiously (Ganz 2005; Moenssens 2006).

Law enforcement agencies have used GPS in a variety of situations, including the following: (1) murder investigations, (2) drug investigations, (3) robbery investigations, (4) public corruption, (5) parole violations, (6) hostage situations (Ganz 2005), (7) sting operations, and (8) remote surveillance (Dowdell 2005). Despite the successful applications of GPS evidence, the technology is prone to problems. In the high-profile investigation of the murder of Stacy Peterson, a GPS attached to the vehicle of husband and suspect Drew Peterson showed him traveling at speeds of

38,000 miles per hour, while a second GPS attached to the vehicle of his missing wife's car did not work during a 3-week period (Ganz 2005).

In many cases, locational information is obtained from GPS units possessed by a suspect. Stacy (2008a, b) lists several instances in which GPS was used as evidence to assist in providing evidence to law enforcement agencies and prosecutors by contradicting a defendant's alibis. Using a suspect's own vehicle navigation GPS device to establish location and time, prosecutors obtained a conviction in the murder of the suspect's parents and sister in 2005. The forensic value of vehicle navigation GPS devices also served a critical role in a Montana rape case that showed the suspect traveled through town looking for a victim. In Pennsylvania, GPS records showed a suspect parked near his home prior to committing arson on his own home.

In other cases, law enforcement agencies attached a GPS to the personal belongings of a suspect in an effort to track movement. Ganz (2005) summarized criminal investigations reliant on GPS tracking initiated by law enforcement. In an investigation of a suspected murderer in Washington, police placed a GPS tracking device on the vehicle of the suspect. After analyzing the track log, investigators were able to locate the buried remains of the victim at one of the stops the suspect made. In another investigation, a GPS unit was attached to a dog that repeatedly brought home the remains of a human skeleton. Investigators were able to locate and recover the remains after tracking the movement of the dog. In an investigation of public corruption, tracking devices were placed on police patrol vehicles after numerous reports of officers loitering in parking lots. Officers were subsequently barred from working in law enforcement after filing false patrol reports in which they reported patrolling activities contrary to the collected GPS data. In another internal affairs investigation, a GPS was attached to the vehicle of an officer to determine where he lived because there was a local law requirement for officers to live within the city limits. GPS was also cited as instrumental in two robbery investigations and a parole violation investigation after law enforcement agencies were able to track and locate suspects.

The versatility and effectiveness of GPS tracking has enabled law enforcement agencies to become increasingly innovative in applications (Mackey 2011). According to a "News Brief" article in Geography and Public Safety (2012), patrol officers keep track of specific locations where suspects dispose of evidence while being pursued. The "geo-tagging" of evidence assists in recovering small items that are traditionally difficult to find after a long chase over a large geographic area. GPS technology has also been used by university law enforcement agencies in innovative ways (Bowman 2011). At the University of California, Santa Barbara, and the University of Wisconsin-Madison, police equipped bicycles with GPS units to deter and apprehend bicycle thieves. National Guard personnel also routinely use advanced GPS equipment (NLECTC 2004). With the aid of a personal digital assistant (PDA) and laser range-finder binoculars, National Guard personnel aboard helicopters are capable of pinpointing the exact location of marijuana fields in remote areas, enabling eradication teams to return and destroy the fields. The coordinates are additionally saved to a database for reporting and analysis.

Perhaps the most common and proliferate use of GPS tracking is associated with monitoring sex offender parolees. Although state laws vary (e.g., based on specific offense or age of victim), many states mandate continuous and lifelong tracking of convicted sex offender parolees by using GPS anklets. In general, sex offender parolees are prohibited from traveling within specific distances of certain areas (e.g., where children play) and are not allowed to have contact with certain people (e.g., victims) (Mielke 2009). Additionally, many states also monitor parolees convicted of other offenses. For example, California has attached GPS anklets to gang members (Wilkinson 2010), and Massachusetts allows courts to impose GPS tracking of stalkers and domestic abusers who violate restraining orders (Hinson 2008). As of 2009, twelve additional states used GPS anklets to monitor domestic abusers (Green 2009).

In an effort to actively enforce exclusion zone and time restrictions, events may be closely monitored for potential violations made apparent through GPS anklets. In California, for example, summertime events such as the California State Fair monitor sex offenders to track their proximity to fair grounds, and gang members who are restricted to entering the fair only during certain times are monitored to ensure compliance with designated hours (Wilkinson 2010). Data collected from GPS anklets not only allow parole agents to monitor violations but are also used by law enforcement agencies to investigate offenses that may occur in a spatial and temporal proximity to the parolee. Preston (2009) reported an example of how the spatiotemporal analysis of GPS data helped absolve a suspect. After an 11-year-old boy disappeared from his Florida home for 5 days, his mother used a website listing the geographic locations of registered sex offenders to locate a sex offender who lived in the vicinity of her son. After reporting the disappearance of her son and the name and location of the sex offender to law enforcement officials, a review of the GPS data established that the suspect was never in the vicinity of the boy.

3.3 Cellular Phones

In 2005, the FCC mandated that all cell phones be capable of extremely precise geolocation by 2012 to support emergency calls (Ball 2008). In an effort to meet FCC mandates, GPS technology has been integrated into cell phones (Barnes and Williamson 2009; Werbach 2007). The purpose of the FCC mandate was to enable emergency responders to determine a 911 caller's location in order to assist the caller during an emergency (Edmundson 2005). Similar to GPS units, cell phones acquire and store positional and temporal data. Unlike most GPS units, locational cell phone data is transmitted to a network (Ball 2008), allowing the data to become available without physical possession of the cell phone.

Cell phone tracking has also experienced a great deal of use for investigating and solving crime. Schmitz et al. (2013) provided four case examples utilizing cell phone records to track and monitor suspects to create a storyboard of activity.

The storyboards Schmitz created for trial helped the court in understanding events surrounding the crimes and aided in the convictions of the accused offenders. Schmitz credited cell phone tracking as a neutral tool for investigations which can provide valuable data leading to a conviction while also excluding innocent suspects. With limited resources available for investigations, the ability to exclude possible suspects permits investigators to optimize their time and resources in pursuit of the guilty party.

Similarly to GPS devices discussed previously, GPS-enabled cell phones have aided law enforcement agencies in numerous ways. Barnes and Williamson (2009) reported the safe recovery of a 9-year-old girl who was allegedly kidnapped by her grandmother. Massachusetts police officers were able to track the girl's cell phone to a motel several hundred miles away, where the grandmother was apprehended by Virginia State Police. In a separate case reported by Stacy (2008a, b), law enforcement officers tracked the missing phone of a homicide victim. The phone, along with the victim's purse, was recovered in the trash at the residence of the offender. Cell phone tracking has also been used proactively to reduce crime. According to Napier (2013), the Dayton Police Department has placed cell phones in vehicles in an effort to catch thieves breaking into vehicles. The theft of the phone automatically communicates the detection of movement to law enforcement officers, who are then able to track the movement of the phone and suspect. Utilizing the camera abilities of the phone, photographs are also taken automatically, providing prosecutors with the date and time of the theft, the movement of the bait phone, and photographic evidence of the suspect.

Despite the increasing use of cell phone tracking by law enforcement agencies, personal surveillance by the aggregation of the billions of individual cell phone users exceeds the efforts of officials (Werbach 2007). In an increasing number of cases, individuals are collecting data about what others are doing and where they are doing it. In effect, the watched have become watchers empowered by cell phone technology. The "smartphones" currently in use enable individuals to activate cell phone carrier location-based tracking options or install third-party tracking applications (e.g., Google Latitude) that incorporate and exploit the built-in GPS abilities, automatically sharing locational information with other cell phone users. Additionally, easy-to-use spyware software is widely available, enabling a third party to surreptitiously install tracking software on someone else's phone. Whether monitoring the location of family members or following the movements of friends through space and time, cell phone tracking by individuals has become a common occurrence. However, cell phone tracking has also provided a means for stalking. Southworth et al. (2005) published a report describing the increased use of technology to assist in intimate partner stalking. Uhler and Keefe (2010) reported on the ability of abusive husbands locate women's shelter after tracking their partner's cell phone to the secret locations. On rare but increasingly frequent occasions, the tracking of spousal cell phones has resulted in tragedy. Uhler and Keefe further described two incidents of murder-suicide committed after estranged husbands tracked their wives' cell phones.

3.4 Unmanned Aerial Vehicles (UAV)

The domestic use of unmanned aerial vehicles (UAV), also known as unmanned aircraft systems (UAS) and drones, has emerged as a recent controversial issue, unrivaled in effectiveness by any other police tactic (Clark 2013b, c). While public perception often associates drones with weapon of war used extensively during operations in Iraq and Afghanistan, the greatest concern is for individual privacy (Wood 2013). A poll conducted by Monmouth University Polling Institute revealed 60 % of American citizens expressed serious concern about their personal privacy if law enforcement agencies began using unmanned drones (Clark 2013c). Equipped with surveillance video and infrared cameras, a major concern is how the video collected from drones will be collected and stored. However, when used responsibly, drones have the potential to make the public safer. Law enforcement could utilize drones to keep officers and civilians safe and provide increased situational awareness during dangerous operations, such as drug busts or hostage situations (Wood 2013). Citing the worldwide recognition of drones as a valuable policing tool, Crouch and Hunt (2013) discussed the significant advantages over traditional methods employing fixed-wing aircraft during incidents and operations. Among the advantages, drones are more cost-effective, less dangerous, and rapidly deployable.

Even as many states are debating bills to curtail and prohibit drone use (Clark 2013c), privacy advocates and the public have acknowledged the usefulness of drones following two high-profile events. Subsequent to the bombing incident at the 2013 Boston Marathon, discussions of domestic drone use emerged as potential surveillance techniques for public events (Clark 2013a). Although privacy advocates and the public favored the use of drones in manhunt cases and for crime scene documentation to search for evidence, the constant surveillance use of drones during public events was debated. Despite the discomfort with drone use for monitoring public events, the use of drones by the Federal Bureau of Investigation (FBI) to monitoring a hostage situation in Alabama in 2013 received little criticism (Clark 2013c).

3.5 Automated License Plate Readers

One of the most widespread, yet least known location tracking technologies is automated license plate readers (Crump 2013). Capable of populating a massive database with the locations of millions of vehicles over extensive time periods, license plate readers have emerged as extremely powerful tracking tools. Commonly deployed by law enforcement agencies to aid in the recovery of stolen cars and license plates, locate wanted felons, monitor for "Amber Alerts," aid vehicle and traffic enforcement, and assist general investigations, automated license plate readers (commonly referred to as an ALPR in the United States) have become an increasingly widespread surveillance and intelligence tool blanketing streets and highways across the country (Roberts and Casanova 2012). ALPRs are often mounted in parking lots, including all airport parking lots following September 11, 2001 (Benzel 2012).

According to a 2012 report by the Police Executive Research Forum (2012), approximately 70 % of all law enforcement agencies utilize ALPRs. Mounted on patrol vehicles and on structures at fixed locations, an ALPR uses high-speed cameras to capture license plate numbers, date/time, and GPS location of every recorded vehicle (Roberts and Casanova 2012). Data captured from vehicles is processed instantaneously, immediately alerting law enforcement of potential matches with wanted vehicles in law enforcement databases. The process of scanning, recording, processing, matching, and alerting is done automatically (Benzel 2012). Conversely, the databases created from ALPR records can be queried by law enforcement agencies to generate a list of times and locations specific vehicles were previously captured on camera.

The deployment of ALPR technology can significantly reduce vehicle theft (Ferraresi 2008). Citing a New York Police spokesman, Baker (2011) reported vehicle theft statistics which showed significant decreases in reported thefts and significant increases in arrests and vehicle recovery. Serving the majority of suburban New Orleans, LA, the Jefferson Parish Sherriff's Office had an average recovery rate of three to four stolen vehicles per month. After deploying fixed and mobile ALPR throughout the Parish, an immediate and noticeable increase was observed, with a recovery of 23 stolen vehicles in less than a month (Hunter 2008). The Philadelphia Police Department experienced similar results with the recovery of 50 stolen vehicles in a single month after deploying ALPR (Steele 2012). Recovering 76 stolen vehicles in 2011, law enforcement agencies in Colorado Springs were also able to make 129 felony arrests, including a suspect accused of abduction and strangulation who was found within hours of stealing his victim's car. Additionally, ALPR automatically identified 14,745 other crimes (Benzel 2012). The Gwinnett County Sherriff Department in Georgia has also benefited from the use of ALPR. In one incident, a stolen license plate identified through an ALPR match resulted in the arrest of a convicted felon who was found with nine weapons and over 1,200 rounds of ammunition (Morris 2009). Deputies also routinely patrol public parks, monitoring compliance of paroled sex offenders prohibited from areas frequented by children.

3.6 Radio Frequency Identification

Radio frequency identification (RFID) technology is an enabling technology for automatic identification based on radio waves (Lieshout et al. 2007). RFID is the generic term for technologies using radio waves to automatically detect people or objects. Although the RFID market is only beginning to develop, widespread adoption of RFID is already a reality (Mackey 2011). Many people already possess RFID-tagged items or are familiar with them. Automatic tollbooth transponders are equipped with RFID as well as many newer credit (e.g., contactless touch-and-go cards) and identification cards (e.g., E-passport) and vehicle anti-theft tracking devices (e.g., LoJack®). An increasingly common practice among pet owners is to embed RFID microchips into their pet to aid in recovery in the event the pet becomes lost. Additionally, merchants commonly use RFID to monitor inventory and prevent theft (Clarke and Petrossian 2013). As RFID become more ubiquitous in society, the potential to utilize captured data for investigation and prosecution will likely increase as well.

An RFID system consists of an antenna and a microchip which is collectively termed an RFID "tag." There are two types of RFID tags: active and passive (Hunt et al. 2007; Lieshout et al. 2007; Mackey 2011). Active tags contain their own internal power source, while passive tags rely upon an external source of power to function. In general, active tags utilize batteries to boost transmission range and increase available memory or data logging abilities. Each RFID tag contains a unique identification chip, allowing individual items to be tagged and tracked when the data is transmitted through the antenna. An RFID reader is used to detect the tiny RFID tags which can be embedded into an endless range of products (Mackey 2011). The RFID reader can identify the specific tagged item, as well as when and where the read was made, and automatically add the information to a database. Although active tags generally have a longer read range (up to 300 ft), many factors affect read range, including the radio frequency used and interference from outside sources. For a complete discussion of RFID technologies, see the RFID guides by Hunt et al. (2007) or Lieshout et al. (2007).

The adoption of RFID-enabled technology by US law enforcement agencies has been slow (Hunt et al. 2007). Applications of RFID technologies can be categorized into three dimensions: improving efficiency, promoting officer safety, and as a crime-fighting tool. In addition to being much smaller and therefore more concealable, RFID tags offer two less obvious, though still considerable, advantages over other GPS- or cell phone-based tracking devices (Edmunds Inc. 2008). First, RFID does not require a clear line of sight to satellites in order to determine a position. This is significant for tracking within buildings and other features that block satellite views. Second, RFID is unaffected by cellular "dead spots," areas with poor cellular reception or completely without cellular service. Areas that may experience cellular dead spots include, for example, interior spaces of large buildings, low-lying valleys, and rural areas. Lieshout et al. (2007) discussed further advantages of RFID over barcodes, a complimentary technology, including the following six points: tags can be read rapidly in bulk to provide a simultaneous reading of all items in a particular location; RFID tags are more heat and chemical resistant; RFID tags are able to store more data; data on RFID chips can be changed; RFID can be located inside items; and RFID tags do not require human intervention for data transmission.

One area of efficiency that has benefited from RFID technology has been evidence tagging. In addition to tracking evidence, RFID can also be used to track paper files (Wethal 2009). The judicial system has strict requirements pertaining to the tracking and accountability of evidence from initial seizure to introduction at trial (Hunt et al. 2007). Successful introduction of credible evidence in court is dependent upon the police knowing exactly where it has been at all times within the chain of custody (see Sect. 2.2.5 for a discussion of chain of custody). Baber et al. (2005) demonstrated through experimentation that RFID tagging of evidence was faster and more efficient for crime scene investigators to catalog and track evidence than traditional paper-based methods involving continual handwritten notes. Bacheldor (2008) further discussed improvements in evidence location and access as well as a more thorough documentation of an items chain of custody when compared to barcode readers or paper labels. Once all evidence is tagged, tracking is performed automatically through RFID readers within each room of the crime lab facility. Personnel removing items are required to scan their identification badges; if they forget, an alarm will sound.

Focusing on safety, law enforcement agencies and correctional facilities have applied RFID technology in a variety of ways. In a study funded by the National Institute of Justice, the use of RFID to develop a reliable and effective "smart gun" was identified as feasible for police force adoption (Hunt et al. 2007). Recognizing that officers were often assaulted and killed by their own weapon, "smart gun" research was initiated as a means of only allowing a weapon to be fired by recognized and authorized users. Through the use of an RFID scanner placed inside a weapon, a "smart gun" would only unlock the trigger and enable firing if the user was tagged and authorized. The same RFID tag could conceptually be used to track the location of officers, further promoting safety. However, a pilot project examining RFID tracking was cancelled after firm resistance by officers who considered the tracking to be intrusive.

Although initial efforts to track law enforcement movement were resisted, correctional facilities were able to implement RFID tracking of guards and inmates successfully (Hunt et al. 2007; Wethal 2009). Implemented as a way to reduce expenses associated with labor-intensive tasks, RFID began to be utilized in correctional facilities in 2002. Hunt et al. (2007) detailed the application of RFID tracking in corrections. The RFID tracking system allowed continual tracking of inmates via a tamper-resistant wristwatch and tracking of guards via a belt-worn device. Capable of pinpointing the exact location of all RFID devices every two seconds, the system database records the movement of inmates. Useful for headcounts and alerting guards when inmates enter restricted areas, the database archive can also be queried after an incident to determine which inmates were involved and recreate an event. With inmates unable to deny their presence at certain incidents, the RFID systems have been responsible for a significant decline in violence. As for the guards, the system automatically alerts officials when the device is removed, when the device determines a guard is knocked down, or when the guard presses a panic button. Wethal (2009) described the implementation of a similar RFID system within six adult and three juvenile correctional facilities. The facilities similarly experienced a decline in inmate violence and increase in prison guard safety.

The crime solving and crime reduction potential of RFID technologies is only beginning to emerge (Hunt et al. 2007). One of the most promising applications is addressing property crimes. RFID technologies have successfully assisted law enforcement agencies in identifying and recovering stolen merchandise as well as deterring theft by increasing the chance of arrest and decreasing the ability of thieves to sell stolen merchandise. Furthermore, recovered RFID-tagged items have been used as evidence, resulting in convictions. For example, the LoJack® vehicle anti-theft device has demonstrated a high rate of success (over 90 %) in identifying and recovering stolen vehicles (Christiana et al. 2007; Edmunds Inc. 2008). Frequently, thieves are still in or near stolen vehicles, resulting in an apprehension in addition to recovery. In many situations, the recovery of an RFID-tagged vehicle has resulted in the recovery of additional stolen untagged vehicles when the destination has been the same (e.g., a "chop shop").

Law enforcement agencies have recognized the advantages of RFID as a cost-effective solution to deterring theft and more easily recovering stolen items. Christiana et al. (2007) cited the demonstrable importance of RFID tracking of valuable items to deter theft or identify and apprehend offenders after they have committed a crime. In Oklahoma, the Rogers County Sheriff's Office received an "excellence in problem-oriented policing" award after they utilized RFID to solve a growing theft problem (Police Executive Research Forum 2001). During the mid-1990s, trailer theft had become a serious problem for the rural community which depended on livestock and utility trailers for business, recreation, and livelihoods. Recovery of stolen trailers was very difficult since most trailers were not registered and were often repainted after they were stolen. The Sheriff's Office implemented an RFID tracking program which resulted in a reduction in the market for stolen trailers and an almost immediate elimination of trailer theft once the program was made public. Working with residents, the Sheriff's Office tagged trailers and created an ownership database enabling the monitoring and tracking of trailers. A similar project was undertaken at Ohio State University in an effort to deter bicycle theft (Johnson et al. 2008). Student support for the project resulted in hundreds of bicycles being tagged and registered, which resulted in the recovery of stolen tagged bicycles and a significant reduction in bicycle theft.

3.7 Summary

Tracking people and objects through space and time with tracking technologies has emerged as an important means of crime prevention and investigation, allowing law enforcement agencies to deter and solve crime. The effectiveness and versatility of GPS and cell phone tracking have proven to be vital tools for investigations which previously required enormous investments of time and resources. The detailed and surreptitiously collected digital data on locations and movement patterns of individuals and objects has enabled law enforcement agencies to identify and track offenders while also excluding innocent suspects.

Unmanned aerial vehicles have the potential to make the public safer, but are currently a controversial tactic. The public concern over drones has resulted in only a few cases of domestic use. Automated license plate readers, however, are actively used by a large percentage of law enforcement agencies. Mounted on vehicles and structures at fixed locations, license plate readers are among the most widespread tracking technologies. Capable of instantaneously processing data, ALPRs have recovered stolen vehicles and led to the arrest of wanted felons. The population of massive databases filled with locational information of millions of vehicles has furthermore provided law enforcement agencies with an additional tool that allows records to be stored and queried for future analyses.

Radio frequency identification is an emerging technology that is also proving to be a valuable tracking tool. Used to track evidence from a crime scene to trial, RFID is an efficient method of cataloging and tracking evidence within the chain of custody. In addition to improving efficiency, RFID has improved officer safety and assisted in crime prevention and investigation.

The spatial tracking technologies discussed in this chapter have illustrated the value for both reducing and solving crimes. From the investigative standpoint, spatial tracking technologies have generated enormous volumes of data which have allowed law enforcement agencies to solve crimes effectively and apprehend offenders. Although the use of spatial tracking technologies in litigation has only been discussed in a limited manner in this chapter, it stands to reason that the data collected have served pivotal roles in prosecution. Of course, one can also reason that the role of spatial tracking technologies in prosecution is dependent upon the technology itself. For example, a GPS data log placing an offender in spatial and temporal proximity to a crime scene would necessitate the admissibility of the data log at trial. However, an offender apprehended in a stolen vehicle through the use of ALPR or RFID may not require the admissibility of the technology since possession of the stolen vehicle establishes guilt. Nevertheless, the data captured from spatial tracking technologies are digital and should be readily available if needed during trial.

References

Baber C, Smith P, Cross J, Zasikowski D, Hunter J (2005) Wearable technology for crime scene investigation. In ISWC 2005, Osaka, Japan, 18–21 October 2005. IEEE Computer Society Press: Washington, DC, pp 138–141. doi:10.1109/ISWC.2005.58

Bacheldor B (2008) Santa Clara crime lab turns to RFID. http://www.rfidjournal.com/articles/view?4434. Accessed 24 Oct 2013

Baker A (2011) Camera scans of car plates are reshaping police inquiries. http://www.nytimes.com/2011/04/12/nyregion/12plates.html?pagewanted=all. Accessed 29 Oct 2013

Ball C (2008) GPS evidence might drive your case home. http://www.law.com/jsp/legaltechnology/PubArticleFriendlyLT.jsp?id=1202425606808. Accessed 14 Jan 2009

Barnes G, Williamson D (2009) Mass. officers use GPS to find kidnapped girl. PoliceOne.com News

Benzel L (2012) License-plate readers create 'massive intelligence database'. http://gazette.com/article/144053#ixzz25i0FgC41. Accessed 13 Oct 2013

Bowman M (2011) Baiting a bicycle thief. TechBeat, pp 1–2

Christiana J, Cirella N, Lee B, Sun J, Werner-Allen G (2007) Law enforcement response to concerns regarding RFID technology. Harvard School of Engineering and Applied Science, Cambridge, MA, p 6

Clark M (2013a) Boston bombings show future use for police drones. Digital Communities. http://www.digitalcommunities.com/articles/Boston-Bombings-Show-Future-Use-for-Police-Drones.html#. Accessed 30 May 2013

Clark M (2013b) Drone limits become law. Digital Communities. http://www.digitalcommunities.com/articles/Drone-Limits-Become-Law.html#. Accessed 13 May 2013

Clark M (2013c) States seek legal limits on domestic drones. Digital Communities. http://www.digitalcommunities.com/articles/States-Seek-Legal-Limits-on-Domestic-Drones-GT.html. Accessed 30 May 2013

Clarke RV, Petrossian G (2013) Shoplifting, 2nd edn, Problem-oriented guides for police problem-specific guides series. U.S. Department of Justice, Office of Community Oriented Policing Services, Washington, DC

Crouch B, Hunt N (2013) SA police to use unmanned drones to spy on criminals. http://www.theaustralian.com.au/news/sa-police-to-use-unmanned-drones-to-spy-on-criminals/story-e6frg6n6-1226671865697. Accessed 15 July 2013

Crump C (2013) Police documents on license plate scanners reveal mass tracking. Free future, vol. 2013. ACLU. http://www.aclu.org/blog/technology-and-liberty-national-security/police-documents-license-plate-scanners-reveal-mass. Accessed 18 July 2013

Dowdell EM (2005) You are here! Mapping the boundaries of the fourth amendment with GPS technology. Rutgers Comput Technol Law J 32:109

Edmunds Inc (2008) Evaluating stolen vehicle recovery systems. http://www.edmunds.com/car-technology/evaluating-stolen-vehicle-recovery-systems.html. Accessed 18 Oct 2013

Edmundson KE (2005) Global positioning system implants: must consumer privacy be lost in order for people to be found? Indiana Law Rev 38:207

Ferraresi M (2008) Police cameras scan plates for stolen cars. http://www.azcentral.com/news/articles/2008/08/14/20080814phx-autotheft0815.html?nclick_check=1. Accessed 17 Oct 2013

Fulton S (2013) Congress must act to stop unwarranted tracking by law enforcement. Technology and Liberty. ACLU. http://www.aclu.org/blog/technology-and-liberty/congress-must-act-stop-unwarranted-tracking-law-enforcement. Accessed 21 June 2013

Ganz J (2005) It's already public: why federal officers should not need warrants to use GPS vehicle tracking devices. J Crim Law Criminol 95(04):1325–1362

Geography and Public Safety News Briefs (2012) Law enforcement use new tagging system to track evidence. Geogr Public Saf 3(2):17

Green A (2009) More states use GPS to track abusers. http://www.nytimes.com/2009/05/09/us/09gps.html. Accessed 13 Oct 2013

Hinson Z (2008) GPS monitoring and constitutional rights. Harv Law Rev 43(1):285–288

Hunt VD, Puglia A, Puglia M (2007) RFID: a guide to radio frequency identification. Wiley, Hoboken

Hunter M (2008) License-plate cameras boost arrests. http://www.nola.com/timespic/stories/index.ssf?/base/news-6/1229581831162320.xml&coll=1. Accessed 17 Oct 2013

Johnson SD, Sidebottom A, Thorpe A (2008) Bicycle theft, vol 52, Problem-oriented guides for police problem-specific guides series. U.S. Department of Justice, Office of Community Oriented Policing Services, Washington, DC

Lieshout MV, Grossi L, Spinelli G, Helmus S, Kool L, Pennings L, Stap R, Veugen T, Van der Waaij B, Borean C (2007) RFID technologies: emerging issues, challenges and policy options. Institute for Prospective Technological Studies, Seville

Mackey DA (2011) Employing surveillance in situational crime prevention. In: Mackey DA, Levan K (eds) Crime prevention. Jones & Bartlett Publishers, Burlington, pp 209–225

Mielke PD (2009) Determining anchor points for sex offenders using GPS data. Geogr Public Saf 2(1):9–10

Moenssens AA (2006) Attaching a GPS locator system to a car: is it a "search"? http://forensic-evidence.com/site/Police/GPS_onCar.html. Accessed 29 Jan 2009

Morris M (2009) Nine guns found in Ga. man's car after alert from license plate reader. http://www.policeone.com/police-products/traffic-enforcement/license-plate-readers/articles/1849696-Nine-guns-found-in-Ga-mans-car-after-alert-from-license-plate-reader/. Accessed 17 Oct 2013

Napier JR (2013) Dayton police bait criminals with mobile technology. Digital Communities. http://www.digitalcommunities.com/articles/Dayton-Police-Bait-Criminals-with-Mobile-Technology.html#. Accessed 22 Aug 2013

NLECTC (2004) BorderTrack. [quarterly newsmagazine of the National Law Enforcement and Corrections Technology Center system]. TechBeat Fall, p 2

Pease K (2005) Science in the service of crime reduction. In: Tilley N (ed) Handbook of crime prevention and community safety. Willan Publishing, Portland, pp 171–197

Police Executive Research Forum (2001) Excellence in problem-oriented policing. Police Executive Research Forum, Washington, DC, p 65

Police Executive Research Forum (2012) How are innovations in technology transforming policing? Police Executive Research Forum, Washington, DC

Preston K (2009) Right place, right time: GPS monitoring in Pinellas County. Geogr Public Saf 2(1):3–5

Roberts DJ, Casanova M (2012) Automated license plate recognition (ALPR) systems: policy and operational guidance for law enforcement. U.S. Department of Justice, National Institute of Justice, Washington, DC

Schmitz P, Eloff C, Talmakkies R, Linnen C, Lourens R (2013) Forensic mapping in South Africa: four examples. Cartogr Geogr Info Sci 40(3):238–247. doi:10.1080/15230406.2013.800273

Southworth C, Dawson S, Fraser C, Tucker S (2005) A high-tech twist on abuse: technology, intimate partner stalking, and advocacy. In Safety net: national safe & strategic technology project of national network to end domestic violence, p 16

Stacy M (2008a) Small GPS devices help prosecutors win convictions. http://www.newsvine.com/_news/2008/08/28/1797896-small-gps-devices-help-prosecutors-win-convictions. Accessed 28 Aug 2008

Stacy M (2008b) Small GPS devices help prosecutors win convictions. PoliceOne.com News. http://www.policeone.com/police-technology/articles/1730192-Small-GPS-devices-help-prosecutors-win-convictions/. Accessed 26 Feb 2009

Steele A (2012) Philadelphia police now have 24-hour real-time surveillance center. http://articles.philly.com/2012-02-23/news/31091391_1_surveillance-network-patrol-officers-cameras. Accessed 17 Oct 2013

Uhler J, Keefe B (2010) GPS technology in cell phones enable abusers to stalk sheltered victims. http://www.abcactionnews.com/dpp/news/region_south_pinellas/st_petersburg/GPS-technology-in-cell-phones-enable-abusers-to-stalk-sheltered-victims. Accessed 16 Oct 2013

Werbach K (2007) Sensors and sensibilities. Cardozo Law Rev 28:2321–2371

Wethal T (2009) Under constant watch. [Studies/research reports]. Law Enforc Technol 36(3):22–28

Wilkinson K (2010) Parolees monitored by GPS don't get free ride at state fair. Government Technology. http://www.govtech.com/gt/articles/766738. Accessed 17 Aug 2010

Wood C (2013) The case for drones. Government Technology. http://www.govtech.com/The-Case-for-Drones.html#. Accessed 31 July 2013

Chapter 4
Spatial Technology Applications

George Roedl, Gregory A. Elmes, and Jamison Conley

Abstract This chapter presents an overview of the various spatial technologies that are utilized by law enforcement agencies to document evidence for the preservation of crime scenes which can be used for further investigations or as evidence during trial. The first technology presented is remote sensing. Four of the more common types of remote sensing are discussed: aerial photography, satellite imagery, ground-penetrating radar, and thermal imaging. The chapter continues with a discussion of geographic information systems and their application for crime mapping and analysis as well as geographic profiling. A review of the innovative uses of laser scanning technologies to document crime and accident scenes concludes the discussion.

Keywords Remote sensing • Geographic information systems (GIS) • Thermal imaging • IR technology • Forward looking infrared (FLIR) • Laser scanning • High-density laser scanning (HDLS) • Geographic profiling

4.1 Introduction

Although the use of geographic data by law enforcement agencies is not new, agencies of all types from enforcement to judiciary are increasingly using contemporary digital geospatial technologies (Peet 2012). The successful applications of geospatial technologies have been well documented in the literature (Goodchild and Janelle 2010; Milla et al. 2005; Sui 2008). Technologies, such as remote sensing and GIS, are currently well established in law enforcement agencies throughout the United States (Markowitz 2002; Ratcliffe 2004). One of the advantages of these

G. Roedl (✉) • G.A. Elmes • J. Conley
Department of Geology and Geography, West Virginia University,
330 Brooks Hall, Morgantown, WV 26506, USA
e-mail: groedl@gmail.com; greg.elmes@mail.wvu.edu; jamison.conley@mail.wvu.edu

G.A. Elmes et al. (eds.), *Forensic GIS: The Role of Geospatial Technologies for Investigating Crime and Providing Evidence*, Geotechnologies and the Environment 11,
DOI 10.1007/978-94-017-8757-4_4,

technologies is their multiuse and multiuser capability. Whether the primary function is to increase efficiency, analyze crime incidents, investigate crimes, share information, or provide evidence in court, the spatial technologies discussed in the chapter are unrivaled solutions. Remote sensing and GIS have numerous well-established roles within crime analysis and law enforcement and are generally considered credible sources of data in courts.

4.2 Remote Sensing

Remote sensing is defined as the technique of obtaining information about objects through the analysis of data collected by special instruments that are not in physical contact with the objects of investigation (Avery and Berlin 1992). The absence of contact ensures the object is not disturbed, a desirable characteristic for preserving evidence at a crime scene and for admitting remote sensing data into evidence. Remotely sensed data has been widely applied to civil and criminal investigations and frequently admitted into evidence during trial (Slonecker et al. 1998). Law enforcement agencies are increasingly applying remotely sensed imagery on a routine basis (Messina and May 2003). Within the judicial system, remote sensing data has been deemed acceptable as both scientific and demonstrative evidence (Hodge 1997; Krouse et al. 2000).

One of the initial applications developed within the remote sensing discipline was for environmental monitoring (e.g., of deforestation and desertification). Therefore, it seems reasonable that environmental forensics was among the first investigative applications of remote sensing (Brilis et al. 2000; de Leeuw et al. 2010; Morrison 2000). As an example, a specialist journal, *Environmental Forensics*, "provides a forum for scientific investigations that address contamination within the environmental media of air, water, soil and biota, and is subject to law court, arbitration, public debate, or formal argumentation" (http://www.environmentalforensics.org/journal.htm). The evolution and availability of sensors and technology have since enabled a range of applications to become common for investigations and for providing substantive and persuasive evidence during litigation. Advantageous as a visualization and analysis tool for investigators and legal practitioners (Crowsey 2002; Markowitz 2002; Grip et al. 2000), remote sensing has also empowered citizens with the capability to monitor and conduct investigations of distant places (Macauley and Brennan 1998).

Aerial photography is considered the oldest and most traditional type of remote sensing (Brilis et al. 2000; Macauley and Brennan 1998). Other types of remote sensing commonly used for investigations and litigation include (1) satellite imagery, (2) ground-penetrating radar, and (3) thermal imaging. There are several additional types of remote sensing less commonly used for forensics which will not be discussed here (e.g., sonar LiDAR and radar). The types of remote sensing discussed in this chapter have well-established methodologies and have been successfully applied to investigative and judicial applications for many years by countless practitioners.

4.2.1 Aerial Photography

Aerial photographs have been available since the nineteenth century (Grip et al. 2000). Broadly defined as any photograph taken from the air, aerial photographs are among the most available and most commonly utilized sources of remotely sensed data and can be taken from a variety of platforms, including fixed-wing aircraft, helicopters, UAVs, balloons, and even kites. Integrated with GPS, modern high-resolution digital cameras are capable of providing extremely detailed images of the surface with locational precision, which may be georeferenced to provide orthophotographs rectified to a planar surface. Aerial photographs provide unbiased, detailed information of a coverage area at the time of capture and are particularly valuable as a nonintrusive means of observation (Brilis et al. 2000). Aerial photographs are commonly used to create maps, a practice known as photogrammetry (Lillesand et al. 2008). A series of aerial photographs taken over time provide an effective and reliable means of displaying, monitoring, recording, and documenting change (Brilis et al. 2000). Among the primary advantages of aerial photographs, compared to other types of remote sensing, are flexibility and the spatial proximity to the area being photographed. Various platforms can be used to either increase or decrease distance from the observed surface. An increase in distance permits photography of a wider field of view, while a decreased distance provides higher spatial resolution.

Aerial photographs are particularly valuable for investigations when private property owners do not permit intrusive observations (Brilis et al. 2000). According to the US Supreme Court decision in California v. Ciraolo (1986), anything capable of being viewed from a public space is not protected by privacy laws in the United States. Therefore, aerial observations and photographs are legally permitted to monitor and document areas considered private property and may be presented as evidence during trial. Although California v. Ciraolo has established legal precedent, it also demonstrates a particular application of aerial photographs by law enforcement officers to investigate and apprehend a suspected marijuana grower. An airplane was used to fly over the suspect's property and take aerial photographs which conclusively showed the presence of marijuana. Typically, aerial photographs are used to confirm or deny the presence of an activity or operation (Grip et al. 2000). On a much larger scale, the US Department of Agriculture (USDA) actively takes aerial photographs of farm property across the United States in an effort to reduce crop fraud (Charles 2006; Marshall 2013). Intended to monitor farmers receiving government subsidies, the USDA program helps maintain current records while also catching violators growing crops or altering the land (e.g., draining wetlands or plowing native prairie) when they should be engaged in conservation activities (e.g., reducing soil erosion).

The potential applications of aerial photography by law enforcement agencies are numerous (Harries 1999). They have been used in a variety of litigations over the last several decades (Slonecker et al. 1998). Aerial photographs have proven to be one of the most valuable tools available to an investigator for documenting evidence and illustrating the layout of a certain crime scenes, such as large outdoor scenes or vehicle accident scenes (Langer and Dunaski 2007). Within such scenes, aerial photographs can provide a different perspective while more accurately recording facts and conditions. For example, France et al. (1992) determined that aerial photography was ideally suited for the detection and delineation of grave sites based on surrounding environmental factors (e.g., changes in vegetation patterns).

The integration of aerial photographs within a GIS provides greater flexibility for forensic analysis, for the management of information, and for the presentation of facts in an easily understood manner (Brilis et al. 2000; Grip et al. 2000). Additional spatial data relevant to the needs of the investigation, such as zoning codes, injury locations, or offender/victim residences, can be superimposed on the photographs (Harries 1999). Messina and May (2003) provided a case study in which detectives used aerial photographs to record the events of a carjacking as a visualization tool for the jury during trial. The creation of a timeline of events helped the prosecution re-create the scene and was a key tool in proving the sequence of events occurred the way law enforcement officers stated, not how the offender alleged. In short, aerial photographs are a powerful tool capable of being utilized in a variety of applications; moreover, when integrated with GIS and additional spatial data, aerial photographs become an even more invaluable investigative and forensic tool.

4.2.2 Satellite Imagery

Satellite imagery is an alternative to aerial photography, capable of collecting large amounts of spatial data in a relatively small amount of time. Satellite imagery has been used as an alternative to aerial photographs to provide nonintrusive, unbiased, and reliable data useful for observations during an investigation or documentation of evidence for litigation. Relative to aerial photographs, satellite imagery can capture larger areas using fewer images. Larger spatial coverage within fewer images makes satellite imagery a faster and more efficient alternative. However, the major advantage of satellite imagery is the fact that satellites can capture imagery from anywhere on the planet. Many modern satellite sensors are directable, capable of providing specific images with a very high level of spatial resolution (detail) and at greater frequencies (temporal resolutions). Additionally, archived data from satellite imagery dating back several decades can be obtained if needed to analyze and document long-term changes.

While one component of the USDA crop fraud program discussed previously utilizes aerial photographs to identify violators receiving subsidies but not complying with requirements, the USDA crop insurance program relies upon satellite imagery to verify crop loss (e.g., due to disease or weather-related events) and investigate suspected fraudulent insurance claims (Marshall 2013; Rocchio 2006). Satellite imagery is used to determine if the imagery of the ground matches the farmers claim. Farmers who file false claims have had to reimburse payments, pay fines, and have become ineligible for future benefits (USA v. Torlai 2013). Covering over 7.6 million acres, improper payments continue to be made despite current efforts.

However, while crop fraud has not been eliminated, it has decreased. The arguable success of the claims monitoring has resulted in fewer claims, a drop in the number of fraud cases, and even admissions of guilt by farmers who received prior settlements but who were not investigated.

An innovative and controversial application of satellite imagery was implemented by town officials in Riverside, NY. Using Google Earth, a free Internet-based application, code officials identified swimming pools that were installed without permits and without the safety inspections which accompany the permit process (Nichols 2010). Identifying 250 illegally installed swimming pools, the town was able to recover $75,000 in permit fees. Because of public concern for privacy, the town has subsequently abandoned the Google Earth method and currently tracks violations through the use of its own GIS database. Craig (2007) provides a thorough review of online imagery providers and associated controversies.

4.2.3 Ground-Penetrating Radar

While the most common application of ground-penetrating radar (GPR) is to detect buried remains of homicide victims, GPR has been used successfully by law enforcement agencies to locate a variety of forensic evidence (Schultz 2007, 2012). GPR effectiveness is based on the active and continuous transmission of high-frequency electromagnetic signals into the ground. The signals experience electrical impedance which can result either in attenuation or reflection. Based on the impedance differences, GPR creates a subsurface profile which can be interpreted by the operator. Images of buried objects are not displayed, but the presence of an anomaly within the subsurface profile indicates prior disturbance (see Schultz 2012 for a complete review of GPR technology). GPR is a nondestructive, nonintrusive means of locating buried objects. Forensic evidence can be located under concrete surfaces without the need to damage the surface in the search, and evidence can be found without being damaged in the process. However, GPR is an expensive technology that requires an experienced operator with specialized training; consequently, GPR is not a tool commonly possessed by law enforcement agencies, but is often secured through other parties when needed. When utilized appropriately, GPR can save time and expense compared with traditional search methods.

Given the nature of GPR, applications are almost entirely confined to the search for buried evidence. Davenport (2001) discussed a case in which GPR was used to locate a body of a missing woman that was buried beneath a concrete slab nearly three decades prior to the search. The GPR survey identified five anomalies, which were likely caused by plumbing. After prioritizing the anomalies, the body was found, and a subsequent autopsy revealed the cause of death was strangulation. Schultz (2007) similarly described a GPR search for a missing body presumed to be buried beneath a concrete slab. In this case, there were several concrete slabs that needed to be surveyed. The results of the GPR survey eliminated slabs without anomalies, which allowed law enforcement to focus on the remaining slab that contained the missing body.

4.2.4 Thermal Imaging

Thermal imaging is also known as thermography. Thermal imaging technologies create an image using temperature (Asano et al. 2012; Laba 1996). More precisely, thermal imaging devices detect and produce images of radiation in the infrared range of the electromagnetic spectrum. Since thermal imaging technology detects infrared radiation (IR), they are also commonly referred to as IR systems. A digital camera capable of capturing IR may be called an IR camera, and a digital video likewise may be called video IR. The images produced are visual displays of the amount of infrared energy emitted, transmitted, and reflected by an object. Developed initially for military applications, thermal imaging devices are capable of sensing small temperature differences and are commonly used during limited visibility (e.g., at night or in fog) to detect objects. Applications of thermal imaging fall into three main categories: military, law enforcement, and commercial (Davenport 2001). Within law enforcement agencies, IR devices are often mounted to aircraft or vehicles and are referred to as forward looking infrared (FLIR). Data from FLIR is continuously recorded in digital format and often integrated with GPS to provide locational information. As with each of the previously discussed types of remote sensing, thermal imaging is nonintrusive (arguably: see Laba 1996), nondestructive, and a reliable means of observation.

The more common law enforcement applications of thermal imaging technologies are for night surveillance operations; security operations; search and rescue operations; the location of fleeing suspects (individuals or vehicles), children, and escaped convicts; and primarily to locate indoor drug laboratories and marijuana-growing operations (Davenport 2001; Laba 1996; Schreiber 2009; Schultz 2008). Indoor marijuana-growing operations require powerful heat lamps for the marijuana plants to flourish. As a result, structures for indoor marijuana cultivation are significantly warmer (Lawyers USA 2009). Law enforcement relies on this detection of thermal difference for obtaining a search warrant of the property (U.S. v. Kattaria 2009). However, it should be noted that a search warrant is currently required to use thermal imaging devices to detect heat emanating from a private residence (Kyllo v. United States 2001). Schultz (2008) provided a list of supporting information that can assist in obtaining a warrant to use thermal imaging, including prior history of cultivation by an individual, information on supplies purchased to support cultivation, records of utility use, and the detection of marijuana odor in the vicinity of the residence.

Cited as a force multiplier, the advantages of thermal imaging devices are priceless (Schreiber 2009). They have the potential to save lives, find lost children, minimize officer danger, and provide efficiency while simultaneously reducing expenses. Law enforcement agencies are constantly seeking new and innovative ways to utilize thermal imaging devices; however, the use of thermal imaging technology is limited by two significant factors: it requires specialized and experienced operators and it is affected by prevailing environmental conditions – especially wind (Davenport 2001; France et al. 1992). Two previous limitations of cost and quality (ability to detect temperature variations) have become less significant as thermal imaging technology has advanced and has become more accessible to law enforcement agencies and the general public.

4.3 Geographic Information Systems

Geographic information systems (GIS) have already enjoyed a long and rich history in civil and criminal investigations, facilitating a diverse range of applications (Dischinger and Wallace 2005). Chamberlayne (2001) considered GIS an essential tool to the mainstream office functions of a crime analyst. A non-exhaustive list of investigative applications includes environmental protection, civil tort lawsuits, forensic engineering, crime scene documentation, arson investigations, forensic research applications, and record investigations. The following section will illustrate the role of a forensic GIS in these roles.

GIS has experienced considerable use in the field of environmental forensics. Describing GIS as software growing in popularity in the courtroom, Reis (2007) also identified the frequent use of GIS in environmental practices. GIS serves as a presentation medium which assists judges, juries, and litigators in understanding multifaceted environmental relationships presented during trial. The reason, as Markowitz (2002) explained, is that GIS provides a powerful tool for visualizing and resolving complex environmental and legal problems. While establishing environmental crime prosecution guidelines, Suggs et al. (2002) drew attention to the difficulty in following the course of an environmental crime, which often involves a series of different, independent activities by numerous people at different times. Aschenback (1991) described GIS as enhancing litigation through the availability of new information capable of providing answers to previously unanswerable questions. Furthermore, Grip et al. (2000) noted that, although historical aerial photographs have proven to be powerful tools in environmental forensic investigations, they become even more powerful when integrated in a GIS.

Citizens involved in the detection and enforcement of environmental laws have also turned to GIS (Thompson 2000). Many citizen monitoring groups have established sophisticated monitoring systems that rival and often exceed the capabilities of public enforcement agencies; for example, the Virginia Water Monitoring Council (VWMC) is developing a GIS application in partnership with the State of Virginia Department of Environmental Quality to show where citizen groups, private industry, government agencies, and other organizations are monitoring for water quality. Environmental data is collected and analyzed by these and other groups to ensure compliance with federal regulations pertaining to air and water quality, wildlife protection, and habitat degradation. Macauley and Brennan (1998) acknowledged that technologies that make data available to citizens can facilitate more efficient enforcement of environmental regulations. Environmental justice issues have also benefited from the use of GIS in litigation. Lewis (2006) recommended identifying and analytically mapping minority population distributions whenever environmental justice concerns are suspected to exist. GIS maps demonstrating high densities of minority residents within close proximity to a new or existing environmental hazard may demonstrate the potential for environmental discrimination, although intent to discriminate, would still need to be established, a difficult task.

GIS modeling of incidents, such as plume and flow models of pollutants, is a common type of evidence produced with GIS. Referring to the ability to identifying exact locations of people, property, and substances in relation to releases or spills under litigation, Wilson et al. (1997) found GIS models extremely effective in mass torts and class actions. Furthermore, superimposing modeled data on an aerial photograph or satellite image promotes realism, aiding in effectiveness as well as conveying a sense of authenticity. GIS further assists in class-action lawsuits by providing scientific boundaries of affected individuals. In the case of Gwendolyn Guillory, et al. v. Union Pacific Railroad (2005), GIS and an expert witness sufficiently defined the geographic boundaries in a mass tort case based on a spatial analysis of individuals affected by a chemical spill near a rail yard.

Additionally, GIS has been utilized successfully for forensic investigations of tragic events. When the TWA flight 800 crashed into the Atlantic Ocean near New York in 1996, the National Transportation Safety Board (NTSB 1996) used GIS to aid in creating boundaries within which to search for victims and to correlate injuries with other physical evidence to assist in identifying the cause of the accident. Investigators matched the location of recovered bodies to seating charts in an effort to identify patterns which would enable them to locate missing victims. The same seating charts were used to determine if physical injuries sustained by the victims were consistent with theories pertaining to the possible cause of the crash, namely, structural failure, explosive detonation/missile impact, and fuel tank explosion. While additional evidence ruled out a structural failure and an explosive detonation, a fuel tank explosion was eventually deemed to be the most probable, but non-conclusive, cause of the disaster, which was substantiated by the GIS maps used in the investigation.

GIS was also used in the reconstruction of the Space Shuttle "Columbia" disaster in 2003 (Brown et al. 2003). Based on the GPS location of found debris, GIS was employed to predict the location of other undiscovered debris. Of particular concern was the location of toxic debris, which posed significant health and environmental concerns, and of human remains of the crew. The GIS-produced maps were distributed to searchers with indications of high probability debris areas and of locations already searched; the final debris-field maps aided in reconstructing the sequence of events which occurred during the disaster. Along with the TWA disaster, determining the most likely sequence of events has contributed to new measures for preventing future disasters. In a more recent example, Graettinger et al. (2012) documented their efforts to develop a GIS analysis of tornado damage in Tuscaloosa, Alabama, which destroyed 4,700 homes while damaging an additional 7,000 homes. The results helped capture the overall distribution of damage and were critical to analyzing damage and obtaining knowledge of potential ways to prevent future devastations, while at the same time providing evidence for claims.

Forensic investigations depend on a well-documented crime scene for continued investigation (Platt 2006). GIS and other spatially enabled technologies have been explored as additional ways to capture and record crime scene information. For example, hikers at Lake Powell in Glen Canyon National Recreation Area discovered human remains, which authorities were able to identify as a man who reportedly drowned 20 years earlier. However, foul play was never ruled out. Faced with rising lake levels that threatened to submerge the area where the remains were found, Park Service officials documented the 3-acre scene with a GPS-enabled digital camera (Corbley 2008). Georeferenced digital photos were then uploaded into a GIS for further examination by the coroner, who was able to ascertain that there was no evident trauma to the body remains. In this particular investigation, limited time prevented more detailed examination of evidence that was scattered over a large area. Park officials credited GPS photomapping and GIS as extremely useful tools for preserving the scene and spatially piecing together clues in an investigation. The GIS map permitted additional investigative analysis long after rising water levels consumed the investigation scene.

Wilson (2003) demonstrated the usefulness of GIS as an investigative analysis tool not only within a single crime scene, but between multiple scenes, citing the high-profile Washington, DC, sniper case as an example. To illustrate the potential of GIS within and between crime scenes, Smith (2001) described the application of GIS in Kosovo. Initially, the GIS was established to identify and inventory damaged buildings, which provided evidence of the homes of ethnic Albanians being targeted. Subsequently, this same GIS was used to map the known locations of landmines, and it was eventually used in the forensic investigation of sites of alleged atrocities.

Using GIS to examine records for patterns and anomalies has proven highly useful as starting points for further in-depth investigations. Frequently, crimes are recorded in official records, but yet go completely unnoticed. In one case study, Morrow-Jones et al. (2005) used GIS to identify potential house-flipping schemes and illicit over-appraisals of property. Citing these practices as money laundering methods, the analysis was also intended to protect unsuspecting buyers from purchasing overvalued property, which might result in foreclosure and neighborhood blight. By applying structured queries of more than 13,000 real estate transaction records, they identified 110 suspicious transactions.

With an estimated $25 million loss to food stamp fraud each year, Louisiana's Department of Social Services used GIS to detect food stamp fraud by investigating anomalous data (ESRI 2004; Schwart 2005; Walsh 2004). Investigators cited the ability of GIS to plot suspect transactions on a map as a way to discern potential fraud in less than a day, in contrast to a 2-week process without GIS. By identifying fraudulent activities, such as high-volume transactions of whole dollar amounts from retailers (an indication of retailers buying food stamps from recipients at reduced value and redeeming them for full value), investigators were able to identify and prosecute fraud cases. The system developed for Louisiana was subsequently deployed in Mississippi with a similar rate of success in detecting food stamp fraud (Raths 2007). The utility of GIS to detect fraud has also attracted the attention of the US Department of Justice. With an estimated $47 billion worth of Medicare fraud occurring every year, the government's attempt to prevent Medicare fraud is becoming more reliant on GIS and geographic data (Westly 2010).

As an emerging technology within the field of forensic investigations, GIS is currently being explored for a variety of potential uses as researchers seek to apply GIS to new domains of knowledge. In one such research avenue, Manheim et al. (2006) examined the utility of GIS as a forensic taphonomy tool for predicting the locations of dumped bodies and understanding the dispersal pattern of scattered remains. Although no dispersal pattern was identified, the ability to predict probable locations of remains represents an important research agenda for forensic investigations. Along a different line of research, Burdette et al. (2007) applied GIS to create maps and a database of marine mammal habitats and fishing gear to identify mammals caught and injured/killed by fisheries. Rope marks were taken from photographs and compared to specific fishery gear to narrow down the list of possible offenders, as different fisheries and fishing methods utilized different types of gear. Morton et al. (2005) designed a poster presentation demonstrating the ways in which the Virginia Department of Game and Inland Fisheries (VDGIF) used geospatial technology to support wildlife law enforcement activities. Realizing the difficulties in mapping incidents occurring in wilderness areas, game wardens were assigned GPS units to routinely record coordinates which were added to a GIS. Authors cited one example incident in which game wardens recorded the location of a hunter suspected of hunting within 500 ft of a duck blind. GIS was used to create a 500 ft buffer around the duck blind to show conclusively the hunter was in violation of the law. VDGIF credited geospatial technologies as being powerful tools for wildlife law enforcement and for communicating details to administrators and within court proceedings.

4.3.1 Crime Mapping and Analysis

Crime mapping and analysis has experienced tremendous growth over the last decade with law enforcement agencies using it to aid intelligence, support criminal investigations, reduce and prevent crime, and increase performance and efficiency (Leitner 2013; Chainey and Tompson 2008). Crime mapping and analysis is not a substitute for, but conversely is complementary to, other forms of crime analysis (Boba 2005). Crime mapping assists crime analysts in analyzing and presenting information (Bruce 2001). Chainey and Ratcliffe (2005) described crime mapping and analysis as a combination of practical criminal justice issues with GIS. Boba (2005: 37) defined crime mapping as "the process of using a geographic information system to conduct spatial analysis of crime problems and other police-related issues." Crime mapping and analysis explores the geography of crime through the identification of patterns and trends in crime data. Theories useful to examining spatial relationships between crime, offenders, victims, and the environment are offered through a subset of mainstream criminology theories known as environmental criminology (see Boba 2005: 59–71; Brantingham and Brantingham 1981; Chainey and Ratcliffe 2005: 79–113; Wortley and Mazerolle 2008). There is currently an abundant and expanding source of literature documenting the successful applications of crime mapping and analysis within the criminology field (Wolff and Asche 2009). For example, Chainey and Tompson (2008) and Leitner (2013) edited collections of case studies, the National Institute of Justice hosts a multiday crime mapping and analysis conference and publishes an online bulletin, and specialized journals (e.g., *Crime Mapping: A Journal of Research and Practice*) provide peer-reviewed articles.

As Bruce (2001) stated, mapping data indicates who, what, when, and where, but analysis helps determine why and what it all means. One of the leading methods of analyzing crime patterns and trends is by identifying clusters (Tompson and Townsley 2010). Also known as hot spots, clusters highlight geographic areas in which crime occurs at a rate higher than average within close spatial or temporal proximity, relative to the distribution of crime across the entire area of interest. Analysts use a variety of methods (e.g., density mapping, spatial distribution statistics, nearest neighbor clustering) to identify hot spots (Boba 2005). For further information, Eck et al. (2005) provided a detailed overview of hot spot analysis methods. The identification of crime hot spots allows law enforcement to focus resources to areas most needed to reduce offenses and is considered a very effective means of preventing crime and reducing disorder (Braga 2007, 2008; Caplan et al. 2011). The use of hot spot mapping to predict future crime locations is common in the crime mapping field (Chainey and Ratcliffe 2005). Tompson and Townsley (2010) demonstrated that temporally sensitive hot spot maps were highly effective for predicting when and where future offenses were likely to occur. Additional research (Bowers and Johnson 2005; Johnson et al. 2005; Lockwood 2012; Rey et al. 2012; Townsley et al. 2000; Youstin et al. 2011) on repeat victimization (locations with repeat offenses occurring) also demonstrates the effectiveness of crime forecasting.

4.3.2 Geographic Profiling

An increasingly common investigative approach and one of the more specialized methods of crime mapping and analysis is geographic profiling (Canter and Hammond 2006, 2007). Warren et al. (1998) argue that according to Zipf's principle of least effort (also known as Zipf's Law) (1949), offenders typically travel the shortest possible distance from their anchor points to their criminal task. This principle generally holds true for most individuals and is commonly applied to investigations of serial crimes (Canter 2009). The approach integrates theory from criminal investigative analysis, environmental criminology, ethnographic geography, journey to crime research, and criminal geographic targeting to aid law enforcement in investigating and solving difficult serial crimes. Environmental criminology provides a general framework for addressing questions related to offender spatial behavior, and crime pattern theory suggests a specific method for determining probable area of criminal residence. Geographic profiling is an example of the practical application of criminological theory to the real world of police investigation. "Some of our [offender profiling] hypotheses … seem now to have passed into the general realm of established detective knowledge … It is this gradual building of elements of certainty by scientific rigour that is the object of the researchers" (Copson 1993, pp. 20–21).

Geographic profiling is a decision support tool for criminal inquiries. It does not solve cases; rather, it focuses on an investigation by providing both an optimal search strategy and a means of managing large volumes of information (Block and Bernasco 2009). Crime locations and their patterns provide clues that, when properly interpreted, can be used to help find the offender. Young (2004) reported a successful application of geographic profiling which was used in Raleigh, NC. The analysis predicted the residential location of a serial rapist/murder within a single block.

Linking unsolved serial crimes to a single or multiple offenders is particularly relevant to police investigations (Bennell and Jones 2005; Bennell et al. 2009). Like all police tactics, geographic profiling reaches its potential when employed as part of a package of techniques. Because address information is so commonly available, a number of different strategies have been developed that can be integrated with other investigative approaches, which results in the more effective and efficient use of limited police resources (Rossmo 2000).

Taylor (1997: 5) suggested that "The technique [of geographic profiling] holds the strongest promise for assisting police in investigations of serial murderers, serial rapists, and serial arsonists." Its full potential, however, results from its integration with other behavioral science techniques, forensic analysis, eyewitness reports, and investigative information (Rossmo 2000). While geographic profiling is primarily an investigative tool, it also has a role in the courtroom. In addition to analyzing the geographic patterns of unsolved crimes for investigative insights, the spatial relationship between the locations of a crime series and an accused offender's activity sites can be assessed in terms of the probability of their congruence (Rossmo 2006; Rossmo et al. 2005). When combined with other forensic identification findings (e.g., a DNA profile), such information increases evidential strength and likelihood of guilt. The question of how to quantify most appropriately the weight of forensic identification evidence and rare trait possession is called the generalized island problem (Balding and Donnelly 1994). Geographic profiles can also be used as supporting grounds for search warrant affidavits (Rossmo 2000).

4.4 3D Laser Scanning

3D laser scanning provides police, the FBI, and criminal defense teams with the ability to rapidly document a crime scene and reconstruct the crime scene for additional investigative purposes or for presentation to a jury during trial (Agosto et al. 2008; Headrick 2013). 3D laser scanning technology provides more useful data and is considered more reliable and accurate than traditional methods (e.g., sketches and photographs). Through the use of a laser scanner, investigators are able to rapidly document a crime scene or accident site with extreme detail and precision. Laser scanners are not constrained by day or night (Foster 2004). Newer high-definition laser scanners (HDLS) are able to capture objects at the crime scene in full color with high resolution (Dabrowski 2010; DeLaurentis 2009; Frei et al. 2004). The results of the scan may be used to generate an interactive 3-dimensional display within which additional evidence, such as photographs and witness statements, can be embedded to provide immediate access with the click of a button. The investigation and any analyses or updates can be performed using a computer, which reduces time and expense (Fries 2006). Furthermore, investigators are capable of moving within the virtual crime scene, unlike the actual physical scene which may need to be preserved to avoid evidence being contaminated or is otherwise inaccessible. Investigators can change perspectives, explore for additional evidence, determine whether a witness could actually see what they claim to have seen and show a jury what a victim may or may not have seen (Foster 2004, 2005). While 3D laser scanning is not a new technology, it has become more available and more affordable for widespread use (DeLaurentis 2009). 3D reconstructions are frequently used for investigations today (Agosto et al. 2008; Headrick 2013).

A variety of investigations have benefited from 3D laser scanning. Cited as the most abundant form of evidence at a crime scene, DeLaurentis (2009) discussed the use of 3D laser scanning to capture shoeprint impressions. Agosto et al. (2008) reconstructed a crime scene to determine suspect's possible walking paths based on blood patterns. Cramilt and Grissim (2010) provided several case examples, including the reconstruction of a helicopter crash and the documentation of an airplane crash into the Austin Internal Revenue Service building in Texas. Craig Fries, CEO of Precision Simulations, Inc., provided several case studies demonstrating the effectiveness of 3D laser scanning for investigating and reconstructing events. Fries (2007) presented a case study in which a police officer fatally shot a suspect driving a stolen vehicle in a high speed pursuit (see also PSI 2012b). Family members of the suspect sued the police department, alleging excessive use of force. The scene of the event was captured with a laser scanner, the physical evidence was examined, and an accurate time-synchronized animated re-creation of the events conclusively proved that the use of deadly force was reasonable given the circumstances. Fries (2006) discussed another instance in which 3D laser scanning was utilized in an innovative way to settle a construction accident case alleging negligence (see also PSI 2012c). 3D reconstructions provided evidence of what the accident victim could see and what the operator of a front loader could see. The evidence resulted in the defendants of the case saving over $2.3 million. Another accident involving a pedestrian being struck by a municipal bus determined that the bus driver could not see the pedestrian, but that the pedestrian could see the bus (PSI 2012d). In a reconstruction of a shooting that killed a deputy sheriff, 3D laser scanning documented the crime scene and analyzed bullet trajectories to determine if the shooting was intentional and premeditated (PSI 2012a). And finally, the reconstruction of an accident scene helped to determine which driver was at fault (PSI 2012e).

4.5 Summary

This chapter has provided an overview of three spatial technologies that are used by law enforcement agencies for investigation and provide evidence. Aerial photography is the most widely used type of remotely sensed data for forensics. Used as a source for the creation of base maps, aerial photographs are effective and reliable means of displaying, monitoring, recording, and documenting change. Satellite imagery is an alternative to aerial photographs that also provides an unbiased, nonintrusive means of observation. Ground-penetrating radar offers an effective solution to locating buried objects more efficiently than traditional search methods. While thermal imaging is most often used to detect indoor marijuana operations, it can be used for a large variety of applications. Geographic information systems are essential tools to assist crime analysts and increasing for presenting evidence. One of the more fundamental applications of geographic information systems is for crime mapping and analysis, which has been proven to increase performance and efficiency while helping to reduce crime and support criminal investigations. One highly specialized area of crime mapping and analysis is geographic profiling, an investigative tool which assists in solving difficult serial crimes. 3D laser scanning allows investigators to document a crime scene or accident site rapidly and accurately with extreme detail and precision and digitally reconstruct the scene. While law enforcement agencies are able to benefit from any of the spatial technologies discussed within this chapter, the use of multiple technologies offers tremendous potential for increasing efficiency, investigating and solving crimes, and providing compelling evidence in court.

References

Agosto E, Ajmar A, Boccardo P, Tonolo FG, Lingua A (2008) Crime scene reconstruction using a fully geomatic approach. Sensors 8(10):6280–6302. doi:10.3390/s8106280

Asano CH, Rogers T, McGuire G, Juck D, Jacobs J (2012) Application of infrared photography for the detection of hidden evidence at the crime scene: passive and active thermography techniques through drywall. [Studies/research reports]. Identif Can 35(2):40–55

Aschenbach RJ (1991) Geographic information systems as a decision making tool. Ohio State Law J 52:351–368

Avery T, Berlin G (1992) Fundamentals of remote sensing and air photo interpretation. Prentice Hall, Upper Saddle River

Balding DJ, Donnelly P (1994) Inference in forensic identification. J R Stat Soc 157(3):1–20

Bennell C, Jones NJ (2005) Between a ROC and a hard place: a method for linking serial burglaries by *modus operandi*. J Invest Psychol Offender Profiling 2(1):23–41

Bennell C, Jones NJ, Melnyk T (2009) Addressing problems with traditional crime linking methods using receiver operating characteristic analysis [Theoretical research]. Legal Criminol Psychol 14(2):293–310

Block R, Bernasco W (2009) Finding a serial burglar's home using distance decay and conditional origin-destination patterns: a test of empirical Bayes journey-to-crime estimation in The Hague. J Invest Psychol Offender Profiling 6(3):187–211

Boba R (2005) Crime analysis and crime mapping, 1st edn. Sage Publications, Thousand Oaks, 301pp
Bowers KJ, Johnson SD (2005) Domestic burglary repeats and space-time clusters. Eur J Criminol 2(1):67–92. doi:10.1177/1477370805048631
Braga AA (2007) Effects of hot spots policing on crime. A campbell collaboration systematic review, available at http://www.aic.gov.au/campbellcj/reviews/titles.html
Braga AA (2008) Crime prevention research review no. 2: police enforcement strategies to prevent crime in hot spot areas. U.S. Department of Justice Office of Community Oriented Policing Services, Washington, DC, p 36
Brantingham P, Brantingham P (1981) Environmental criminology. Sage, Thousand Oaks
Brilis GM, Gerlach CL, van Waasbergen RJ (2000) Remote sensing tools assist in environmental forensics. Part I: traditional methods. J Environ Forensics 1:63–67. doi:10.1006/enfo.2000.0009
Brown SC, Crum S, Foote VS (2003) GIS and GPS emergency response lessons learned from the space shuttle Columbia disaster. J Ext 41(4). http://www.joe.org/joe/2003august/iw1.php. Accessed 1 Apr 2011
Bruce CW (2001) A thousand words for a picture. Is the overvaluation of GIS disrupting a critical balance in crime analysis? Massachusetts Association of Crime Analysts, Crime Analysts' Round Table
Burdett LG, Adams JD, McFee WE (2007) The use of geographic information systems as a forensic tool to investigate sources of marine mammal entanglement in fisheries. J Forensic Sci 52(4):904–908. doi:10.1111/j.1556-4029.2007.00466.x
California v. Ciraolo (1986). 476 U.S. 207
Canter D (2009) Developments in geographical offender profiling: commentary on Bayesian journey-to-crime modelling. J Invest Psychol Offender Profiling 6(3):161–166
Canter D, Hammond L (2006) A comparison of the efficacy of different decay functions in geographical profiling for a sample of US serial killers. J Invest Psychol Offender Profiling 3(2):91–103
Canter D, Hammond L (2007) Prioritizing burglars: comparing the effectiveness of geographical profiling methods. [Article]. Police Pract Res 8(4):371–384. doi:10.1080/15614260701615086
Caplan J, Kennedy L, Miller J (2011) Risk terrain modeling: brokering criminological theory and GIS methods for crime forecasting [Applied research]. Justice Q 28(2):360–381. doi:10.1080/07418825.2010.486037
Chainey S, Ratcliffe J (2005) GIS and crime mapping, 1st edn. Wiley, Hoboken
Chainey S, Tompson L (2008) Crime mapping case studies: practice and research. Wiley, Chichester
Chamberlayne D (2001) Functions of crime analysis and the roles crime analysts play. http://www.macrimeanalysts.com/rt/f1dc1.htm
Charles D (2006) Crop cops take to the sky. NPR Morning Edition. http://www.npr.org/templates/story/story.php?storyId=5710986. Accessed 24 Sept 2008
Copson G (1993) Offender profiling. Paper presented at the presentation to the Association of Chief Police Officers Crime Sub-Committee on Offender Profiling, London, England, May 1993
Corbley K (2008) GPS photo-mapping in law enforcement. Law Enforcement Technology, January. http://www.officer.com/print/Law-Enforcement-Technology/GPS-photo-mapping-in-law-enforcement/1$40229. Accessed 4 Feb 2009
Craig BJ (2007) Online satellite and aerial images: issues and analysis [Minnesota legal studies research paper no. 08–11]. North Dakota Law Rev 83:547–578
Cramblitt B, Grissim T (2010) 3D scanning: visualizing scenes in another dimension. Evid Technol Mag 8(5):18–21
Crowsey R (2002) A legal assistant's guide to legal applications of geospatial information. http://www.crowsey.com/pdf/spatialInformation.pdf. Accessed 30 Sept 2008
Dabrowski J (2010) High definition laser scanning takes forensic geosciences to a new level. http://www.forensicmag.com/articles/2010/06/high-definition-laser-scanning-takes-forensic-geosciences-new-level. Accessed 31 Oct 2013
Davenport GC (2001) Archaeologists as forensic investigators: defining the role. Hist Archaeol 35(1):87–100

de Leeuw J, Georgiadou Y, Kerle N, de Gier A, Inoue Y, Ferwerda J et al (2010) The function of remote sensing in support of environmental policy. Remote Sens 2010(2):1731–1750. doi:10.3390/rs2071731

DeLaurentis P (2009) Three D scanning: a new tool for cracking tough cases. Forensic Mag 6(1):37–40

Dischinger SS, Wallace LA (2005) Geographic information systems: coming to a courtroom near you. Colo Lawyer 34(4):11–23

Eck JE, Chainey S, Cameron JG, Leitner M, Wilson RE (2005) Mapping crime: understanding hot spots. Office of Justice Programs, Partnerships for Safer Communities. US Department of Justice: Washington, DC, p 72

ESRI (2004) Taking a bite out of illegal food stamp traffic. ArcUser, July–September 2004

Foster REL, LAPD (ret.), MPA (2004) High definition surveying gives law enforcement a whole new picture of a crime scene. http://www.hitechcj.com/id193.html. Accessed 12 Sept 2005

Foster REL, LAPD (ret.), MPA (2005) Crime scene investigation. [e-magazine]. Government Technology

France D, Griffin T, Swanburg J, Lindemann J, Davenport G, Trammell V et al (1992) A multidisciplinary approach to the detection of clandestine graves. J Forensic Sci 37(6):1445–1458. doi:10.1520/JFS13337J

Frei E, Kung J, Bukowski R (2004) High-definition surveying (HDS): a new era in reality capture. In NATSCAN, 2004, International archives of photogrammetry remote sensing and spatial information science. Leica Geosystems HDS, vol. XXXVI, 8 W2 ed, pp 262–271

Fries C (2006) New tools for reconstruction. http://www.precisionsim.com/files/1513/2285/6604/NewToolsClaimsArticleLetterSize.pdf. Accessed 21 Feb 2013

Fries C (2007) New tools for crime scene reconstruction. Prosec Brief 28(4):28–30

Goodchild MF, Janelle DG (2010) Toward critical spatial thinking in the social sciences and humanities. GeoJournal 75(1):3–13

Graettinger AJ, Grau D, Van de Lindt J, Prevatt DO (2012) GIS for the geo-referenced analysis and rapid dissemination of forensic evidence collected in the aftermath of the Tuscaloosa tornado. Construction Research Congress 2012 (May 21–23, 2012). Purdue University, West Lafayette, IN, pp 2170–2179. http://rebar.ecn.purdue.edu/crc2012/papers/pdfs/-306.pdf. Accessed 24 Jan 2013

Grip WM, Grip RW, Morrison RD (2000) Application of aerial photography and photogrammetry in environmental forensic investigations. Environ Forensics 1(3):121–129

Gwendolyn Guillory et al. vs. Union Pacific Railroad Company, et al. (2005). 915 So. 2d 1034, Court of Appeal of Louisiana, Third Circuit

Harries K (1999) Mapping crime: principle and practice. Office of Justice Programs, National Institute of Justice, Washington, DC, p 206

Headrick D (2013) 3D laser scanning gives law enforcement the upper hand in crime scene investigation and recreation. http://3dforensicscans.com/2013/3d-laser-scanning-gives-law-enforcement-the-upper-hand-in-crime-scene-investigation-and-recreation. Accessed 8 Feb 2013

Hodge SH (1997) Satellite data and environmental law: technology ripe for litigation application. Pace Environ Law Rev 14:691–731

Johnson SD, Bowers KJ, Pease K (2005) Predicting the future or summarising the past? Crime mapping as anticipation. In: Smith MJ, Tilley N (eds) Crime science: new approaches to preventing and detecting crime. Willan Publishing, Cullompton, pp 145–163

Krouse AJ, Ferry MM, Crowsey RC (2000) Satellite imagery: the space Odyssey arrives in the courtroom. http://www.crowsey.com/pdf/spaceOdyssey.pdf. Accessed 30 Sept 2008

Kyllo v. United States (2001). 533 U.S. 27

Laba JT (1996) If you can't stand the heat, get out of the drug business: thermal imagers, emerging technologies, and the fourth amendment. Calif Law Rev 84(5):1437–1486

Langer M, Dunaski M (2007) Aerial photography: an elevated perspective. Evid Technol Mag 5(4):28–33

Lawyers USA (2009) Court says thermal imaging search supported by probable cause. PoliceOne.com News

Leitner M (ed) (2013) Crime modeling and mapping using geospatial technologies. Springer, Heidelberg

Lewis BC (2006) Changing the bathwater and keeping the baby: exploring new ways of evaluating intent in environmental discrimination cases. Saint Louis Univ Law J 50:459–516

Lillesand TM, Kiefer RW, Chipman JW (2008) Remote sensing and image interpretation, 6th edn. Wiley, Hoboken

Lockwood B (2012) The presence and nature of a near-repeat pattern of motor vehicle theft. Secur J 25(1):38–56, http://dx.doi.org/10.1057/sj.2011.5

Macauley MK, Brennan TJ (1998) Enforcing environmental regulation: implications of remote sensing technology. Resources for the Future (RFF), Washington, DC

Manhein MH, Listi GA, Leitner M (2006) The application of geographic information systems and spatial analysis to assess dumped and subsequently scattered human remains. J Forensic Sci 51(3):469–474. doi:10.1111/j.1556-4029.2006.00108.x

Markowitz KJ (2002) Legal challenges and market rewards to the use and acceptance of remote sensing and digital information as evidence. Duke Environ Law Policy Forum 12(2):219–264

Marshall P (2013) USDA's high-res view of fraud. http://gcn.com/articles/2013/10/09/gcn-award-usda-fraud-detection.aspx. Accessed 14 Oct 2013

Messina J, May J (2003) Aerial photography and remote sensing for solving crimes. In: Leipnik MR, Albert DP (eds) GIS in law enforcement: implementation issues and case studies. Taylor & Francis, New York

Milla KA, Lorenzo A, Brown C (2005) GIS, GPS, and remote sensing technologies in extension services: where to start, what to know. J Ext 43(3). Online at http://www.joe.org/joe/2005june/a6.shtml. Accessed 9 Nov 2013

Morrison RD (2000) Critical review of environmental forensic techniques: part I. Environ Forensics 1:157–173

Morrow-Jones HA, Anacker KB, Chen W, Horner MW, Malecki EJ (2005) Rapid property transactions and price changes: an exploratory study in forensic geographic information system (GIS). FOCUS Geogr 48(3):19–25

Morton DD, Harner LF, Shires SM (2005) Geospatial technology in support of wildlife law enforcement. In 61st annual Northeast fish and wildlife conference, Virginia Beach, 17–20 April 2005

Nichols R (2010) Is Google Earth eyeing your pool? Government Technology. http://www.govtech.com/gt/articles/768352. Accessed 20 Aug 2010

NTSB (1996) Accident report: in-flight breakup over the Atlantic Ocean Trans World Airlines Flight 800 Boeing 747–131, N93119 near East Moriches, New York July 17, 1996. Washington, DC, p 341

Peet D (2012) Fighting crime using geospatial analytics. Police Chief 79(6):46

Platt R (2006) Crime scene: definitive guide to the world behind the tape. DK Publishing, New York

PSI (2012a) 3-D laser scanning and ballistic trajectory analysis. http://www.precisionsim.com/index.php/news-and-articles/. Accessed 21 Feb 2013

PSI (2012b) 3-D laser scanning and photogrammetry. http://www.precisionsim.com/index.php/news-and-articles/. Accessed 21 Feb 2013

PSI (2012c) Reconstruction of construction accidents. http://www.precisionsim.com/index.php/news-and-articles/. Accessed 21 Feb 2013

PSI (2012d) Reconstruction of municipal bus accidents. http://www.precisionsim.com/index.php/news-and-articles/. Accessed 21 Feb 2013

PSI (2012e) The working model. http://www.precisionsim.com/index.php/news-and-articles/. Accessed 21 Feb 2013

Ratcliffe JH (2004) Crime mapping and the training needs of law enforcement. Eur J Crim Policy Res 10:65–83

Raths D (2007) GIS integration. GovTech. http://www.govtech.com/geospatial/GIS-Integration.html. Accessed 19 Dec 2012

Rey S, Mack E, Koschinsky J (2012) Exploratory space–time analysis of burglary patterns. J Quant Criminol 28(3):509–531. doi:10.1007/s10940-011-9151-9

Ries DG (2007) Computer presentations by lawyers in the conference room, classroom, and courts. Pa Bar Assoc 78:56–70

Rocchio L (2006) Fighting crop insurance fraud with Landsat. Photogramm Eng Remote Sensing 27(7):725–727

Rossmo DK (2000) Geographic profiling. CRC Press, Boca Raton
Rossmo DK (2006) Geographic profiling in cold case investigations. In: Walton RH (ed) Cold case homicides: practical investigative techniques. CRC Press, Boca Raton, pp 537–560
Rossmo DK, Laverty I, Moore B (2005) Geographic profiling for serial crime investigation. In: Wang F (ed) Geographic information systems and crime analysis. Idea Group Publishing, Hershey, pp 102–117
Schreiber S (2009) Seeing through to the hot spots: amid falling costs and improved features, law enforcement may be able to find a few more reasons to pick up a thermal imaging camera. [Technical reports]. Law Enforc Technol 36(7):35–40
Schultz JJ (2007) Using ground-penetrating radar to locate clandestine graves of homicide victims: forming forensic archaeology partnerships with law enforcement. Homicide Stud 11(1):15–29. doi:10.1177/1088767906296234
Schultz PD (2008) The future is here: technology in police departments. Police Chief 75(6):19–27
Schultz JJ (2012) Detecting buried remains using ground penetrating radar. Department of Anthropology, University of Central Florida, p 235
Schwartz M (2005) Louisiana battles food stamp fraud. Bus Intell J 10(4):5
Slonecker ET, Shaw DM, Lillesand TM (1998) Emerging legal and ethical issues in advanced remote sensing technology. Photogramm Eng Remote Sensing 64(6):589–595
Smith DG (2001) Kosovo: applying GIS in an International Humanitarian Crisis. ArcUser, July–September 2001
Suggs JA, Beam EW, Biggs DE, Collins W Jr, Dusenbury MR, MacLeish PP et al (2002) Guidelines and resources for conducting an environmental crime investigation in the United States [Guest editorial]. Environ Forensics 3(2):91–113
Sui DZ (ed) (2008) Geospatial technologies and homeland security: research frontiers and future. Springer, Dordrecht
Taylor RB (1997) Crime and place: what we know, what we can prevent, and what else we need to know. Paper presented at the National Institute of Justice Annual Research and Evaluation Conference, Washington, DC, July 1997
Thompson BHJ (2000) Innovations in environmental policy: the continuing innovation of citizen enforcement. Univ Illinois Law Rev 2000(1):185–236
Tompson L, Townsley M (2010) (Looking) back to the future: using space-time patterns to better predict the location of street crime. Int J Police Sci Manage 12(1):23–40
Townsley M, Homel R, Chaseling J (2000) Repeat burglary victimisation spatial and temporal patterns. Aust N Z J Criminol 33(1):37–63
U.S. v. Kattaria (2009). 553 F.3d 1171, U.S. Court of Appeals, 8th Circuit
United States of America v. Gregory Peter Torlai Jr (2013). D.C. No. 2:08-cr-00329-JAM-1
Walsh T (2004) Louisiana takes a bite out of food stamp fraud. Gov Comput News 23(23):2
Warren J, Reboussin R, Hazelwood RR, Cummings A, Gibbs N, Trumbetta S (1998) Crime scene and distance correlates of serial rape. J Quant Criminol 14(1):35–59
Westly E (2010) Computers key to fighting Medicare fraud. Institute of Electrical and Electronics Engineers. http://spectrum.ieee.org/biomedical/ethics/computers-key-to-fighting-medicare-fraud. Accessed 19 Dec 2012
Wilson C (2003) Mapping the criminal mind. New Sci 178(2392):46–50
Wilson AC, Jones SG, Smith MA, Liles R (1997) Tracking spills and releases: high-tech in the courtroom. Tulane Environ Law J 10:371–394
Wolff M, Asche H (2009) Towards geovisual analysis of crime scenes – a 3D crime mapping approach. In: Sester M, Bernard L, Paelke V (eds) Advances in GIScience, Lecture notes in geoinformation and cartography. Springer, Berlin/Heidelberg, pp 429–448
Wortley RK, Mazerolle LG (2008) Environmental criminology and crime analysis, Crime science series. Willan Publishing, Portland
Young G (2004) Mapping mayhem: the geography of crime. J Prof Invest 71:2
Youstin TJ, Nobles MR, Ward JT, Cook CL (2011) Assessing the generalizability of the near repeat phenomenon. Crim Justice Behav 38(10):1042–1063. doi:10.1177/0093854811417551
Zipf GK (1949) Human behavior and the principle of least effort. Addison-Wesley, Reading

Part II
Case Studies

Chapter 5
Using Near Repeat Analysis for Investigating Mortgage Fraud and Predatory Lending

Ronald E. Wilson and Ann D. Fulmer

Abstract Mortgage fraud and predatory lending are pervasive white-collar crimes that are difficult to investigate. As such, fraud investigators must use analytical techniques that can identify loans with the greatest potential to show systematic irregularities in the lending process. Foreclosures from the 2000 decade housing crisis are widely known to have occurred because of large-scale fraud, to which many were clustered. Despite displaying geographic patterns, an approach that has rarely been considered for fraud investigations from foreclosures is to use spatial analysis to identify geographic patterns of these crimes. We demonstrate in this chapter a spatial analysis method that can be applied more widely to help fraud investigators identify loans for scrutiny that show geographically systematic patterns of foreclosure.

Keywords Mortgage fraud • Spatial dependence • CrimeStat • Near repeat analysis

5.1 Introduction

The housing crisis of the mid-2000 decade led to the largest volume of foreclosures ever experienced (Bianco 2008; Crump et al. 2008; Wilson and Paulsen 2009) with a rate about four times (400 %) the historical average (Wachter 2009). The fallout of the crisis left a landscape of concentrated foreclosures with a lasting negative effect on many neighborhoods. Foreclosed houses can exert a direct negative economic impact on nearby properties because properties share vicinity-based

R.E. Wilson (✉)
U.S. Department of Housing and Urban Development, University of Maryland, Baltimore County, Washington, DC, USA
e-mail: rwilson@umbc.edu

A.D. Fulmer
Interthinx Corporation, 30005 Ladyface Court, Agoura Hills, CA 91301, USA

G.A. Elmes et al. (eds.), *Forensic GIS: The Role of Geospatial Technologies for Investigating Crime and Providing Evidence*, Geotechnologies and the Environment 11, DOI 10.1007/978-94-017-8757-4_5,

monetary value from having similar characteristics. Foreclosed properties transfer negative value through price devaluation, or "discounting," that reduces the value of the original and nearby properties simultaneously (Scott 2010). Neighboring properties can lose between 0.9 and 8.7 % in value after another property is vacated from foreclosure (Pennington-Cross 2004; Immergluck and Smith 2005; Lee 2008; Lin et al. 2009). Losses can cumulatively mount when multiple foreclosures are concentrated (Simons et al. 1998; Schuetz et al. 2008; Leonard and Murdoch 2009), which can lead to social and economic decline (Baxter and Lauria 2000; Metzger 2000; Wyly et al. 2004; Galster et al. 2006; Kaplan and Sommers 2009; Wilson and Paulsen 2009; Scott 2010; Fisher et al. 2013).

Mortgage fraud[1] has grown considerably over the past decade and predatory lending[2] is still pervasive. The number of mortgage-related Suspicious Activity Reports (SAR) submitted to the Financial Crimes Enforcement Network (FinCEN) of the US Department of Treasury is higher than ever. SAR submissions increased 2,071 % between 2000 and 2010 (James and Butts 2011) and grew 31 % in 2011 alone (FinCEN 2012). It has been estimated that about 38 % of foreclosures were the result of mortgage fraud[3] and a high proportion from predatory lending schemes. Mortgage fraud alone is straining the Federal Bureau of Investigation (FBI) resources.[4] James and Butts (2011) show that the number of cases to which the FBI issued indictments doubled from 1,531 in 2010 to 3,020 by February 2011. These trends signal that mortgage fraud and predatory lending are not abating and that fraud investigations must remain steadfast in identifying, preventing, and prosecuting fraudsters. Investigators must consider new tools and methods for detecting problems to identify trends not seen in other approaches. With mortgage fraud and predatory lending contributing to the concentration and spread of foreclosures, a spatial analysis to detect mortgage fraud and predatory can be a useful approach for fraud

[1] Mortgage fraud involves deceptive practices that use intentional misstatements, misrepresentations, or outright omissions in the process of making a loan. Following the FBI's definition of activities, mortgage fraud is comprised of the following: (1) encouraging a borrow to lie about identity, (2) equity theft, (3) inflating borrower income, (4) falsifying appraisals, and (5) encouraging borrows to lie about income. These activities are generally classified as either fraud for profit or fraud for housing and target anyone who has the ability to qualify for a loan.

[2] Predatory lending involves illegal practices that apply exorbitant fees, rates, and unnecessary terms to loans to strip housing wealth from the borrower. Predatory practices are not defined in federal law and states differ in their definitions, but Freddie Mac identifies the following as general practices: (1) charging interest rates and/or fees that far exceed reasonable lender compensation, (2) repeatedly refinancing (flipping) at a lower interest rate after originally lending at a high rate to strip the borrower's equity in order to pay new points and fees, (3) failing to report borrower credit information to limit borrowers to obtain the lowest interest rate available, (4) steering to higher-cost mortgages when they are eligible for lower-cost financing, and (5) requiring credit insurance products to be financed upfront.

[3] In testimony to the Financial Crisis Inquiry Commission, Ann Fulmer of Interthinx Corp. stated that about 38 percent of foreclosures were the results of mortgage fraud, September 21, 2010, http://fcic-static.law.stanford.edu/cdn_media/fcic-docs/2010-09-21%20Ann%20Fulmer%20Supplemental%20Written%20Testimony.pdf (accessed August 17, 2012).

[4] Statement of Assistant Director Chris Swecker before the House Financial Services Subcommittee on Housing and Community Opportunity, October 7, 2004. http://www.fbi.gov/congress/congress04/swecker100704.htm (accessed March 29, 2009)

investigations. A near repeat analysis, as one approach, can help investigators identify loans for patterns of fraud they would not have been detected otherwise.

Near repeat analysis is an emerging approach in the spatial and temporal measurement of crime. When a crime occurs, there is a high probability that, in a relatively short period of time, other crimes will occur at the same location (repeat) or within close proximity (near repeat) (Johnson et al. 1997; Townsley et al. 2000; Bowers and Johnson 2004; Johnson et al. 2007; Ratcliffe and Rengert 2008; Short et al. 2009; Wells et al. 2011; Haberman and Ratcliffe 2012). Repeat and near repeat crimes occur because incident sites share favorable location-related commonalities (similar social, physical, and economic features) conducive to other crime opportunities. A primary theoretical underpinning of repeat/near repeat research is rooted in the First Law of Geography (Miller 2004; Tobler 1970). This law purports that everything in geographic space is related, but closer things are more related because they share stronger commonalities, i.e., the near repeat concept is based on *spatial dependence*. This geographic principle permits the repeat/near repeat approach to be applied to many housing-related problems because geography frames the locational and temporal occurrence of those problems. Analyzing foreclosures, as one example, with a near repeat analysis can provide an alternative way to detect systematic mortgage or predatory lending fraud cases.

We demonstrate in this chapter a near repeat analysis method to measure the concentration and spread of foreclosures to identify loans to scrutinize for mortgage fraud and predatory lending in Charlotte-Mecklenburg County, NC. We highlight the direct connection with geographic concepts that provide the foundation for understanding why spatial analysis can be of value to fraud investigations. Our method incorporates a series of spatial analysis techniques that identifies parameter settings, tests for near repeats, creates near repeat geographies, and facilitates more robust geographic analysis from the near repeat results. Our intent is to demonstrate different ways of examining the near repeat geography to identify potential connections between foreclosures. We also demonstrate how to partition data to analyze foreclosures for near repeat patterns across different social, economic, or ecological dimensions. We finish our chapter by discussing a number of analytical avenues to further fraud investigations.

5.2 The Concentration and Spread of Foreclosures: A Change in Routine Activities

Foreclosures regularly occur from life-related events, such as high medical bills, job loss, divorce, family death, or other household events that rapidly exhaust income that lead missed mortgage payments. These events typically lead to isolated and randomly scattered foreclosures across a metropolitan area. During the housing collapse of the mid-2000s, though, foreclosures concentrated unlike any previous time – many the result of mortgage fraud or predatory lending. The concentration occurred from four non-mutually exclusive events that created a negative geography, which were the (1) financial industry deregulation, (2) emergence of new lending practices and technologies, (3) purchasing behavior changes, and (4) territorial expansion of metropolitan areas (Wilson and Paulsen 2009).

5.2.1 Financial Deregulation at the Federal Level

Rooted in the idea of American exceptionalism, the US government has historically pushed for increased homeownership, especially for minorities in the latter decades, because it is believed to be the most reliable way for average citizens to build wealth, as well as stabilize communities (Wyly et al. 2004; Olson 2005; Bullard 2009). To maximize obtaining a home, in the mid-1990s, the federal government deregulated the mortgage market and created federal agencies to help potential buyers more easily obtain loans including many previously unqualified borrowers (Doms and Motika 2006; Chambers et al. 2007; Bianco 2008).

5.2.2 New Lending Practices and Technologies

With the advent of mortgage-backed securities,[5] banks offered loans with historically low interest rates and new flexible loan terms that let almost anyone qualify because these securities reduced lending risk (Immergluck and Wiles 1999; Apgar and Calder 2005; Wachter et al. 2006; Crump et al. 2008; Fulmer 2009; Kaplan and Sommers 2009; Pavlov and Wachter 2009; Smith 2010). Borrowers from all economic backgrounds now qualified for loans and were intensively targeted, in particular minority and low-income populations – aka "greenlining" (Immergluck and Wiles 1999, Apgar and Calder 2005; Wachter et al. 2006; Crump et al. 2008; Welsch 2008; Fulmer 2009; Kaplan and Sommers 2009; Pavlov and Wachter 2009; Crossney 2010).

New technologies also facilitated loan growth. Underwriting software helped stream line the loan process and minimized the human bias that might impede loan originations. Large databases and data mining techniques helped to identify burrowers to target as well as find the right loan terms to help them make a purchase.

5.2.3 Frenzied Purchasing Behavior

Borrowers took swift advantage of subprime loans and engaged in an irrational purchasing frenzy in the belief that if they did not purchase soon, they would never be able to afford a house (Rohe et al. 2000; O'Sullivan 2003) or unable to get into a

[5] Mortgage-backed securities allowed banks to reduce the risk of any one loan, even a group of loans, from defaulting because they bundle all these loans together as a single product. As a bundle, the impact of any loan going into default is minimized for investors because the loss is minimal due to the other loans being stable.

trendy market (Crump et al. 2008). As such, many borrowers accepted lending terms that later resulted in untenable payments (Fulmer 2009; Collins et al. 2009; Kaplan and Sommers 2009; Crossney 2010).

5.2.4 *Metropolitan Growth and Territorial Expansion*

Housing demand surged with the growth in lending between the mid-1990s and mid-2000s, which led many metropolitan areas to territorially expand through building out large plots of land, revitalize older neighborhoods, or fill in empty space. This expansion created a supply of new houses that were quickly filled with borrowers who purchase them with subprime mortgages (Immergluck and Smith 2005; Crump et al. 2008).

For changes in these events to lead to crimes, motivations must exist and barriers be removed for opportunity to emerge. Prior to the four events listed above, routine activities of the mortgage industry were nominal and governed the lending process in which foreclosures occurred from regular life events and be randomly scattered across the landscape. More so, while there were motivated offenders and suitable targets in the mortgage industry prior to the housing crisis of the 2000 decade, there were capable guardians in the form of federal regulations, qualification limits on borrowers, and lack of technologies to facilitate the process. As the regulations and qualifications relaxed and as technologies became ubiquitous, the routine activities of actors in the lending process changed, effectively removing the capable guardians to allow motivated offenders to target large volumes of suitable targets. These elements and changes are the hallmarks of Routine Activities Theory (Cohen and Felson 1979; Gottfredson and Hirschi 1990),[6] which led to the convergence of fraud incidents in time and space to form "risky loan" geographies, collectively ready to default once the market collapsed.

5.3 The Consummation of Risky Loan Geographies: Formation from Spatial Relationships

The volume of subprime mortgages left borrowers collectively sharing the same geographic space of default risk. These "risky loan" geographies led to neighborhoods full of abandoned properties that exacerbate the spatial variation – i.e., spatial heterogeneity – of the metropolitan landscape. A single foreclosure has only a small impact on a neighborhood. When concentrated, foreclosures can have a cumulative negative impact on a neighborhood and can reverberate across a metropolitan area

[6]The General Theory of Crime allows the application of Routine Activities Theory – typically applied to street crime – to white-collar crime due to attitude and behavioral characteristics underpinning the motivations for the perpetration of mortgage-related crimes.

and impinge the government's ability to provide services and maintain the quality of life (Simons et al. 1998; Schuetz et al. 2008; Welsch 2008; Harding et al. 2009). The fallout can be long lasting in which the true damage is not seen until the many years later. At lower loan volumes, households with "at-risk" loans rarely co-occurred within proximity to each other, but because of high volumes, borrowers with high-risk loans were living in the same neighborhoods near each other, en masse. The suddenness, magnitude, and geographic scale of foreclosures in the mid-2000s geography became affixed to these loans because these loans formed a spatially and temporally dependent group that would later manifest into a spatial pattern of systematic defaults when the subprime market collapsed.

This spatial connection between houses is manifested from the positive or negative transfer of economic value to other nearby houses based on property condition, occupant behavior (Galster 1982; Pennington-Cross 2004; Harding et al. 2009; Lin et al. 2009), proximity to built or natural amenities (Gibbons and Machin 2008; Mendelsohn and Olmstead 2009), or market supply principles (Whitaker and Fitzpatrick IV 2011). Foreclosures primarily transfer negative value, or "devaluation," to other properties from the house being vacated and added to the local supply, often steeply discounted. A single foreclosure can devalue nearby properties between 0.9 and 22 %, and their spatial impact can range from an eighth to half a mile in distance (Pennington-Cross 2004; Immergluck and Smith 2005; Lee 2008; Campbell et al. 2009; Lin et al. 2009).

Mortgage fraud and predatory lending also geographically connected the houses through inflated sales prices. Once the first fraudulently inflated sales price is recorded in the deed, realtor, commercial appraisal, and tax databases, it can be referenced as a comparable sale to support an artificially inflated appraisal for other nearby properties. As inflated values were recorded, fraud began to geographically corrupt the tax digests leading to "official" records of "legitimate" market values. The spatial heterogeneity of neighborhoods across a metropolitan area becomes amplified from the database recording that geographically propagates the effects from fraud. As dispersed buyers located into these properties with high-cost loans, they create a new geography of highly leveraged buyers living in overinflated properties. Unable to recognize fraudulent transactions, county appraisers overlooked the early inflated sales as signals of a growing market; additional sales at fraudulently inflated prices created the appearance of price appreciation from market forces as neighborhoods across the areas grew as well. When a critical mass of sales is reached, the neighborhoods are reassessed leading to a geographic range of overinflated house values – by as much as 30 % – and extend several miles (American City & County 2009).[7] In the immediate aftermath of the first wave of fraud, the sales prices of all properties within a quarter mile rise by as much as 4 % (Carswell 2009). The appreciation is illusory and increases the housing costs for new buyers, and new construction will be overpriced and exacerbate the spatially heterogeneity

[7] Tax officials in Boulder County, Colorado, reported tax assessment inflation up to six miles away from the affected neighborhoods.

of housing submarkets across the metropolitan landscape through speculation and overbuilding (Crump et al. 2008).

The wolf finally came[8] in 2006 with the housing market collapse. The spatial aspects of these four events that led up to the concentration of high-risk loans gave way to a concentration of foreclosures, grossly altered settlement patterns and geographically spreading associated social and economic problems. The market began correcting itself with a flood of loans entering foreclosure, taking neighborhood values down across an area. With the sheer amount of loans leading to foreclosure, and only about a third of them resulting from fraud, identifying which ones to investigate becomes challenging. If mortgage fraud or predatory lending schemes have any systematic geographic pattern, then using spatial analysis may help to reveal those loans for instigation.

5.4 Geography and Data

We used Charlotte-Mecklenburg, NC, for our analysis because the local market did not experience the hyperinflated price increases seen in other metropolitan markets, which allowed lower- and middle-income borrowers more opportunity to purchase a house without a reduced likelihood of overextending their finances. Unlike cities in rapidly appreciating housing markets, the Housing Price Index for the greater Charlotte area rose 32 % between 2000 and 2006 (peak of the housing boom) before declining, compared to increases of 66 % for Chicago, IL; 125 % for Las Vegas, NV; and 180 % for Tampa, FL (Office of Federal Housing Enterprise Oversight 2008). Because Charlotte-Mecklenburg did not experience price escalations seen in other housing markets (see Fig. 5.1), housing will be more affordable to a larger number of families.

Lack of rapidly escalating house prices and increased access to credit make housing more affordable to lower- and middle-income populations across a wider set of local markets. North Carolina is a nonjudicial foreclosure state, which means the state does not intervene in the process and allows it to proceed according to typical monthly administrative notices and filings.

We used parcel data from the Charlotte-Mecklenburg Tax Assessors Office to pinpoint the actual location of the foreclosed houses. We used all residential properties[9] that passed through the foreclosure process under North Carolina state law without short sale transfer or other pre-foreclosure process preventing vacancy.

[8]This phrase was borrowed from John Hoerr's 1988 book *And the Wolf Finally Came: The Decline and Fall of the American Steel Industry* that described the collapse of the steel industry and the devastating geographic impact on the Pittsburgh metropolitan region and connected geographies. This phase is aptly appropriate here because of the geographic magnitude of the fallout from one industry.

[9]Property types were condominium, manufactured, and single- and multifamily dwellings of all types. Condominiums and multifamily dwellings will appear in the same results cell in the results because they are at the same location, i.e., they are vertically integrated in this analysis. Single-family

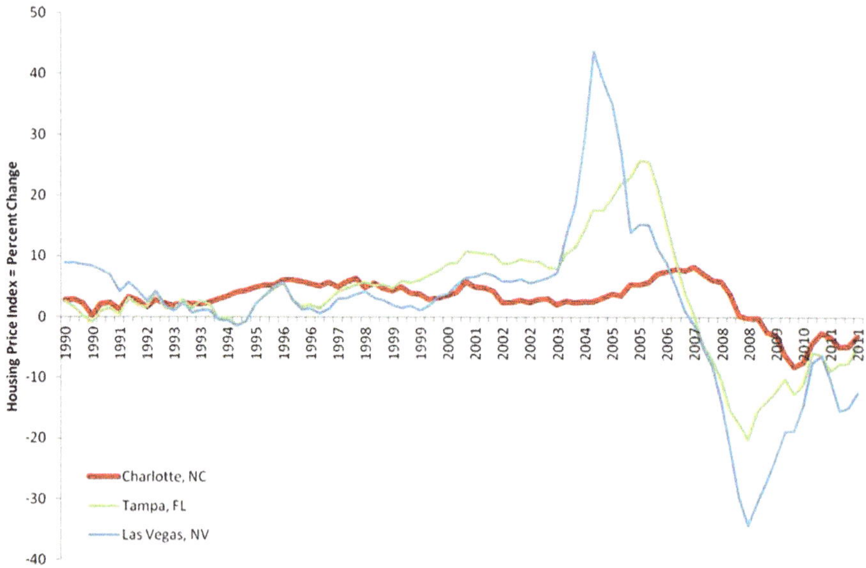

Fig. 5.1 Housing price index change for Charlotte-Mecklenburg and two other major metropolitan areas

Table 5.1 Foreclosure dwelling-type descriptives for Charlotte-Mecklenburg parcel data, 2003–2008

Dwelling types – foreclosed parcels				
	Frequency	Percent	Valid percent	Cumulative
Condominium	1,704	12.7	12.7	12.7
Manufactured	46	0.3	0.3	13.0
Multifamily	200	1.5	1.5	14.5
Single family	11,511	85.5	85.5	100.0
Total	13,461	100	100	

Foreclosure, here, is defined as the state by which a lender repossesses a property after the mortgage delinquency process has come to an eviction conclusion. We identified 13,461 foreclosed properties[10] with transfer dates between 2003 and 2008 indicating loan originator repossession (Table 5.1).

There were 5,097 forecloses in which no original sales price was reflected and 134 outlier properties (prices ≥ 99th percentile). These anomalous properties were temporarily removed in order to evaluate their price distribution to determine how affordable most dwellings were in the Charlotte-Mecklenburg housing market (see Fig. 5.2).

and manufactured homes will be spatially dispersed and appear in the distance interval cells of the results.

[10] Parcel data were provided by the Geography Department at the University of North Carolina, Charlotte.

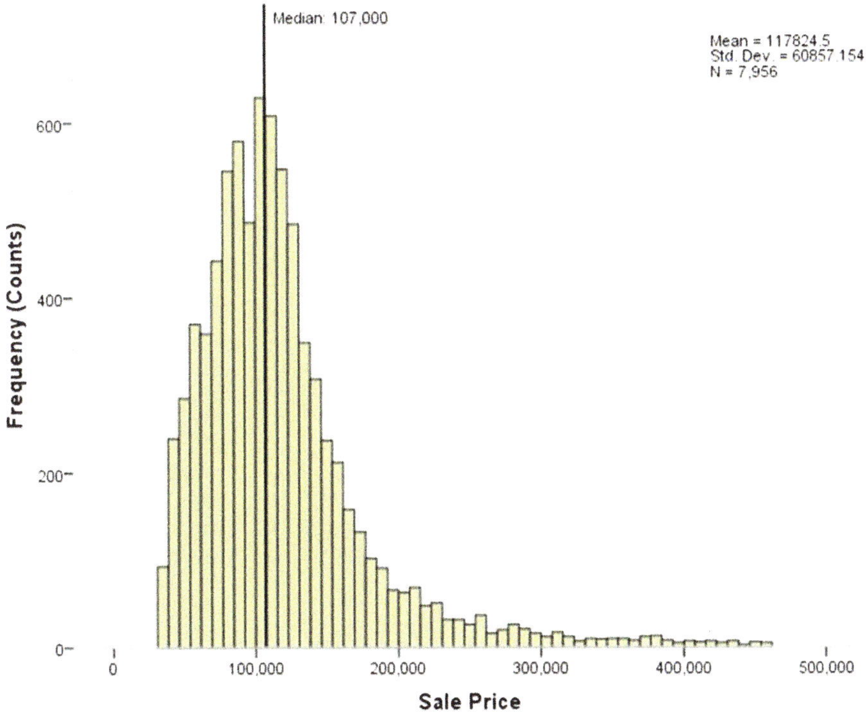

Fig. 5.2 Price distribution for Charlotte-Mecklenburg on foreclosed houses

The distribution in Fig. 5.2 suggests that many properties were well within reach by most of the population, particularly with the availability of flexible loan terms. Despite the outliers we, nevertheless, used all 13,461 properties for the near repeat analysis because price would not impact a spatial and temporal proximity analysis because the approach relies solely on location and date.

The geographic distribution of foreclosures between 2003 and 2008 is depicted in Fig. 5.3. We aggregated the foreclosed parcels for each year into census block groups and created an average rate across all 6 years.

Figure 5.3 shows a wide crescent pattern that nearly encircles Charlotte city center. The Eastway, Providence, Westover, and the South divisions to the south of the city contain many of the wealthiest neighborhoods[11] in the county and did not appear to have experienced high rates of foreclosures. The geographic range in the crescent covers much of the inner and outer suburbs of the city and shows an uneven, extensive, and cohesive pattern of foreclosures. Figure 5.3 is a limited spatial and temporal picture of foreclosures and provides little information about

[11] An analysis of block group data shows a wedge radiating from center city Charlotte to the south in which there is no concentrated disadvantage and many of the block groups show the top 40 % of median incomes in the county.

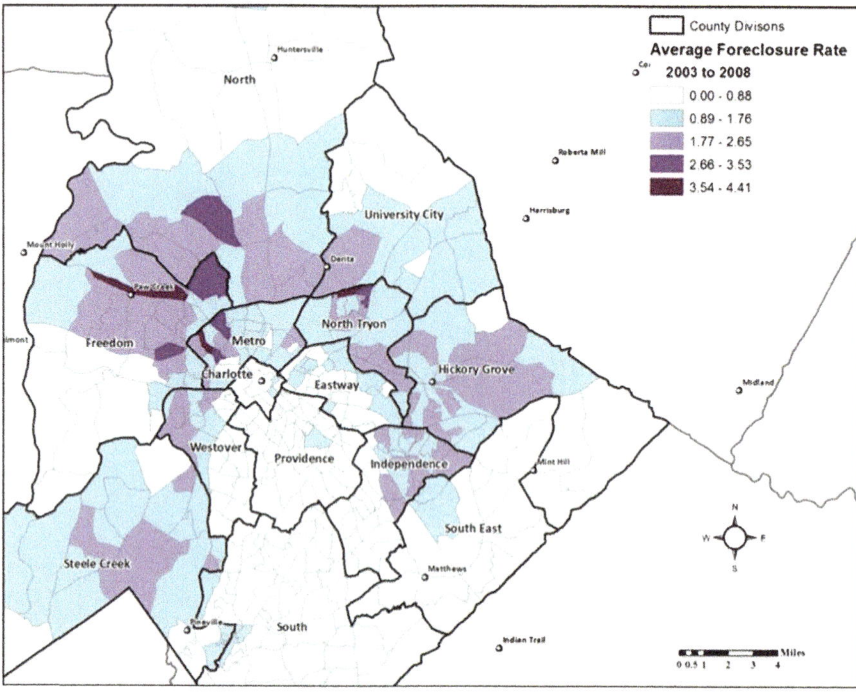

Fig. 5.3 Average foreclosure rate (per block group) of Charlotte-Mecklenburg County – 2003–2008

the geographic patterns of foreclosures in relation to each other. The near repeat process will help to uncover more precise and meaningful patterns for focusing investigations.

5.5 Identifying Geographic and Temporal Intervals to Set Near Repeat Parameters

We first identified meaningful distance and time intervals for the near repeat analysis. Arbitrary selection of either interval will overly bias our results and render them less reliable. Too large or small an interval can impose measurement problems by aggregating either too many or too few of observations within any interval. We identified the spatial interval by conducting a nearest neighbor analysis CrimeStat 3.3 (Levine 2010) on the foreclosed parcels to obtain the average distance between pairs of locations.

The Nearest Neighbor Index (NNI) is a global statistic that produces a single value to indicate whether clustering or dispersion exists among the locations. The

NNI measures the distance of each location to the closest location, which are then summed for all locations and divided by the total number to derive an average minimum distance. A standard deviation distance is also calculated in CrimeStat 3.3 to report the variation of the minimum distances. The average minimum distance is then compared to a derived average minimal distance[12] to produce the NNI ratio, which ranges between 0 and 2.149[13] with a balance point of 1.0. Values closer to 0 indicate clustering. Values around 1.0 indicate a random distribution. Values closer to 2.149 indicate dispersion. A z-score is also calculated to indicate the level of statistical significance that the observed pattern is similar or dissimilar to the randomized pattern.

The NNI results point to a clustering of foreclosures with a value of 0.3835 ($z=-136.92$). The average distance between foreclosures was 264.3 ft with a standard deviation[14] of 374.8 ft, indicating that foreclosed dwellings are very close to each other – either in the same building or on the same/adjacent streets. One standard deviation above and below the mean equates to approximately one-eighth of a mile (639.1 ft), which is what I used for the near repeat distance interval to measure houses within proximity.

Temporal intervals should also be based on a time frame in which proximal repeats are expected to occur. Like the spatial intervals, too large or small time intervals will make measurement difficult and less precise. We used 30 days for the time interval because foreclosures follow a monthly pattern of delinquency stages and default processing procedures. As such, we used a 30-day time interval to account for an abrupt change from occupancy to vacancy from the delinquency the foreclosure stages.

5.6 Identifying Repeat and Near Repeat Foreclosures

We used the Near Repeat Calculator (NRC) (Ratcliffe 2009) software program to measure the spatial and temporal proximity between foreclosed residential parcels. Originally developed for measuring crime, the NRC can be used with any data in which there is an expectation of events exhibiting patterns of reoccurrence in close proximity to each other in short time periods. The results provide statistical evidence about risk levels for proximal events emanating from a catalyst (originating) event. That is, the NRC specifically calculates contagion estimates for the same or nearby areas.

---

[12] The random distance is the expected nearest neighbor distance that represents the distribution of observations if their patterns are completely spatially random. The random distance is calculated with an approximate value, which is $0.5 \times \text{sqrt}(A/N)$ where A is the total area being analyzed and N is the total number of observations in the area.

[13] The value 2.149 is the empirical ceiling of the nearest neighbor index. Theoretically a value could be higher, but none have been observed in previous research.

[14] A standard deviation distance is also calculated in CrimeStat 3.3 to report the variation of the minimum distances.

		TIME: In Proximity	TIME: Not In Proximity
Distance	In Proximity	Distance & Time	Distance Only
	Not In Proximity	Time Only	Neither Distance nor Time

Fig. 5.4 2×2 Knox contingency table to test distance and time relationships (Grubesic and Mack 2008)

DISTANCE INTERVALS (0 → d) / TIME INTERVALS (0 → t)					
D1 & T1	D1 & T2	D1 & T3	...	D1 & TN	~T for D1
D2 & T1	D2 & T2	D2 & T3	...	D2 & TN	~T for D2
D3 & T1	D3 & T2	D3 & T3	...	D3 & TN	~T for D3
...	...	...	...	...	...
DN & T1	DN & T2	DN & T3	...	DN & TN	~T for DN
~D for T1	~D for T2	~D for T3	...	~D for TN	~T & ~D

Fig. 5.5 *N*×*N* Knox contingency table to test distance and time relationships

The NRC expands the George Knox's original concept of simultaneously measuring distance and time aspects (Knox 1964). He derived a statistical test – known as the Knox test – that used a 2×2 contingency table[15] to determine if two events were occurring in close proximity to each other by (a) distance only, (b) time only, (c) both distance and time, or (d) not proximal for either (see Fig. 5.4). Under the Knox test, the event is either in the specified distance (d_1) or not (d_0) and in the designated time period (t_1) or not (t_0).

The NRC extends the Knox test by allowing multiple distance and time intervals instead of just two intervals for each dimension, thus creating an $N \times N$ contingency table (see Fig. 5.5).

The NRC allows for a protracted set of distance and time intervals from d_1 to d_n and t_1 to t_n. The NRC additionally allows for repeat events at the same location, which would now be d_0, and time (t_0) to be detected separately from locations that are in proximity on both dimensions.

The NRC first measures the observed distances between all pairs of event locations and their date stamps to record a count of event pairs that occur within a specified set of distance and time intervals. A Monte Carlo permutation approach is

[15] A contingency table in this context is matrix of two or more categories that depict the relationship between each factor across all rows and columns.

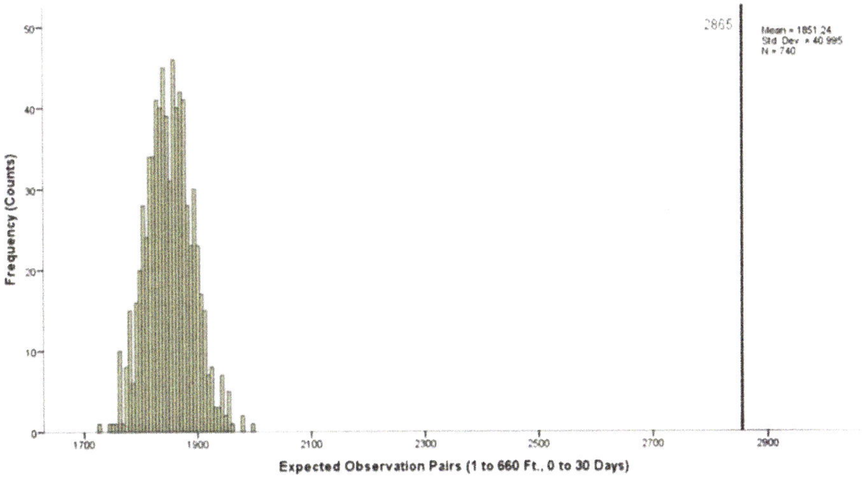

Fig. 5.6 Distribution of expected pairs within first distance and time interval

employed to randomize the event dates.[16] The Monte Carlo approach is fitting for our approach because it represents the random foreclosure process that should occur from the disparate life-related events that lead to mortgage defaults. Locations remain fixed during the permutation process because the interchanging of dates redefines the spatial relationship between locations such that two locations may or may not be in proximity to each other within a time interval. After each randomization trial, the distances[17] and dates for each event pair are remeasured and then combined with the other trial results to create a null empirical (expected) distribution. The observed event pair count is then compared with the mean of the expected distribution to determine if the actual number of nearby event locations is different than the expected within a time interval (see Fig. 5.6).

The permutation process is carried out for each distance and time interval individually so that observed event counts are evaluated with their respective empirical distributions. Statistical significance is determined for each distance and time interval by calculating the number of times the observed event pair count exceeds the mean of the expected distribution. If the observed count is not contained in the expected distribution, then the distance-time relationship between a pair of observations is likely due to some underlying process and not by chance.

Using the results from the nearest neighbor analysis, we set the Near Repeat Calculator to measure foreclosure proximity with eight distance intervals of a standard city street block 660 ft (1/8th of a mile), which is a little under the ±1

---

[16]The number of randomization trials is based on a selected level of significance to be achieved; $p \leq 0.5$: 19 randomizations; $p \leq 0.01$: 99 randomizations; $p \leq 0.001$: 999 randomizations.

[17]The distances are only measured the first time and stored in a matrix. Because the locations remain fixed, there is no need to remeasure them for each permutation run.

Table 5.2 Observed over mean expected values for near repeat foreclosures – 2003–2008

	0–30 days	31–60 days	61–90 days	91–120 days	121–150 days	151–180 days	More than 180 days
Same location	**13.45**	0.88	0.94	0.83	0.77	0.73	0.50
1–660 ft	**1.54**	**1.22**	**1.3**	**1.15**	**1.13**	**1.16**	0.95
661–1,320 ft	**1.3**	**1.12**	**1.12**	**1.07**	**1.06**	**1.08**	0.97
1,321–1,980 ft	**1.04**	**1.04**	1.03	**1.05**	**1.03**	**1.01**	0.99
1,981–2,640 ft	0.98	1	0.98	**1.03**	**0.99**	**1**	1
2,641–3,300 ft	**1.03**	**1.03**	1.01	**1.06**	1	**1.05**	0.99
3,301–3,960 ft	1.01	**1.04**	**1.02**	1	**1.1**	**1.05**	0.99
3,961–4,620 ft	1.01	**1.04**	1.01	**1.03**	1.02	**1.05**	0.99
4,621–5,280 ft	1.02	1.02	1.01	**1.04**	1.01	1.01	1
More than 5,280 ft	1	1	1	1	1	1	**1**

standard deviation distance found in the nearest neighbor analysis. We set the time intervals for six periods of 30 days (6 months), which corresponds to the typical delinquency and default process for a nonjudicial foreclosure state. The NRC offers the selection of three significance levels ($p = 0.05$, 0.01, or 0.001) to determine the level of strength between observed pair counts and the expected distribution. To ensure our results were robust, we used p-values of 0.001 as our statistical threshold for determining relationship strength, as well as to distinguish between levels of statistical strength for each of the distance and time intervals. Finally, distance can be measured in one of two ways in the NRC, which is straight-line distance (Euclidean) or following a simulated street grid (Manhattan). We measured foreclosure proximity with Euclidean distance because houses can go into foreclosure for similar reasons – i.e., borrower or loan characteristics – that are not relative to street patterns.

The NRC produces two contingency tables.[18] One table reports the likelihood ratios (Knox ratios) of a repeat or near repeat occurring. The second table reports the significance levels for each corresponding Knox ratio and is used to identify which ratios are significant and to which level. Knox ratios are derived by dividing the observed pair counts by the mean of the expected distribution for each distance and time interval (Grubesic and Mack 2008). Table 5.2 shows the Knox ratios for repeat and near repeat foreclosures occurring in each distance and time interval. Knox ratios in bold red are significant at 0.01 and bold light red at 0.05 – all light gray ratios are not significant.

Ratios less than 1 indicate that foreclosures are less likely to occur. Ratios above 1 indicate that foreclosures are more likely to occur. For example, there are 1,850 expected foreclosure event counts to occur within 1–660 ft of each other, between 0 and 30 days, with the observed pair count being 2,856.[19] Dividing the observed pair

---

[18] The likelihood table can present which ratios are significant alone. However, we present both tables so that users understand the output of the Near Repeat Calculator.

[19] The observed and expected counts are reported in another table produced by the NRC, which is entitled "Verbose."

Table 5.3 Significance levels for near repeat foreclosures – 2003–2008

	0–30 days	31–60 days	61–90 days	91–120 days	121–150 days	151–180 days	More than 180 days
Same location	**0.01**	0.95	0.82	1	1	1	1
1–660 ft	**0.01**	**0.01**	**0.01**	**0.01**	**0.01**	**0.01**	1
661–1,320 ft	**0.01**	**0.01**	**0.01**	**0.01**	**0.01**	**0.01**	1
1,321–1,980 ft	**0.01**	**0.01**	0.06	**0.01**	**0.03**	0.26	1
1,981–2,640 ft	0.85	0.61	0.91	**0.01**	0.79	0.52	0.2
2,641–3,300 ft	**0.02**	**0.02**	0.32	**0.01**	0.46	**0.01**	1
3,301–3,960 ft	0.33	**0.01**	**0.04**	0.44	**0.01**	**0.01**	1
3,961–4,620 ft	0.31	**0.01**	0.18	**0.01**	0.06	**0.01**	1
4,621–5,280 ft	0.16	**0.03**	0.17	**0.01**	0.19	0.16	1
More than 5,280 ft	1	1	1	1	1	1	**0.01**

count by the expected creates a Knox ratio of 1.54 and indicates a 54 % (Knox ratio – 1) greater chance that another foreclosure will occur within an eighth of a mile and in a month's time.

Table 5.3 reports the corresponding significance levels for each Knox ratio. Statistical significance is determined by calculating *p*-values from the observed and expected pair count comparison.

P-values are calculated for each distance and time interval with $p = 1 - n_e/(n_s + 1)$, where n_e is the Knox ratio and n_s is the number of permutations. For example, with foreclosures that occur within 1–660 ft, between 0 and 30 days, the *p*-value for the 1.54 Knox ratio is $p = 1 - 1.54/(999 + 1)$, which is $p = 0.0054$. This *p*-value is statistically significant at the 0.01 level, and the likelihood of the 2,856 foreclosures being in the closest proximity to each other within the first 0–30 days has less than a 1 % likelihood being due to chance. Knox ratios in bold red are significant at 0.01 and bold light red at 0.05.

The overall results in Table 5.3 show an extensive pattern of repeat and near repeat foreclosures across both distance and time. Foreclosures at the same location – a repeat event – within 30 days have a 1,245 (13.45–1) percent chance of reoccurring; these are likely condominiums or other multifamily dwellings that share the same physical location. At 0–30-, 91–120-, and 121–150-day intervals, near repeat foreclosures extend up to three-eighths of a mile (1,980 ft). The spatial effect for near repeat foreclosures at both 31–60 and 91–120 days extends up to 1 mile (5,280 ft). A distance decay pattern is present in each time interval with the Knox ratios generally decreasing rapidly across the distance intervals.[20]

[20] If further spatial analysis is to be conducted, plotting the values in a graph will show a distance decay curve that can be be used to guide the selection of a mathematical function to produce a similar distance decay curve for weighting distance.

5.7 Creating Near Repeat Geographies

The Knox tables, though, only tell investigators that repeat and near repeat foreclosures are occurring, which only offers limited information. To conduct any spatial analyses, though, the near repeat results need to be converted into geographic objects for further analysis. We use additional output from the NRC to visually identify geographic patterns of repeat and near repeat foreclosures to focus investigative efforts on the repeat and near repeat clusters in relation to their local geographies. The NRC allows for the selection of locations that fall within a range of distance and time intervals identified as significant in the Knox contingency tables to import into a geographic information system (GIS) for visualization or spatial analysis. For each location in that distance and time range, the NRC appends the number of times each location was a catalyst (originator) to other locations, as well as how many times that location was a near repeat to a catalyst. The locations themselves can be analyzed alone or in conjunction with the corresponding catalyst or near repeat counts. We used the NRC output to examine the locations in combination with their corresponding local geographies to potentially identify other factors that might underlie rapidly concentrating foreclosures in Charlotte-Mecklenburg. We then imported the results into the ESRI ArcGIS mapping software to overlay of the identified repeat and near repeat locations with the average foreclosure rates to examine the patterns in the context of the average foreclosure rates by block group (see Fig. 5.7).

We classified the repeat and near repeat foreclosures based on quartiles derived from Open GeoDa 1.40 (https://geodacenter.asu.edu/projects/opengeoda) analytical software. The circle sizes are based on the calculation of the interquartile range of the ratio between originators and near repeats. This depiction helps to visualize the dual relationship a house can have as the initial foreclosure as well as a near repeat to other foreclosures. The larger the circle, the stronger the connection the originator has with a chain of near repeats.

The repeat and near repeat patterns in Fig. 5.7 refine the analysis in three ways that are much more revealing about location than the average rates aggregated to block groups shown in Fig. 5.3. First, the results specifically show which foreclosed properties were the most problematic with large dark red circles. Their classification as outliers reveals their particular influence as originating events and how strong the connection is with other foreclosures nearby. Second, the results show more specifically where problem houses are and reveal that they are in specific areas of a block group and not distributed across block groups. Third, the near repeat patterns are not in the tracts that had the highest average rates of foreclosures as might be expected. If investigators where to use the simple rate, they might look to the wrong block groups for sample loans. It is one thing for a block group to have the highest rate, but it is another to have a systematic pattern, the latter being what investigators more likely want to follow.

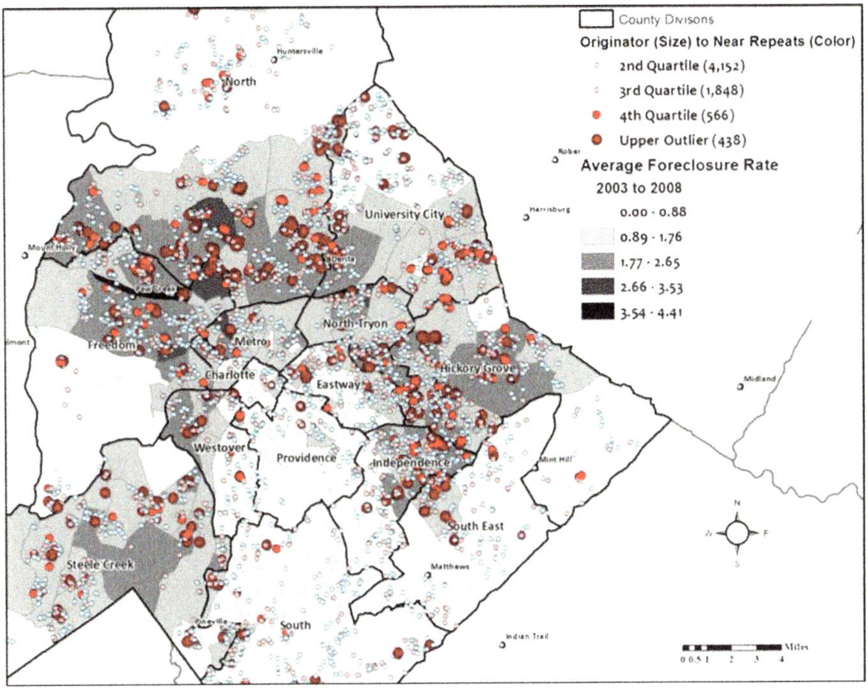

Fig. 5.7 Originator to near repeat foreclosure relationships over the average foreclosure rate – 2003–2008

We next used the Knox test in the ClusterSeer 2.0 (BioMedware 2002) analytical software to identify linkages between foreclosures based on the results from Table 5.2 to reveal the distance and time connections between the originator and near repeat foreclosures in Fig. 5.2. We identified foreclosure linkages that were within a distance of 1,980 ft and within a 90-day time frame because the results showed consistent pattern of significant Knox ratios. Figure 5.8 reveals the direct relationships between the foreclosure originators and near repeats not produced by the NRC.

The northwest of Charlotte-Mecklenburg County highlighted in Fig. 5.8 shows numerous clusters of originator and near repeat foreclosures but now includes links to between near repeats. This visual now allows for directly identifying which originators and near repeats are associated with each other. With these links, groups of foreclosures and their corresponding loans can be examined for commonalities between or across them. Figures 5.7 and 5.8 both direct the investigator's attention to the most problematic houses and their relationships for focused attention.

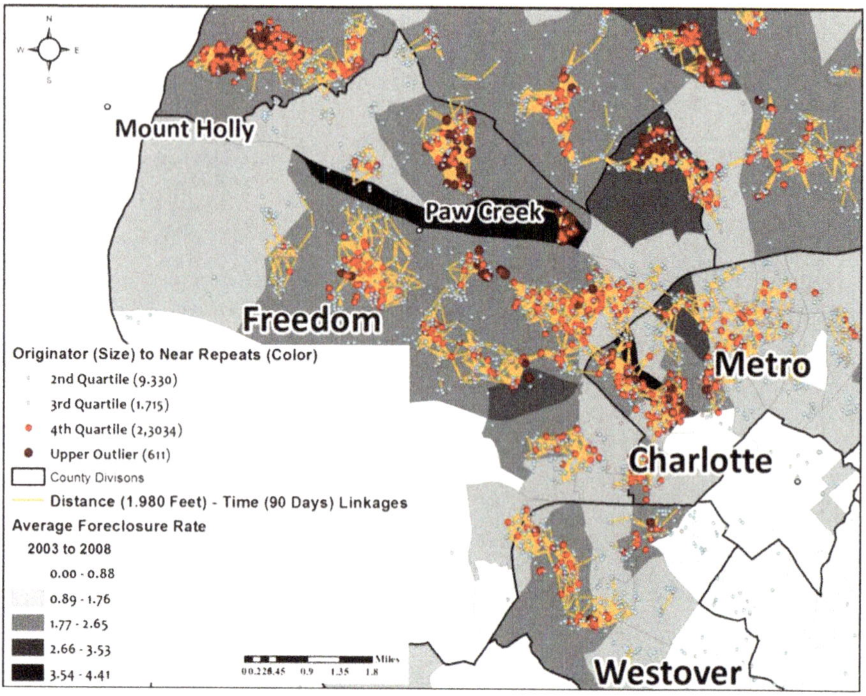

Fig. 5.8 Repeat to near repeat foreclosure locations with distance and time linkages – 2003–2008

5.8 Examining Differing Foreclosure Characteristics with Near Repeat Analysis

The near repeat approach allows input data to be partitioned in any number of ways. Quercia and Stegman (1992) identified that loan, borrower, or property characteristics have the potential to drive specific foreclosure patterns. We demonstrate in this section how splitting foreclosures along economic characteristics can reveal more specific spatial and temporal patterns of foreclosures and associated factors.

Table 5.1 shows an extensive spatial and temporal pattern of near repeat foreclosures across Charlotte-Mecklenburg. This pattern is unlikely because the uneven distribution of social, economic, and housing characteristics across the area. We examined concentrated disadvantage[21] as one neighborhood-level characteristic,

[21] Concentrated disadvantage is comprised of five census variables, which are the number of (1) families below poverty, (2) families receiving public assistance, (3) unemployed individuals in the civilian labor force, (4) families with children that are female headed, (5) residents who are black, and (6) the median family income in 1999 in a census tract. Criminological studies examining neighborhood effects have noted these factors being highly correlated (Morenoff et al. 2001)

Table 5.4 Observed over mean expected values for near repeat foreclosures for houses in disadvantaged block groups – 2003–2008

	0–30 days	31–60 days	61–90 days	91–120 days	121–150 days	151–180 days	More than 180 days
Same location	22.45	0..5	0.37	0.46	0.16	0.05	0.31
1–660 ft	1.35	1.23	1.2	1.15	1.24	1.19	0.95
661–1,320 ft	1.1	1.17	1.05	1.09	1.12	1.11	0.98
1,321–1,980 ft	1.06	1.08	1.04	1.1	1.04	1.06	0.99
1,981–2,640 ft	1.03	1	1	1.07	1.09	1.04	0.99
2,641–3,300 ft	1.08	1.03	1.01	1.04	0.97	1.04	0.99
3,301–3,960 ft	1.09	1.06	1.07	1.07	1.06	1.06	0.99
3,961–4,620 ft	1.06	1.1	1.01	1.07	1.05	1	0.99
4,621–5,280 ft	1.05	1.1	1	1.05	1.03	1.07	0.99
More than 5,280 ft	0.99	0.99	1	1	1	1	1

Table 5.5 Observed over mean expected values for near repeat foreclosures for houses not in disadvantaged block groups – 2003–2008

	0–30 days	31–60 days	61–90 days	91–120 days	121–150 days	151–180 days	More than 180 days
Same location	9.58	0.93	1.08	0.9	0.85	0.87	0.69
1–660 ft	1.5	1.22	1.33	1.15	1.11	1.15	0.95
661–1,320 ft	1.37	1.13	1.16	1.07	1.06	1.08	0.97
1,321–1,980 ft	1.04	1.06	1.03	1.05	1.05	0.99	0.99
1,981–2,640 ft	0.97	0.99	0.97	1.03	0.97	1.01	1
2,641–3,300 ft	1.02	1.02	1.03	1.07	1.02	1.04	0.99
3,301–3,960 ft	1	1.04	1.03	1	1.06	1.03	0.99
3,961–4,620 ft	1	1.02	1	1.02	1.02	1.06	1
4,621–5,280 ft	1.02	1.02	1.03	1.04	1.01	1	1
More than 5,280 ft	1	1	1	1	1	1	1

which represents a level of economic status of neighborhood residents. High concentrated disadvantage indicates that neighborhood residents are resource deprived, thus making it difficult to climb the economic ladder and achieve a higher quality of life. Low concentrated disadvantage is the opposite where residents have ample opportunities to live a higher quality of life. Tables 5.4 and 5.5 show the near repeat results for foreclosures in tract of low and high concentrated disadvantage.

Separated by concentrated disadvantage levels, Tables 5.4 (high concentrated disadvantage tracts) and 5.5 (low concentrated disadvantage tracts) show that the distance and time intervals exhibit different patterns of statistical significance than Table 5.1; the significance levels are much more scattered in the former tables. The most striking difference between the tables is the likelihood that a dwelling will go into foreclosure at the same location within 0–30 days. In block groups where concentrated disadvantage is high, there is a 2,145 % chance that a

---

but represent a conceptual idea that represents more than just being economically impoverished. To ensure that any analysis did not suffer from correlation problems, a factor analysis was used with varimax rotation using maximum likelihood estimation to calculate factor loadings.

dwelling will go into foreclosure at the same location, compared to an 858 % chance in a more affluent block group. This result indicates that multifamily complexes have a higher risk for a repeat foreclosure from families in which concentrated disadvantage is more prevalent.

Both low and high concentrated disadvantage geographies show comparable near repeat patterns on continual basis over 6 months at a similar geographic range of about one quarter of a mile (1,320 ft). The near repeat patterns in both Tables 5.4 and 5.5 do diverge, though, from Table 5.1 with patterns being more condensed in the first three time and distance intervals for foreclosures that occurred in block groups of low concentrated disadvantage. Unlike Table 5.1, the spatial effects disappear in Table 5.4 after a quarter of a mile for the 151–180-day interval. Both Tables 5.4 and 5.5 show a continuing pattern of near repeat foreclosures across time for the first two distance intervals.

The separate findings allow for researching the differences and allowing investigation strategies to be refined because the results indicate spatial contagion effect for block groups with higher concentrated disadvantage, but low disadvantaged block groups had a more extensive pattern. The next step is to identify the geographic patterns of the two neighborhood types.

5.8.1 Visualizing Differing Distance and Time Patterns

Given the consistent time and distance patterns in Tables 5.4 and 5.5, we selected those foreclosures that had near repeats across all six time intervals for up to a third of a mile (1,920 ft) or geographic visualization and further spatial analysis. We imported the results into ArcGIS and created a logged ratio[22] of originators to near repeats to visualize the locations as a bivariate relationship (see Fig. 5.9). The logged ratio shows the relationship between the number of times a location is an originator to other near repeats and the number of times it is a near repeat to other originators. Values below 1 indicate a foreclosure is a near repeat to many other originating foreclosures, which indicates a stronger and more clustered relationship to many other nearby foreclosures. Values above 1 indicate that a foreclosure is an originator to a less number of near repeats, meaning it is more isolated with a weaker relationship. As values approach 0, more near repeats are associated with the catalyst and share a stronger relationship. As values increase away from 1, this means the catalyst has less of a relationship with the near repeats.

[22]We took the log of the ratio for displaying the relationship order to remove the distribution asymmetry that ratios can create. In many instances ratio values above 1 stretch to the maximum of their range, whereas values below 1 are confined to a range of 0 and 1. This creates a long tail in the distribution of 1 that can skew distribution and affect thematic mapping for comparing the distribution around the center point of 1. Taking the log of the ration alters the values by pulling in the larger values in the tail creating a range of values above 1 that is similar to the range of values below 1. This makes the values on either side of 1 comparable with each other in thematic mapping.

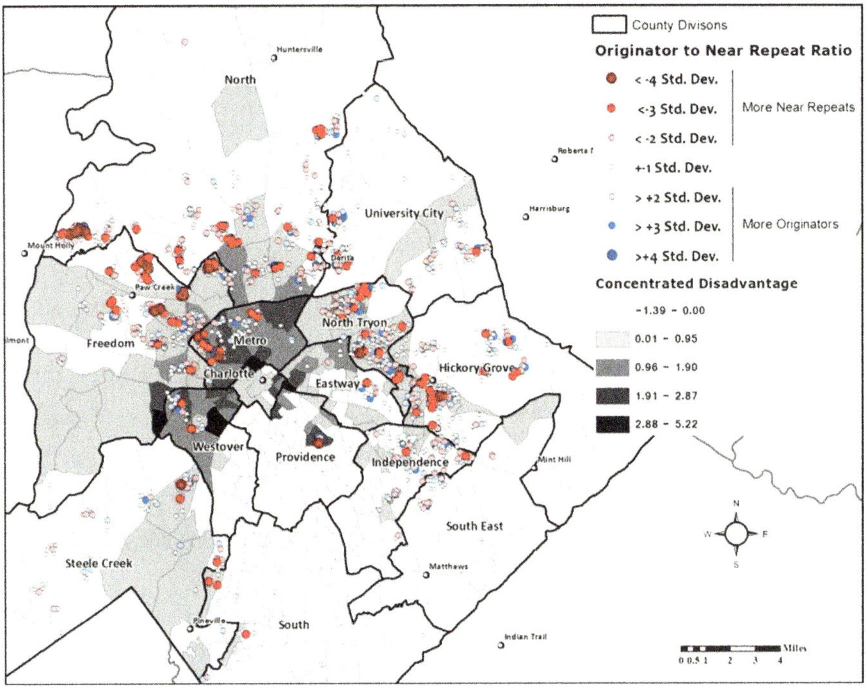

Fig. 5.9 Originating and near repeat foreclosure locations over concentrated disadvantage – 2003–2008

Figure 5.9 first reveals that repeat and near repeat foreclosures were not in the block groups with the highest levels of concentrated disadvantage but also shows that near repeats are more dispersed than in block groups of low concentrated disadvantage. This finding indicates to investigators that those in neighborhoods with the lowest economic status did not suffer the brunt of the foreclosure process in Charlotte-Mecklenburg as has been typically thought or shown in other studies[23] (Calem et al. 2004; Immergluck and Smith 2005; HUD 2010). The results do show that many repeat and near repeat foreclosure clusters are in block groups with higher levels of concentrated disadvantage than not. For example, in the north there were clusters just outside of Mount Holly and Paw Creek, as well as several clusters around and directly to the north of Derita. To the west, there were pockets of clusters in tracts of low concentrated disadvantage in the Eastway, Hickory Grove, and Independence divisions. The southern divisions were nearly devoid of any repeat or near repeat patterns.

[23] Other studies have shown that lower-income populations took advantage of the opportunity to purchase a house in their neighborhood.

The size and the color of the circles indicate where the strongest relationship is between originators and near repeats within a third of a mile from each other. There are several clusters of locations that are −3 and −4 standard deviations below the mean that stretches across the northwest from central city Charlotte, with groups to the east in North Tryon and Hickory Grove divisions and few to the south. These near repeat foreclosures are strongly associated with other near repeats in the immediate area. These clusters are also distinctly separated from each other, signaling to investigators that there might be systematic mechanisms underlying these foreclosures. There are far less clusters of values above one, which indicates that there were many isolated originators in which there were weak associations between the catalyst foreclosure and the other foreclosures nearby. More of these clusters are located in the east with several scattered through the northwest immediately outside city center Charlotte.

5.8.2 Extended Spatial Analysis of the Near Repeat Results

NRC results can guide parameter specification for a subsequent spatial analysis, particularly for establishing a geographic sampling frame. To visualize the spatial concentrations of repeat and near repeat foreclosures, we used the Dual Kernel Density Estimation (DKDE) technique from CrimeStat 3.3 to produce a density surface of repeat and near repeat foreclosures between census tracts that had low and high concentrated disadvantage. The DKDE is ideal for revealing the differing patterns between two different geographic distributions by producing a density surface that reveals the clustering relationships between them with gradients of continuous estimates that summarize clusters of points for more succinct interpretation of patterns. Estimates are calculated by overlaying a grid system across the geography in which the distance from each cell to every observation within a specified distance is measured and weighted based on how close it is to the cell origin and where it corresponds on the distance decay curve. This operation subtracts the density estimates of the houses in non-disadvantage tracts (secondary file) from the estimates for houses in disadvantaged tracts (primary file), producing divergent values above and below 0 to depict areas of concentration of foreclosures based on levels of concentrated disadvantage.

We used the negative exponential function for two reasons. First, the low NNI results indicate close proximity between foreclosure locations. Second, the distance decay patterns in Tables 5.4 and 5.5 show a steep spatial relationship decline in likelihoods across all of the time intervals (see Fig. 5.10), adhering to the empirical research cited above regarding close proximity. Thus, a distance decay function with a steep declining curve that strongly weights close values and rapidly decreases the weights as locations get further away would be representing the spatial relationship between locations.

The negative exponential function creates a steep declining distance decay curve that forms a very narrow peak at the origin that monotonically decreases up to the

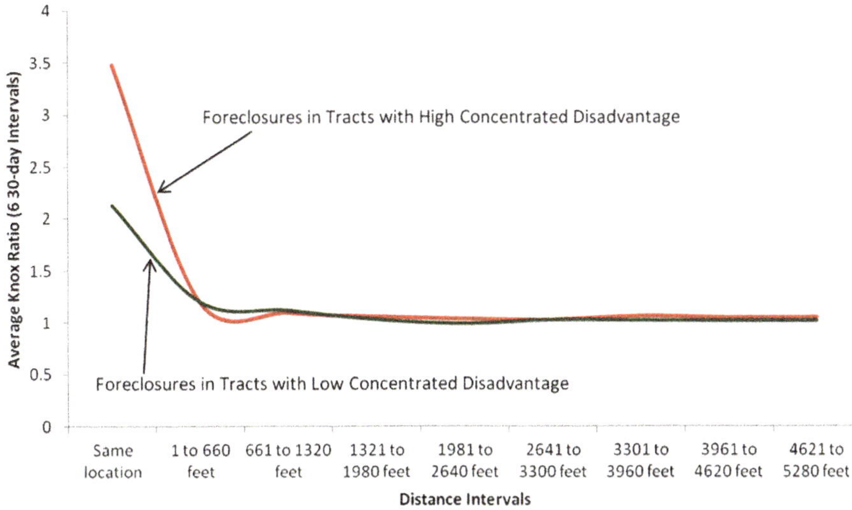

Fig. 5.10 Average Knox ratio decrease along 30-day time intervals – by tract disadvantage level

specified distance –similar to the curves in Fig. 5.9. Locations in very close proximity to the cell are weighted much higher than those further away.[24]

Figure 5.11 shows the results of a DKDE analysis for the geographic concentrations of foreclosures in tracts with lower and higher concentrated disadvantage using a quarter of a mile search distance.

Figure 5.5 supplements the NRC analysis by capitalizing on the appended output to reveal the differing geographic patterns of foreclosures that occurred in tracts with lower and higher concentrated disadvantage. The importance here is that investigations can focus on areas for sampling purposes. The likelihood of other properties within these areas is high, and loans from houses not yet gone into foreclose can be examined and compared to the foreclosed loans. Investigators can oversample in the clusters or create a geographically stratified frame to ensure they get an adequate number of loans from these areas that they would not have otherwise if they just sample from the pool of loans in nonspatial database.

5.9 Implications for Investigating Mortgage Fraud and Predatory Lending

Mortgage fraud and predatory lending cases are extremely complex and are traditionally labor intensive. Fraud cases require the coordinated effort of working groups and task forces from a variety of state and federal regulatory and criminal

[24] Events located outside the specified distance are excluded, which is the case for all the mathematical functions in CrimeStat 3.3 except for the normal function. The normal function includes all observations across the geography.

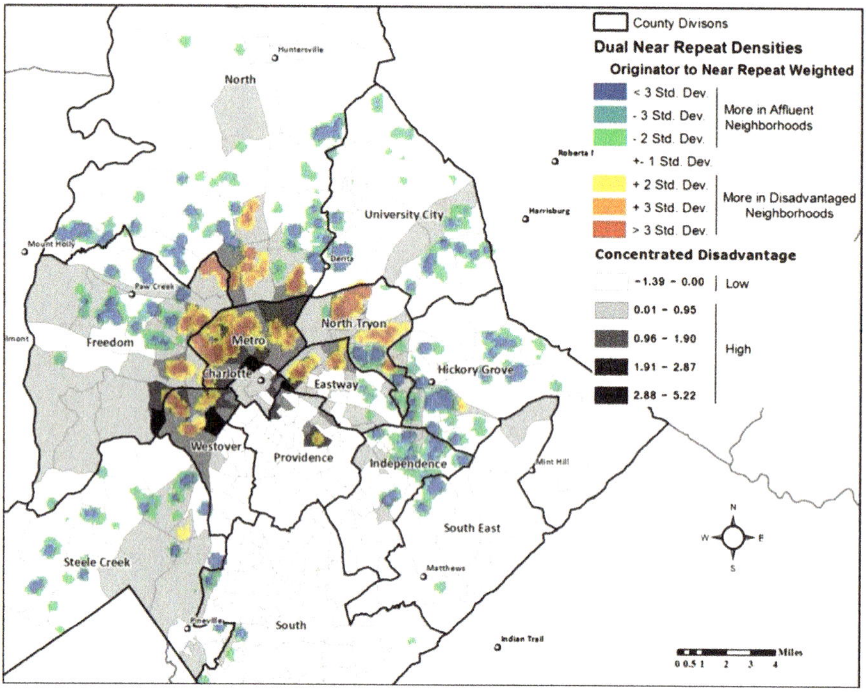

Fig. 5.11 Originating and near repeat foreclosure densities over concentrated disadvantage – 2003–2008

agencies to uncover and prosecute. These investigations take, on average, a minimum of 2 years from initiation to indictment. In many cases, the subjects continue operating their schemes until the day they are indicted. So anything that can help investigators identify systematic fraud should expedite their efforts in making cases against the fraudsters. There are a multitude of implications for investigations from a near repeat analysis, but we list some general ideas below about how this approach offers new avenues for analyzing whether or not foreclosures might have emanated from mortgage fraud or predatory lending.

5.9.1 Enhanced Investigations

Investigations can be enhanced by simply taking advantage of one of the geographic information systems' (GIS) most powerful capabilities, which is linking data attributes by geography when other options are more limited. Since mortgage fraud or predatory lending has been shown to be geographically focused, GIS, then, becomes a valuable tool for analyzing foreclosure data from that perspective and to determine if targeting was or is occurring. Visualizing loan-related data can foster new avenues of investigations that otherwise would not have been possible or generate opportunities from identifying patterns and geographic correlations with other geographic patterns.

5.9.2 Distribution of Other Geographic Factors

The spatial heterogeneity of concept (variation demographic and economic characteristics across the landscape) provides investigators with an opportunity to identify geographic areas to examine more closely. Supplemental geographic data can be used to determine if any visual or statistical correlations exist with the near repeat clusters that might prompt investigation into their relationships. Examining the geographic distributions of demographic or economic factors across neighborhoods can highlight reasons for fraudster targeting or why a clustering of bad loans emerged in those neighborhoods. We used concentrated disadvantage as an example because the research cited above identifies that poorer neighborhoods are usually targets for mortgage fraud and predatory lending for a host of reasons. But because housing is tied to the land and their value is tied to proximity to other geographic entities, factors such as land values, new building permits, income levels, changing neighborhood conditions, or the number of sales can be informative to investigations or research. Most urban environments have huge collections of geographic data that have not been considered for use in understanding the connections between geographic patterns. More so, there are a number of individual- or household-level data sets that can also be mapped and spatially analyzed in this same way.

5.9.3 Time Is a Key Cofactor

The time course of repeats and near repeat adds a new dimension toward understanding the concentration of foreclosures. Knowing how fast foreclosures are spreading, or the temporal extent of the process, can be enlightening about the underlying causes. If time trends are present, this suggests there is something systematic about these loans that investigators should consider. Identifying the pace at which foreclosures are concentrating leads to suspicions of some underlying cause because foreclosures should not be occurring in any systematic way, particularly in conjunction with spatial patterns. Foreclosures can cluster over long periods of time because some neighborhoods will have a demographic that puts them more at risk for defaulting, but those clusters take numerous years to form. Foreclosures reoccurring in the same neighborhood within a month, and across months within a year, signal that these loans are collectively problematic. Time now indicates a more systematic pattern than just space alone because the likelihood of that is statistically improbable, but theoretically. Geography allows a view of patterns that otherwise would have been missed. If investigations can identify clusters of foreclosures as they are emerging, it might be possible to take action before the damage plays out and identify the fraudsters and their plans and reduce damage to the homeowners and their communities.

5.9.4 New Sampling Strategies

The identification of repeat and near repeat foreclosures allows for the development of new sampling frames for scrutinizing loans due to the existence of spatial dependence between the foreclosure locations. With a spatial analysis approach, drawing samples can be based on the originator and near repeat linkages or cluster geographies. With the linkages, specific loans that share the distance and time relationship directly can be examined for their commonalities. With the density surface concentrations, oversampling can be conducted because foreclosures should not normally be clustering, which can ensure enough loans are drawn to yield sought after commonalities if they exist. If there are a large number of clusters, a more balanced sampling frame could be derived to draw more evenly from all of the clusters. Or, a stratified sampling approach might be taken to divide up the larger geography based on the identified patterns in the spatial analysis. If commonalities are found among the loans in the clusters, then investigations can explore, for example, chains (Wells et al. 2011; Haberman and Ratcliffe 2012) of near repeats that reveal connection patterns between the near repeats from one loan to the next. So, a near repeat analysis on the different lenders could be revealing, or not. But that is the test.

5.9.5 Strengthening Prosecutions

Investigation outcomes and research on near repeats can help to show that mortgage fraud and predatory lending are not victimless crimes as is often the perception. As white-collar crimes, these forms of fraud are often seen as victimless because no one is physically harmed. However, these crimes have real consequences beyond the borrower or lender's financial loss and can lead to the destruction of neighborhoods and the emergence of social ills. The ultimate fallout is that the neighborhoods deteriorate to the point that more serious forms of crime emerge, in which people do get physically harmed. Several recent papers have documented that violent and property crime is a growing issue in neighborhoods where foreclosures have concentrated (Immergluck and Smith 2006; Cui 2010; Goodstein and Lee 2010; Ellen et al. 2012; Katz et al. 2012; Arino and Baumer 2012; Arnio et al. 2012; Baumer et al. 2012; Jones and Pridemore 2012; Stucky et al. 2012; Wallace et al. 2012; Kirk and Hyra 2012; Wilson 2015, Forthcoming). Evidence about the long-term damage can be used to bolster cases against the fraudsters and work to increase their sentences, fines, reimbursement requirements, or other penalties by exposing the real damage they cause.

5.9.6 Monitoring Sales for Early Detection

There are few instances in which sales are likely to cluster, such as when new developments are finished or older neighborhoods are gentrified, but like foreclosures, sales should not cluster in short periods of time either. Since we outlined that house sales are the first indication of a geographic pattern, then a near repeat analysis can be employed to detect for clustered sales patterns.

5.10 Conclusion and Further Considerations

Research has concretely identified that foreclosures have spatial patterns, but the detection of those patterns required some form of spatial analysis. We have demonstrated a spatial analysis method to test for repeating patterns of foreclosures across space and time as a means to investigate for mortgage fraud and predatory lending. Because a near repeat analysis only tests for the occurrence of the events themselves, the data can be partitioned for analysis along any number of criteria (see Grubesic and Mack 2008; Levine 2010 for other examples). Regardless of criteria used to partition the data, if a relationship exists in space and time, then the pattern should be detected if parameters are appropriately specified. Analyzing foreclosures with this method offers a new avenue of investigation into the systematic processes of how foreclosures spread and concentrate.

This near repeat method is founded on well-established geographic principles that explain the existence of spatial patterns, including a time dimension. This method provides more credibility to investigations because spatial patterns can simply form if enough time elapses. We first determined that *spatial dependence* was present among foreclosures to form clusters within short periods of time. We then revealed that *spatial heterogeneity* exists in which clusters varied across the landscape. These results present investigators with specific areas to examine a specific group of loans for fraud because they were anomalous along both dimensions of space and time simultaneously, something which is unlikely to occur absent a systematic cause.

We were unable to test *routine activities* directly with regard to the interaction between motivated offenders, suitable targets, and lack of capable guardians because we do not know which foreclosures resulted from fraudulent activity. However, during our analysis we did test the changes in activity with the Near Repeat Calculator by examining patterns in years prior to the collapse (2003–2005) with those at the peak (2006–2008), to which we found a substantial reduction in near repeat patterns. In the latter time frame, the likelihood of repeat and near repeat foreclosures diminished by between 50 and 1,745 % in spatial proximity and the temporal patterns – patterns across multiple intervals all but disappeared. These results suggest the reintroduction of capable guardians in the form of banks being more judicious in lending, a moratorium on foreclosures, and the emergence of fraud investigations as the housing market was collapsing.

Our hope is that this analysis will prompt new thinking about how to examine loan files for evidence of fraud by using geography as a means to focus investigations because there will never be enough time, people, or money to examine all loans for mortgage fraud or predatory lending. A near repeat analysis will not reveal the specific reasons underlying patterns of these crimes, but it will allow investigators or researchers to examine the groups of loans to potentially uncover those reasons. At the very least, this method can help investigators take more immediate actions to thwart fraudsters by identifying the areas where they will likely target in the future. If fraudsters are operating with geography in mind, or at least if their fraudulent activities form geographic patterns, then the use of GIS cannot be ignored as an investigative tool.

Acknowledgments The authors would like to thank Tony Grubesic of Oregon State University, Peter Frandsen of the US Department of Justice, Joe Stachyra of the US Geological Survey, and our blind reviewers for providing comments to refine and improve this chapter. The authors would also like the thank Michael Bess from the Charlotte-Mecklenburg Police Department for providing insight into trends and findings that led to a better understanding of the geography of the city and county.

The views expressed in this chapter are those of the authors and do not represent the official positions or policies of the Office of Policy Development and Research or the US Department of Housing and Urban Development; the University of Maryland, Baltimore County; or Interthinx Corporation.

References

American City & County (2009) Price was not right. http://americancityandcounty.com/technology/mortage-fraud-gis-analysis-200912/index.html. Accessed 11 Jan 2010

Apgar WC, Calder A (2005) The dual mortgage market: the persistence of discrimination in mortgage lending. In: de Souza Briggs X (ed) The geography of opportunity: race and housing choice in metropolitan America. Brooking Institution Press, Washington, DC, pp 101–126

Arnio AN, Baumer EP (2012) Demography, foreclosure, and crime: assessing spatial heterogeneity in contemporary models of neighborhood crime rates. Demogr Res 26(18):449–488

Arnio AN, Baumer EP, Wolff KT (2012) The contemporary foreclosure crisis and U.S. crime rates. Soc Sci Res 41(6):1598–1614

Baumer EP, Wolff KT, Arnio AN (2012) A multicity neighborhood analysis of foreclosure and crime. Soc Sci Q 93(3):577–601

Baxter V, Lauria M (2000) Residential mortgage foreclosure and neighborhood change. Hous Policy Debate 11(3):675–699

Bianco KM (2008) The subprime lending crisis: causes and effects of the mortgage meltdown. CCH Mortgage Compliance Guide and Bank Digest. http://www.consejomexicano.org/Emails/subwprev.pdf

BioMedware (2002) ClusterSeer 2.0. Ann Arbor. http://www.biomedware.com/?module=Page&sID=clusterseer. Accessed 1 Aug 2013

Bowers KJ, Johnson SD (2004) Who commits near repeats? A test of the boost explanation. West Criminol Rev 5(3):12–14

Bullard J (2009) President's message: is the rate of homeownership nearing a bottom? The Regional Economist, October. https://www.stlouisfed.org/publications/re/articles/?id=1715

Calem PS, Hershaff JE, Wachter SW (2004) Neighborhood patterns of subprime lending: evidence from disparate cities. Hous Policy Debate 15(3):603–622

Campbell J, Giglio S, Pathak P (2009) Forced sales and house prices. Working paper. National Bureau of Economic Research, 14866

Carswell AT (2009) Effects of mortgage fraud on property tax assessments. J Prop Tax Assess Adm 6(2):5–17

Chambers M, Garriga C, Schlagenhauf DE (2007) Accounting for changes in homeownership rate. Working paper 21. Federal Reserve Bank of St. Louis

Cohen LE, Felson M (1979) Social change and crime rate trends: a routine activity approach. Am Sociol Rev 44:588–608

Collins CM, Harvey KD, Fulmer A, Nigro PJ (2009) Mortgage Fraud's impact on housing markets: an Atlanta case study in neighborhood collateral damage. White paper submitted to a meeting to Mortgage Fraud, Foreclosures and Neighborhood Decline. National Institute of Justice

Crossney KB (2010) Is predatory mortgage lending activity spatially clustered? Prof Geogr 62(2):153–170

Crump J, Newman K, Belsky ES, Ashton P, Kaplan DH, Hammel DJ, Wyly E (2008) Cities destroyed (again) for cash: forum on the U.S. foreclosure crisis. Urban Geogr 29(8):745–784

Cui L (2010) Foreclosure, vacancy and crime. Working paper. Available at http://papers.ssrn.com/sol3/papers.cfm?abstract_id=1773706. Accessed 1 Aug 2013

Doms M, Motika M (2006) The rise in homeownership. Federal Reserve Bank, San Francisco Economic Letter, 3

Ellen IG, Lens MC, O'Regan K (2012) Memphis murder mystery revisited: do housing voucher households cause crime? Hous Pol Debate 22(4):551–572

Fisher LM, Lambie-Hanson L, Willen PS (2013) The role of proximity in foreclosure externalities: evidence from condominiums. Public Policy Discussion Papers, No. 13–2. Federal Reserve Bank of Boston

Fulmer A (2009) Burning down the house: mortgage fraud and the destruction of residential neighborhoods. White paper submitted for a meeting on Mortgage Fraud, Foreclosures and Neighborhood Decline. U.S. Department of Justice, National Institute of Justice, Washington, DC

Galster G (1982) The social neighborhood: an unspecified factor in homeowner maintenance. Urban Aff Q 15(2):235–254

Galster GC, Cutsinger JM, Malega R (2006) The social costs of concentrated poverty: externalities to neighborhood households and property owners and the dynamics of decline. Revisiting Rental Housing: A National Policy Summit. http://www.npc.umich.edu/publications/working_papers/?publication_id=107&

Gibbons S, Machin S (2008) Valuing school quality, better transport, and lower crime: evidence from house prices. Oxford Rev Econ Policy 24(1):99–119

Goodstein RM, Lee YY (2010) Do foreclosures increase crime? Working paper. Available at http://papers.ssrn.com/sol3/papers.cfm?abstract_id=1670842. Accessed 1 Aug 2013

Gottfredson MR, Hirschi T (1990) A general theory of crime. Stanford University Press, Stanford

Grubesic T, Mack E (2008) Spatial and temporal interaction of crime. J Quant Criminol 24:285–306

Haberman CP, Ratcliffe JH (2012) The predictive challenges of near repeat armed robberies. Policing 6(2):1–16. doi:10.1093/police/pas012

Harding JP, Rosenblatt E, Yao VW (2009) The contagion effect of foreclosed properties. J Urban Econ 66(3):164–178

Immergluck D, Smith G (2005) Measuring the effect of subprime lending on neighborhood foreclosures: evidence from Chicago. Urban Aff Rev 40(3):362–389

Immergluck D, Smith G (2006) The impact of single-family mortgage foreclosures on neighborhood crime. Hous Stud 21:851–866

Immergluck D, Wiles M (1999) Two steps back: the dual mortgage market, predatory lending, and the undoing of community development. Research report. Woodstock Institute, Chicago

James D, Butts J (2011) Thirteenth periodic Mortgage Fraud Case report. LexisNexis® Mortgage Asset Research Institute, May. Last Accessed at: http://nationalmortgageprofessional.com/sites/default/files/LexisNexus_Report_05_09_11.pdf

Johnson SD, Bowers K, Hirshfield A (1997) New insights into the spatial and temporal distribution of repeat victimization. Br J Criminol 37(2):224–241
Johnson SD, Bernasco W, Bowers KJ, Elffers H, Ratcliffe J, Rengert G, Townsley M (2007) Space–time patterns of risk: a cross national assessment of residential burglary victimization. J Quant Criminol 23:201–219
Jones RW, Pridemore WA (2012) The foreclosure crisis and crime: housing-mortgage stress associated with violent and property crime in U.S. metropolitan areas? Soc Sci Q 93(3):671–691. doi:10.1111/j.1540-6237.2012.00887.x
Kaplan DH, Sommers GG (2009) An analysis of the relationship between housing foreclosures, lending practices, and neighborhood ecology: evidence from a distressed county. Prof Geogr 61(1):101–120
Katz CM, Wallace D, Hedberg EC (2012) A longitudinal assessment of the impact of foreclosure on neighborhood crime. J Res Crime Delinq 50(3):359–389. doi:10.1177/0022427811431155
Kirk DS, Hyra DS (2012) Home foreclosures and community crime: causal or spurious association. Soc Sci Q 93(3):648–670. doi:10.1111/j.1540-6237.2012.00891.x
Knox G (1964) Epidemiology of childhood leukemia in Northumberland and Durham. Br J Prev Soc Med 18:17–24
Lee K (2008) Foreclosure's price-depressing spillover effects on local properties: a literature review. Community affairs discussion paper. Federal Reserve Bank of Boston, Boston
Leonard T, Murdoch J (2009) The neighborhood effects of foreclosure. J Geogr Syst 11:317–322
Levine N (2010) *CrimeStat*: a spatial statistics program for the analysis of crime incident locations (v 3.3). User's manual. Ned Levine & Associates, Houston; National Institute of Justice, Washington, DC
Lin Z, Rosenblatt E, Yao VW (2009) Spillover effects of foreclosures on neighborhood property values. J Real Estate Finance Econ 38(4):387–407
Mendelsohn R, Olmstead S (2009) The economic valuation of environmental amenities and disamenities: methods and applications. Annu Rev Environ Resour 34:325–347
Metzger JR (2000) Planned abandonment: the neighborhood life-cycle theory and national urban policy. Hous Policy Debate 11(1):7–39
Miller HJ (2004) Tobler's first law and spatial analysis. Ann Assoc Am Geogr 94(2):284–289
Morenoff JD, Sampson RJ, Raudenbush SW (2001) Neighborhood inequality, collective efficacy, and the spatial dynamics of urban violence. Criminology 39:517–558
Office of Federal Housing Enterprise Oversight (2008) http://www.fhfa.gov/DataTools/Downloads/Pages/House-Price-Index.aspx
O'Sullivan O (2003) New foreclosure phenomenon. ABA Bank J 95(11):77–83
Olson MW (2005) Exploring the benefits and challenges of an ownership society. Remarks by Governor Mark W. Olson at the Community Development Policy Summit: Federal Reserve Bank of Cleveland, Cleveland
Pavlov AD, Wachter SM (2009) Subprime lending and real estate prices. Institute for Law & Economic research paper no. 09–36. University of Pennsylvania, Philadelphia
Pennington-Cross A (2004) The value of foreclosed property. J Real Estate Res 28:193–214
Quercia RG, Stegman MA (1992) Residential mortgage default: a review of the literature. J Hous Res 3(2):341–379
Ratcliffe JH (2009) Near repeat calculator (version 1.3). Temple University, Philadelphia; National Institute of Justice, Washington, DC. http://www.temple.edu/cj/misc/nr/. Accessed 1 Aug 2013
Ratcliffe JH, Rengert G (2008) Near-repeat patterns in Philadelphia shootings. Secur J 21:58–76
Rohe WM, Van Zandt S, McCarthy G (2000) The social benefits and costs of homeownership. Working paper no. 00–01. Research Institute for Housing America, Washington, DC
Schuetz J, Been V, Ellen IG (2008) Neighborhood effects of concentrated mortgage foreclosures. NYU Law and Economics research paper no. 08–41
Scott FW (2010) Estimating the effect of mortgage foreclosures on nearby property values: a critical review of the literature. Fed Reserv Bank Atlanta Econ Rev 95(3):1–9
Short MB, D'Orsogna MR, Brantingham PJ, Tita GE (2009) Measuring and modeling repeat and near-repeat burglary effects. J Quant Criminol 25(3):325–339. doi:10.1007/s10940-009-9068-8

Simons RA, Quercia RG, Maric I (1998) The value impact of new residential construction and neighborhood disinvestment on residential sales price. J Real Estate Res 15(1/2):147–161

Smith JC (2010) The structural causes of mortgage fraud. Syracuse Law Rev 60:433–461

Stucky TD, Ottensmann JR, Payton SB (2012) The effect of foreclosures on crime in Indianapolis. Soc Sci Q 93(3):602–624. doi:10.1111/j.1540-6237.2012.00890.x

Tobler W (1970) A computer movie simulating urban growth in the Detroit region. Econ Geogr 46(2):234–240

Townsley M, Homel R, Chaseling J (2000) Repeat burglary victimization: spatial and temporal patterns. Aust N Z J Criminol 33:37–63

U.S. Department of Housing & Urban Development (2010) Report to congress on the root causes of the foreclosure crisis. U.S. Department of Housing and Urban Development, Washington, DC

Wachter SM (2009) The foreclosures crisis and what is to be done. Testimony to the Joint Economic Committee, Committee on Financial Services, U.S. House of Representatives on Current Trends in Foreclosure and *What More Can be done to Prevent Them*

Wachter SM, Russo K, Hershaff JE (2006) Subprime lending: neighborhood patterns over time in US cities. Institute for Law & Economic Research paper no. 06–19. University of Pennsylvania, Philadelphia

Wallace D, Hedberg EC, Katz CM (2012) The impact of foreclosures on neighborhood disorder before and during the housing crisis: testing the spiral of decay. Soc Sci Q 93(3):625–647. doi:10.1111/j.1540-6237.2012.00886.x

Wells W, Ling W, Ye X (2011) Patterns of near-repeat gun assaults in Houston. J Res Crime Delinq 49(2):186–212

Welsch A (2008) Economic trends, foreclosures and county budgets. Issue brief. National Association of Counties, Washington, DC

Whitaker S, Fitzpatrick IV TJ (2011) The impact of tax-delinquent and foreclosed property on sales prices of neighboring homes. Working paper 11–23. Federal Reserve Bank of Cleveland, Cleveland

Wilson RE (2015) The neighborhood context of foreclosures and crime. Cartogr Geogr Inform Sci 42(1) (Forthcoming)

Wilson RE, Paulsen D (2009) A theoretical underpinning of neighborhood deterioration and the onset of long-term crime problems from foreclosures. White paper submitted for a meeting on Mortgage Fraud, Foreclosures and Neighborhood Decline. U.S. Department of Justice, National Institute of Justice, Washington, DC

Wyly EK, Hammel DJ, Atia M (2004) Capital is the landlord: class monopoly and new geographies of subprime and predatory mortgage lending. Working paper

Chapter 6
State Registration of Sex Offenders: Public Notification, Web Mapping, and Spatial Issues

Mark R. Leipnik and Xinyue Ye

Abstract State governments in the United States register sex offenders and provide to the public notification of the location, identifying information and criminal history of these registered sex offenders (RSOs). Four fifths of US states use web maps to achieve this goal. The design, content, and maintenance of these RSO web maps vary greatly. This chapter examines the history of registration of sex offenders, similarities and differences among RSO web maps and spatial issues related to them, as well as the use and misuse of these web maps.

Keywords Geographic information system (GIS) • Web map • Registered sex offender (RSO) • Public notification • Spatial issues

6.1 Introduction

> …discipline tries to rule a multiplicity of men to the extent that their multiplicity can be dissolved into individual bodies that can be kept under surveillance, trained, used and if need be punished.… Michel Foucault, "Society Must Be Defended," March 17, 1976

Noted philosopher Michel Foucault stated the ideas propounded in the above quotation as part of a series of influential public lectures that focused on state power and the regulation of the individual, in particular their sexuality. However, he was

M.R. Leipnik (✉)
Department of Geography and Geology, Sam Houston State University,
P.O. Box 2148, Huntsville, TX 77341, USA
e-mail: geo_mrl@shsu.edu

X. Ye
Department of Geography, Kent State University,
413 McGilvrey Hall, Kent, OH 44242, USA
e-mail: xye5@kent.edu

G.A. Elmes et al. (eds.), *Forensic GIS: The Role of Geospatial Technologies for Investigating Crime and Providing Evidence*, Geotechnologies and the Environment 11,
DOI 10.1007/978-94-017-8757-4_6,

not thinking of computerized registries of sex offenders with online maps stored in a geographic information system (GIS), global positioning systems (GPS) monitoring, or residential restrictions based on spatial analysis. These were all technologies that were nascent or nonexistent at that time and approaches to management of offenders that did not develop for several decades. He was referring to the eighteenth-century innovation, first systematically implemented in France's Bourbon police state, of using population registries that "were used to ensure the spatial distribution of individual bodies (their separation, their alignment, their serialization and their surveillance) and the organization around those individuals of a whole field of visibility" (Foucault 1976).

The idea of registering population (not merely counting them in a census) existed in Sung Dynasty China before it did in France. Nonetheless, the modern approach was pioneered in France and introduced into other European nations by Napoleon whose invasive police state systematically imposed registration requirements on residents and required workers to obtain identity cards (Broers 1999). It is notable that before the age of computerization and information networks, the French registry system was vulnerable to criminals like Landru, the serial killer active between 1915 and 1920 who simply registered in multiple *communes* under at least ten different aliases (Bardens 1990). Long before the concept of registration of criminal sex offenders per se appeared, most continental European police agencies maintained detailed registries of all criminals, as well as tracked the residency and travels of the general population and they still do.

In contrast to the situation now existing in the United States of America (USA), the information in European registries is confidential. Where the practice in the United States differs from every country on earth is that there is not only a process for registration of criminal sex offenders but that there is a nationwide mandate that each state provide a mechanism for the public to be notified of the names, addresses, descriptions, photographs, and other characteristics of RSO's.

This chapter will examine the history of that requirement, the use of web maps to meet that requirement in 41 of 50 states currently, and the great variations in the design and content of those web maps. The chapter also examines many of the spatial issues which are related to the accuracy and interpretation of the data on those sites. The typical use and the potential for misinterpretation and misuse of the information will also be examined.

6.2 Public Notification Requirements and Methods

6.2.1 Global Context

Public notification of the names, locations, physical characteristics (including photographs), and criminal history of sex offenders is a relatively recent development. It is also a situation that is unique in the United States. Most continental European law enforcement agencies have a centralized authoritarian "police state" tradition in which extensive centralized files are maintained, but information about criminals is not provided to the public. The United Kingdom has a more open approach but also is noted for protecting the rights of suspects and maintaining the confidentiality of investigations and files. The colonial police forces of Great Britain adopted a centralized authoritarian model, as did those of other colonial powers. These colonial police forces were the progenitors of those in most of the developing world, and the centralized model has also been attributed to the police of Japan. The United States has been noted for its exceptionalism with a decentralized and community-oriented policing approach historically (Mawby 2011). Developed countries including the United Kingdom (UK), Canada, Ireland, Australia, New Zealand, Germany, the Czech Republic, France, and Belgium all maintain computerized databases of sex offenders. In these nations, data on sex offenders is maintained in a GIS either by the national police (in France, Belgium, and by the RCMP in Canada) or by a state police agency (like the Bavarian State Police or Victoria State Police in Australia) or by a regional constabulary like the London Metropolitan Police. Germany and the United Kingdom in particular use GIS to study the spatial distribution of RSOs in relation to crime incidents (Leipnik and Albert 2003). In the United Kingdom information about the locations of approximately 32,000 sex offenders is confidential, and only generalized locations of incidents involving sex crimes are available to the public in online maps (Cheney and Ratcliff 2005).

6.2.2 US Situation

In contrast, every state and territory and most Native American tribes in the United States have online web sites that provide access by the public to detailed information about approximately 700,000 RSOs and their locations on approximately 165 official web sites. There are a wide range of estimates of the number of RSOs in the United States ranging from about 325,000 to 747,408 (NCMEC 2013). The main reason for the differences in estimates is that every state, territory, and tribe maintains their own registries with inconsistent criteria, particularly for listing and removing low-level offenders. The double, triple, etc., counting of offenders in the registries of multiple states is a source of overestimation, as is inclusion of multiple aliases and failure to remove offenders after the expiration of registration, exoneration, deportation, or death. All these factors tend to inflate numbers. Conversely, some offenders get omitted from some estimates. This includes juveniles and low-risk offenders. Also many states omit those RSOs that are institutionalized on an original charge, for reoffending, held on nonsexual charges or parole violations. RSO numbers are tending to increase because registration terms often are for life and the process only started in most states between 1996 and 2006. The information available on public web sites about these numerous RSOs generally includes data for those residing in the community and those that are incarcerated, in civil commitment, or those that have absconded or failed to register (Thomas 2011).

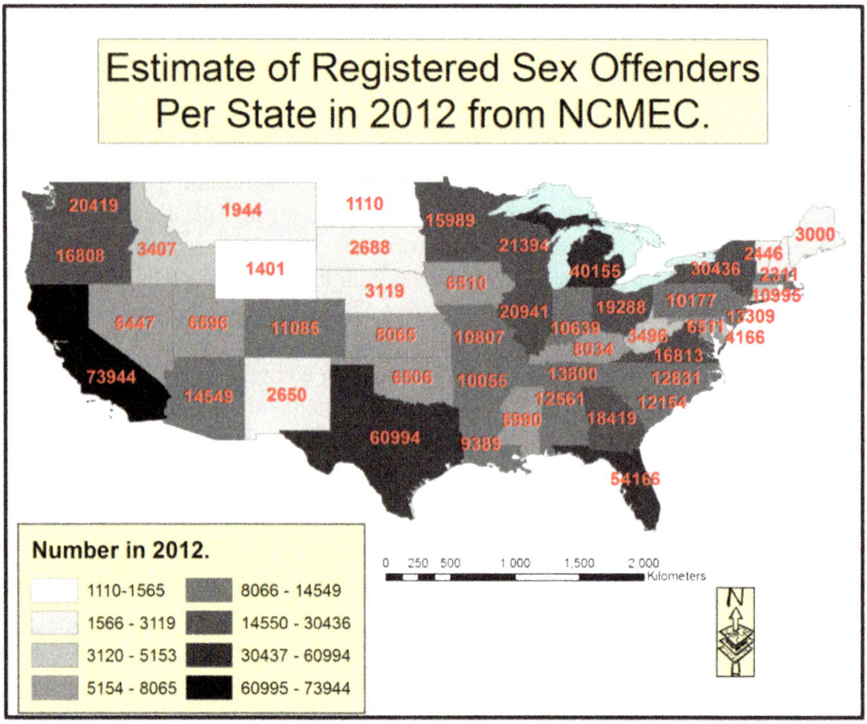

Fig. 6.1 Map of conterminous United States showing the number of registered sex offenders and rate per 100,000 population based on 2012 data from NCMEC and 2010 population estimates (Map prepared by Mark Leipnik, 2013)

State-maintained web sites are the primary means of dissemination of this data, and there are also some private web sites with links to relevant data. An example of a map based on 2012 data from the National Center for Missing and Exploited Children (NCMEC) on the estimated 736,747 RSOs in the United States appears in Fig. 6.1.

It shows that California, large in size and largest in population, had the most RSOs with 73,944. There were great differences in the number of RSOs per 100,000 inhabitants in each state counted by the 2010 census. The national average was 235/100,000, but the small state of Delaware had the highest rate with 527/100,000 and neighboring state of Pennsylvania had the lowest rate with only 69/100,000 (NCMEC 2013). The NCMEC is one of the many organizations with data on the spatial distribution of RSOs; most state sites have web maps available. There is, however, great variability in the design and content of such sites as well as significant deficiencies in them. For example, nine states, most territories, and all tribes lack a web mapping capability. Many of those 41 states, Washington DC, and Guam that do have web maps have issues with missing data, outdated information, and RSO locations that either cannot be geocoded or are geocoded incorrectly. Every state and territory also has a large, difficult-to-quantify, and highly variable issue of offenders who fail to register, abscond, or are in violation of the terms of their parole or of registration requirements (such as providing false information) or of the requirements of local or statewide residential restrictions.

6.2.3 History of Regulation

Public notification came about in the 1980s and 1990s in the United States as a result of growing anger by the families of victims which was amplified by television and online media and which generated a crescendo of public frustration that resonated with state level and then eventually national politicians. The resulting series of laws at the state and federal levels have been generally named for those who have fallen victim to predatory sex offenders. The first of these cases, and the law it inspired, set the pattern for those that followed. It involved Earl Shriner, who was the kidnapper of two teenage girls. He served his entire 10-year sentence in a Washington State prison and was released in 1987. He was not paroled because he informed prison staff about his intention to attack children upon his release. Paradoxically, because he served his entire sentence, there was no legal basis for his supervision after release. Neither was there a registration requirement, so his past crimes and proclivities were not easily accessible information. He then committed two more attempted sexual assaults on girls in 1988 and received short terms in local jails. Then in 1989 he attacked and grievously sexually mutilated a 10-year-old boy. Pressure on the governor and Washington State legislature stemming from public outcry led to three influential developments embodied in two laws: the Community Protection Act of 1990 and the Sexually Violent Predators Act of 1990 (Washington State, 1990). These acts required the registration of all sex offenders including details of their residence and work locations, current photographs and physical descriptions, and details of their crimes and prison records. Another requirement was a mechanism for civil commitment of predatory offenders, so these very rare criminals could be held involuntarily in treatment facilities after release from prison, whether deemed sane or not (Winick 2005). The third, and perhaps the most significant feature of the 1990 laws, was the public notification requirement. This required the "community" into which sex offenders were to be released to be notified via media outreach, public meetings, or some other mechanism. Initially this notification process took the form of evening public meetings in places like high school auditoriums where local officials explained to often disturbed residents that a sex offender was being released back into their community.

The rapid development of the Internet in the 1990s soon led to the obsolescence of public notification via in-person meetings or notices in print or broadcast media with that approach being superseded by web-based methods and in most states (including Washington State) by web maps. Although Washington was the first state to institute a public notification requirement, other states also had incidents that resulted in similar state laws. The Minnesota Predatory Offender Registration Act of 1991 was enacted in response to the abduction of 11-year-old Jacob Wetterling.

This act required a 10-year registration for sexual offenses involving children. In 1992 Louisiana adopted a particularly stringent law that included registration, an option for "chemical castration," extensive public notification, and the optional emplacement of yard signs and placards on automobiles of RSOs (Louisiana State Police 2012). Unlike Washington and Louisiana, Minnesota did not require public notification or provide public access to the registry and in fact as of 2013 Minnesota remained one of only nine states that did not have an official RSO web map. Minnesota today provides a web site without mapping capabilities, but that is publicly accessible in response to the 1996 federal requirements of "Megan's Law."

Megan's Law originated in New Jersey, which along with Tennessee, Alaska, Indiana, and 18 other states had by 1994 adopted some variation of registration of sex offenders. The New Jersey law was in response to the rape and murder of 7-year-old Megan Kanka, a girl that had lived on the same street as her killer, a man with a record of multiple previous child sex offenses. This criminal embodied a combination of a child murderer, a sex criminal, a recidivist, and an unsuspected and apparently harmless neighbor. This man represents the type of offender that the public appears to fear most and demand protection from to the greatest degree.

These fears however are typically overblown since most sexual crimes do not involve children and of the 60,000–70,000 arrests each year in the last decade in the United States for sex crimes against children, only 40–50 each year involve homicide (Finkelhor 2008). Also recidivism among sex offenders is not particularly high. A 1994 Bureau of Justice Statistics (BJS) study indicated that only 5 % of sex offenders were rearrested for another sex crime within 3 years of their release (BJS 1994). Another 2003 study found that only 3.3 % of released child molesters were again arrested for committing another child sex crime within 5 years (BJS 2003). BJS data from 2007 indicate that within 3 years of release, 2.5 % of rapists are rearrested for rape compared with over 70 % of burglars and car thieves rearrested for these other crimes, respectively (BJS 2007). Despite these findings, restrictive, and in some cases severe, laws related to sex offenders proliferated in the 1990s and 2000s. The proliferation of similar, but distinct, state laws proved this issue was politically a highly popular one for state legislators to embrace.

This encouraged US Congress in 1994 to begin the process of enacting seven separate laws on the subject, usually at 2-year intervals in election years (1994, 1996, 1996, 1998, 2000, 2003, and 2006). Five of these laws are important, while two represent minor adjustments to the others. The first federal law was the Jacob Wetterling Crimes against Children and Sexually Violent Offender Registration Act of 1994 (revised in 1998). The law required registration by states, territories, and tribes of those persons who committed violent sex crimes and all sex crimes involving children, even if not violent. The establishment of registries by the states in response to this law was not instantaneous (Stimson and Noronha 2011). Civil liberties groups like the American Civil Liberties Union opposed the registration requirement at state and federal levels and were especially effective in Massachusetts which was the last state to enact RSO registration 4 years after passage of the federal mandate. The federal law did not specify much detail about the information to be maintained in the registry and left it up to states to enact their own laws and develop their own procedures for enforcement which were generally somewhat different in each state. In this it set the pattern for later federal laws relating to registered sex offenders. Since a sex crime is in most cases not a federal offense, it must be punished and felons regulated by the adoption of laws in each state. In this regard, the several states are serving as "laboratories" for the evolution of a range of sex offender registration and public notification approaches, some more effective than others.

The most famous of these series of federal laws was "Megan's Law" of 1996. This is such a well-known statute that the term "Megan's Law" has become synonymous with information about RSOs as in *Megan's Law map* or *Megan's Law list.* It was modeled on the New Jersey statute of the same name. The federal version requires registration by each state of rapists and child sex offenders and "community notification" by every state of the location and characteristics of RSOs, typically by means of state-maintained web sites. But the federal law does not mention the concept of web mapping nor do the state laws enacted to comply with it. Federal sex crimes also incur public notification; this along with web-based coordination among states was required by the Pam Lychner Sexual Offender Tracking and Identification Act of 1996. It mandated the establishment of a federal database on RSOs to help track offenders who moved from one state to another. This law was strengthened in 1998 by the requirement that a link to all the state web sites that were mandated by Megan's Law be maintained by the federal government. The most recent addition to this arsenal of laws was the Adam Walsh Child Protection and Safety Act of 2006. This law addressed a weakness in state notification requirements in that that they often lacked any risk assessment information. This meant that someone guilty of the more common but minor offenses of "statutory rape" or indecent exposure was not easily differentiated from the much rarer but more dangerous violent predator on a state's web site. This law provides a mandate to the states that the risk of RSOs to the community be assessed and made publicly available (Ackerman et al. 2011).

The Adam Walsh Law posed a challenge to state governments since it required that offenders "risk" needed to be assessed in some formal manner, an assessment that is then subject to legal appeal and reassessment at a later time due to changed circumstances. Assessing the risk of RSOs is a costly, complex, and controversial process (Craig et al. 2008). The law mandated a three-tier risk assessment system. Tier III were high risk (usually violent rapists or child sex predators), tier II were moderate risk (often repeat offenders), and tier I were low risk (usually those guilty of nonviolent sex crimes or a single offense involving an adult victim). The law also stipulated that high-risk juvenile offenders continue to be registered after they turn 18.

The federal law sought to impel each state to adopt its own version of this law by withholding 10 % of federal law enforcement grant funds if it failed to comply. Seven years later, only Alabama, Delaware, Florida, Kansas, Louisiana, Maryland, Michigan, Mississippi, Missouri, Nevada, Ohio, Pennsylvania, South Carolina, South Dakota, Tennessee and Wyoming, and Guam, the Northern Mariana Islands, and the US Virgin Islands had complied (National Conference of State Legislatures 2013). It is notable that all the compliant states have web maps of RSOs. But 34 states and the District of Columbia have not complied. That means most states including the three most populous are not in compliance with the law. In fact several states including California and Texas are not complying because they have performed cost-benefit analysis showing that a loss of 10 % of federal grant monies would be less than the cost of complying. Oddly, Texas has risk information about RSOs publicly available in its web maps, but chooses not to comply (U.S. Senate 2009).

Another problem with the system of having the federal government mandate a rather vague set of requirements for registration and public notification but leave it up to states to adopt their own laws to implement the mandate is that there is not a uniform national system of public notification. This lack of uniformity can be rapidly grasped by a close look at the web sites maintained by each of the 50 states. The design, ease of use, and information content of these web sites is markedly different. As the detailed discussion in the next section will show, the design and content of these sites vary widely and some are dysfunctional, making it difficult for users to comprehend their proximity to RSOs and hence take prudent defensive measures or rationally assess risks either individually or as a society (US Senate 2009).

6.3 Web-Based Mapping of Registered Sex Offenders

The biggest difference among states with respect to public notification of the locations and details of RSOs is that some states have web sites with web maps and some lack web maps entirely (see Fig. 6.2 for a map of the conterminous United States highlighting those states that do and do not use web maps).

The 13 states using the *Offender Watch* approach and Tennessee, Maryland, Missouri, and Pennsylvania feature particularly high-quality web maps. See Fig. 6.3 for an example of a map generated from Tennessee's very well-designed web site.

6.3.1 Lead Agency

The question of which agency is responsible for the RSO web site is somewhat simplified by Megan's Law being a mandate at statewide level, so it must cover every jurisdiction within a state. There exist over 18,000 law enforcement jurisdictions in the United States with complex overlapping responsibilities that engender cooperation and confusion (Leipnik et al. 2013). However, at the state level, there are fewer "usual suspects" that could take responsibility for an RSO web site. In states that do not have web maps, it is generally the Department of Corrections or the Board of Paroles or state police that has authority for the maintenance of the tabular registry.

Of states with web maps, only Utah and Wisconsin have the responsibility given to the Department of Corrections. Utah uses the *Offender Watch* approach, while Wisconsin uses the *Family Watchdog* web site. In Maryland, a state with a very good web map, largely developed at Towson University which uses a customized base map from Esri, Inc., the site is a joint effort of the state police and Department of

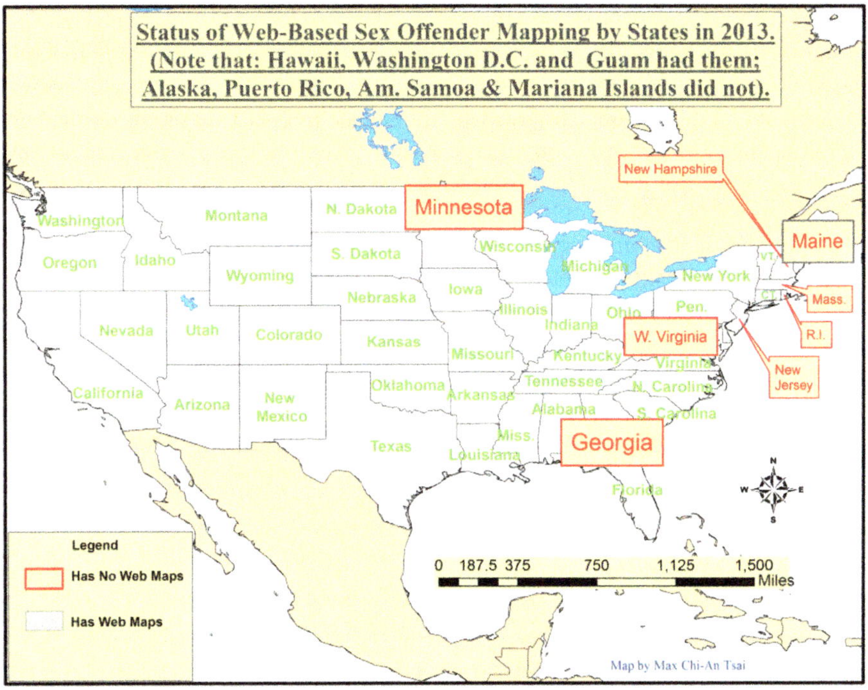

Fig. 6.2 Use of web maps for RSO notification by states in conterminous United States in 2013 Note that in 2014 Alaska also created a RSO web map. (Map prepared by Mark Leipnik, 2013)

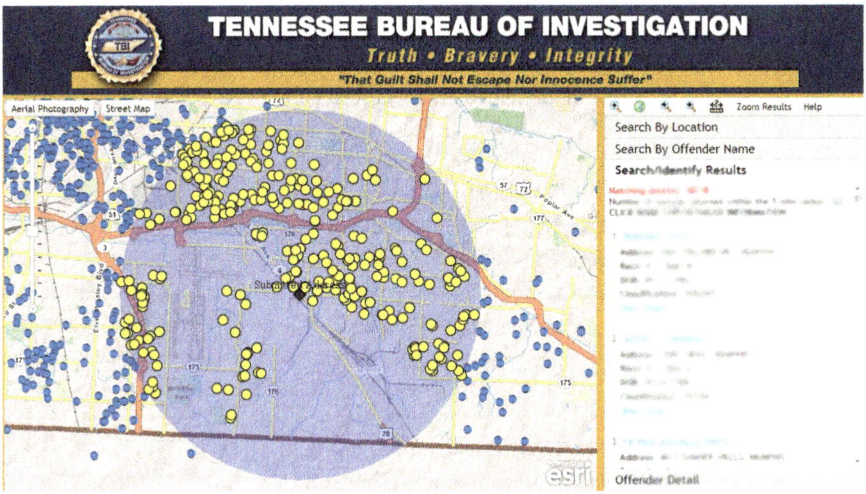

Fig. 6.3 Map of registered sex offender locations using Tennessee's top-quality web site. Tennessee Bureau of Investigation, 2013. http://www.tbi.state.tn.us/sex_ofender_reg/sex_ofender_reg.shtml

Corrections. This is the only site that is a joint undertaking. In most states the state police are responsible for the web site, and states with a powerful state police such as Michigan and Delaware have good sites. Several states have a State Bureau of Investigation which is not a highway patrol or state trooper-type organization, but a specialized crime investigation agency. Kansas, Kentucky, and Tennessee all have these agencies, and they are responsible for the RSO web maps. Alaska has a very active state police but until 2014 no web maps. The new web site is managed by them and has many "unmappable" offenders along with those mapped at locations such as public docks and fish cannries related to the apparent concentration of transient RSO's in that trade. Hawaii, another atypical state, has a state police with few responsibilities beyond guarding state offices, and there the Attorney General is responsible for a limited RSO web, notable for using only one type of symbol for all offenders. Hawaii is like California, Ohio, and North Carolina in that the State Attorney Generals are responsible. In New York State, the Criminal Justice Services Department is responsible. Arkansas is unusual in that they have a State Crime Information Center that developed the web maps.

6.3.2 *Offender Watch*

The states of Alabama, Connecticut, Illinois, Indiana, Kentucky, Louisiana, New Mexico (in 2013), Ohio, South Carolina, Utah, Vermont, and Washington (and in late 2013 Michigan will be added) all share a similar approach that features high-quality online RSO maps and intuitive tools for navigating the data displayed. This approach is *Offender Watch* and is developed and maintained by the firm Watch Systems, Inc., based in Covington, Louisiana. Figures 6.4 and 6.5 contain the Porter County Indiana, gateway page, and the locations and pixelated names of RSOs living in downtown Salt Lake City, Utah. Watch Systems was founded in 2001 and has numerous clients both for mapping RSOs and at county appraisal districts which use web-based GIS tools for public access to property tax-related information.

The first state to use the Offender Watch approach was Ohio, and more have adopted this approach recently including New Mexico and Illinois which both replaced earlier internally developed RSO web maps. Watch Systems hosts the sites on its server which provides reliable access. Access to certain other state-maintained web sites is problematic, for example, Missouri's RSO site had multiple issues with calling geocoding functions and displaying base maps and the Pennsylvania's RSO site was down for extended periods in 2013. Watch Systems has 12 years of experience in RSO web mapping serving 13 states, so clients benefit from economies of scale in use of a common high-quality base map, in design of effective tools and user interfaces, in system administration, in error detection and correction, and in rapid update of information versus developing and maintaining an internal web site and map server in each state with similar capabilities (Roberts 2013).

The Offender Watch approach allows the public to search for information using a map primarily to locate points representing RSO locations or alternatively generate a tabular list of names. A list of names of RSOs is determined based on a

Fig. 6.4 Porter County Indiana's gateway page, served and designed as are all Indiana County Sheriff's RSO pages by Watch Systems. Porter County Indiana Sheriff's Office, 2013. http://www.portercountysheriff.com/main/sexoffender.html

geographically specified area like a county, zip code zone, city, or street address, or alternatively information on a single RSO can be obtained by entering a name (or alias). If a street address is chosen as the basis of a starting location, then a buffer zone of various radii can be generated around that location so RSOs residing in that area can be identified. Once a single RSO of interest has been selected, users then have the option to go back to a map to see the primary and/or secondary residence of the offender. Although the excellent cartographic design of all 13 Offender Watch State's web sites is very similar, there are differences in approach. For example, the size and number of buffer zones and whether or not incarcerated offenders can be excluded at the outset from the search.

It is also interesting to note that many county and local law enforcement agencies use the Offender Watch approach independently of a state agency. Some of these are in states without web maps of RSOs, but many are in states with official state RSO web sites, but where for various reasons local authorities have chosen to adopt what they consider a superior approach. For example, officials with the Saint Lawrence County, New York Sheriff's Office, use Offender Watch to notify the public about offender status as well as submitting the information to the state of New York's official web site. In New York State when an offender moves or changes jobs, the

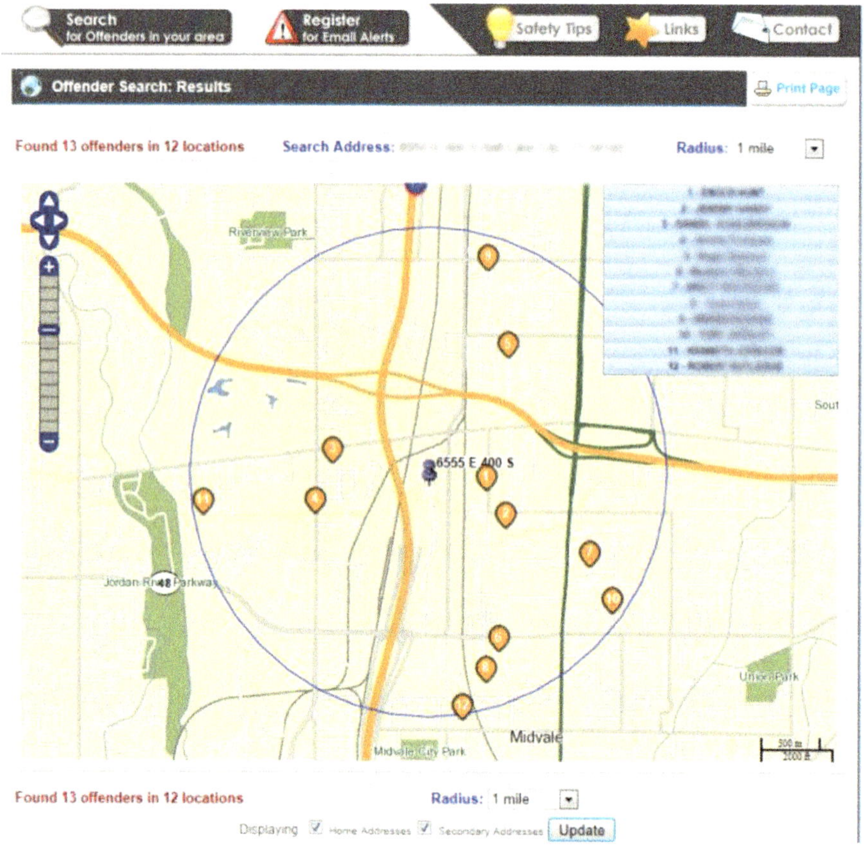

Fig. 6.5 Map from the Utah Sex Offender Registry web site of RSO locations in Salt Lake City Utah using the Offender Watch approach. Utah Department of Corrections, 2013. http://www.communitynotification.com/cap_office_disclaimer.php?office=54438

local agency monitoring them submits a form by surface mail or by fax to a state office in Albany, and it can take a month for the information to be updated and reflected on the state's RSO web map. This paper-based process is antiquated compared to the online submission of updated information to Watch Systems. Local people in Saint Lawrence County were complaining that the state RSO web maps and databases did not reflect changes in offender locations they were aware of (Watch Systems 2013).

6.3.3 *Quality of Web Maps*

The states that use Offender Watch such as South Carolina, Utah, and New Mexico along with Washington DC, Tennessee, Maryland, Missouri, and Pennsylvania have custom-designed interactive web maps of high quality (see Fig. 6.6 for an image

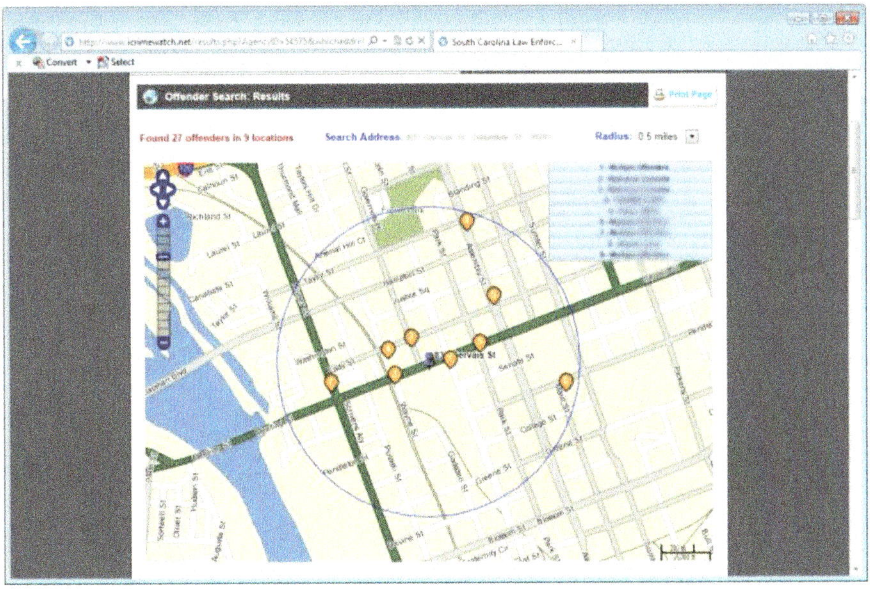

Fig. 6.6 Map of South Carolina RSOs offenders using Offender Watch cartographic design. South Carolina Law Enforcement Division, 2013. http://www.communitynotification.com/cap_main.php?office=54575

from South Carolina's RSO web map). These "good" maps feature many advanced cartographic design concepts that are typical of web maps developed in the period since 2005.

While the evaluation of the quality of cartography used in base maps as well as the ease of use of tools and range of analysis capabilities is a complicated and somewhat subjective topic, there are some basic differences between high-quality web maps and basic ones. For example, if all street centerlines have only one line weight (thickness) and one line color that is worse in terms of design than where an appropriate range of weights and colors are used, if contrasting and connotative colors are chosen that is better in design terms than where the color choice is rather arbitrary, if appropriate cartographic generalization is used, if additional layers are added based on scale ranges, and if label placement is effective—these are all important features of "good" web maps. Well-developed tools for pan and zoom and the ability to generate multiple buffer zones and add or subtract layers are also important attributes of high-quality web maps.

The full range of issues surrounding cartographic design of web maps is outside the scope of this chapter, but it is not difficult to identify advanced maps such as those of states like Tennessee and the Offender Watch states from those of states that rely on the good but not customized maps generated by Google, Bing, and other mapping services and the inferior efforts of some other states such as California (see Fig. 6.7 for an RSO map of Santa Barbara, California). The quality of cartographic design in the content of digital maps is discussed by Slocum et al. (2008), while the

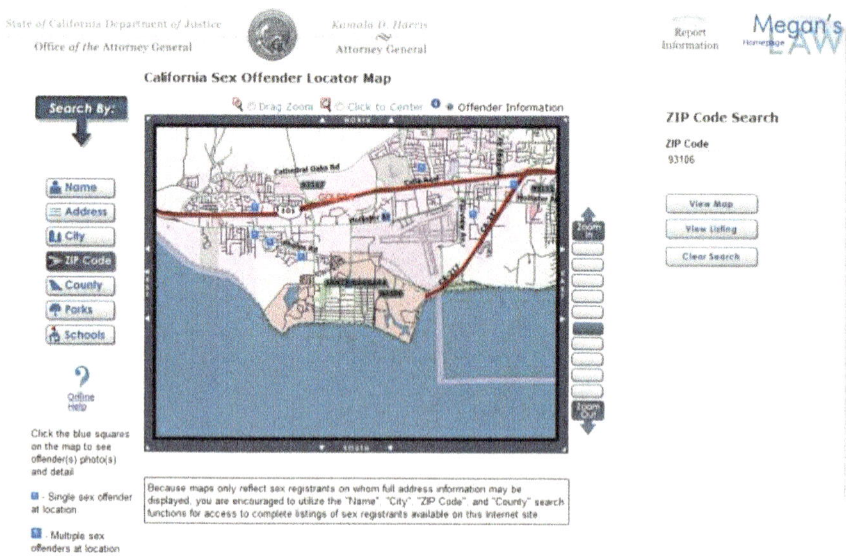

Fig. 6.7 Basic map from California's RSO web site of Santa Barbara area. State of California Office of the Attorney General, 2013. http://www.meganslaw.ca.gov/

more specialized issue of the quality of cartographic design of web maps is further discussed by Aileen Buckley of ESRI, Inc., in an online article (Buckley 2011).

6.3.4 Interactive Features

Until the recent adoption of the Offender Watch approach, Illinois had an early version of the web mapping approach which featured quite generalized base maps and limited pan and zoom functions, and California shares some of these characteristics which is a reflection of early adoption of the approach but also a lack of revision of the web site design as newer approaches have become the standard. This might reflect budget woes which are particularly acute in the most populous state (Leiphon 2008). One redeeming feature on the California web site is that once a specific RSO has been identified, there is an option to send the state a text message with a report on this specific offender. Such a function is an important aspect of a proactive approach to managing sex offenders. Every RSO web site has contact information, and many have features where the users can obtain additional information or view material on how best to use the information or learn more about protecting themselves from sex offenders. Some sites allow users to receive email alerts if an offender moves into proximity to them. The Offender Watch approach lets users register for email alerts as well as submit either corrections or "tips" related to specific offenders. For example, Mercer County Ohio's Sheriff's Department web site has a tab for "sex offenders" which opens another web page that lets users receive email alerts if an RSO locates within a 1-mile buffer zone of an address they have designated and/or moves into the county. This site also features an imbedded video of Shaquille O'Neal discussing how Internet predators can be avoided and suggesting users down load software to shield children browsing the web from predators. All counties in Ohio have similar capabilities.

Whether the maps are sophisticated in design like those of Ohio or more basic like those in California, the maps are essential to navigating the information spatially. However, an easy-to-use means of sending information back to the agency about a specific sex offender is a very valuable truly interactive public participation feature (Sieber 2006). Of course the value of tips can be questionable, and as in all communications to law enforcement particularly those protected by the apparent anonymity of the Internet, the proportion of spurious and duplicative information may be quite high. The ability to send a "tip" may however reassure the public and can lead to the prevention of crime as well.

6.3.5 *Map Mash-Ups*

There are a large number of states that use Google Maps mash-ups, including such diverse places as Hawaii, Iowa, Arizona, and Mississippi. A Bing Maps mash-up is used by Michigan, and in North Dakota, Yahoo Maps are used. Some maps feature sophisticated designs such as that of Montana, and some are much cruder cartographically such as those of Oregon.

Once an RSO has been identified, his or her address information is submitted to these cloud-based sites and geocoded. It can take a while for reported errors in geocoding to be addressed by firms like Google, and most errors are probably never reported by users. If the location can be successfully geocoded, then it can then be displayed on a user-selected type of base map. The choice of base map ranges from just the use of a conventional street map alone as in Iowa and Kansas (perhaps due to limited topographic relief) to one with hill-shaded terrain in the background as in Arizona and Arkansas to a satellite view as in Colorado and to an option for a hybrid of streets and a satellite view as in Montana and Idaho. Only in New York does there seem to be an option for seeing RSO locations displayed atop a "bird's-eye view" using high-resolution oblique aerial photography as a base map, perhaps because of the preponderance of high-rise urban structures in the places with the most RSOs in New York State. Two states (Nevada and Wisconsin) have only a link to a relatively poor cartographic design private web site called Family Watchdog which features context-related pop-up ads such as offers to expunge ones criminal record.

6.3.6 *Symbolization of Offender Locations*

The manner in which offender locations are symbolized varies greatly as does the provision of information about risk, status (i.e. "compliant", "incarcerated", etc.), and whether the location portrayed is only the primary residence or (in some states) a secondary residence; place of work and place of education (if the RSO is in school) are included. The Offender Watch states allow users to select "other addresses" which include secondary addresses for a specific offender in some states, but in other states like New Mexico, only information on a single primary residence is available.

The offender locations on every RSO web map are symbolized with point marker symbols only, although often a range of three or more primary colors and or a range of up to six different symbols are employed. New Mexico until 2013 used small 3D human figures, and Virginia uses a 3D symbol for a home or office or school location, but most commonly a drop point (an inverted teardrop symbol) is used. Push pin symbols and lozenge-shaped symbols are used as well. One can conceive of a situation where an area fill might be appropriate, for example, if an RSO owned, lived on, or worked on a large property. There are certainly cases of RSOs living in rural areas were a point symbol does not capture their location very effectively. Of course the disclaimers on most state RSO web maps indicate that symbol placement is approximate.

6.3.7 *Attribute Data*

Every state allows the public to obtain information on the web about the first and last name, age, gender, eye color, hair color, and addresses of offenders. Louisiana, which has a severe sex offender registration law, has in addition to this information data about any aliases used, current and past primary and secondary residences, place(s) of work, place(s) of schooling, make of car(s), model of car(s) and license plate number(s), home phone number, IP address, scars, tattoos, and very detailed information about the crime as well as level (tier I, II, or III) information.

Most states have color photographs of all offenders on their web sites. However, other states, notably New York, have only one that is often grainy and frequently in black and white in some cases it appears to be a decade or more out of date. However, an analysis of the spatial location of offenders with poor-quality photographs posted on the New York State RSO web site indicates that most of these are in cases of incarcerated offenders based on the associated addresses being an institutional setting. Conversely, most states have high-resolution color photographs, and some states have multiple angles and different dates displayed. In states such as Florida, these images and other summary details like name and tier information pop up on the map when a point location associated with an offender is queried (see Fig. 6.8 for an example of a Gainesville, Florida, RSO map).

In many states, the detailed attribute data appears in a summary form displayed below the map for all the offenders whose locations are displayed on the map. This approach is taken by the "Offender Watch" states. For these states, a sequential number starting at 1 appears in the symbols on the map of RSO locations, and this number relates to a number in the list which appears below the map of detailed attribute data for the offenders displayed on the map, and thus the information can be correlated. Arkansas uses a letter inside the point location symbol with the same intent, but if more than 26 RSOs are on the map, this approach is problematic.

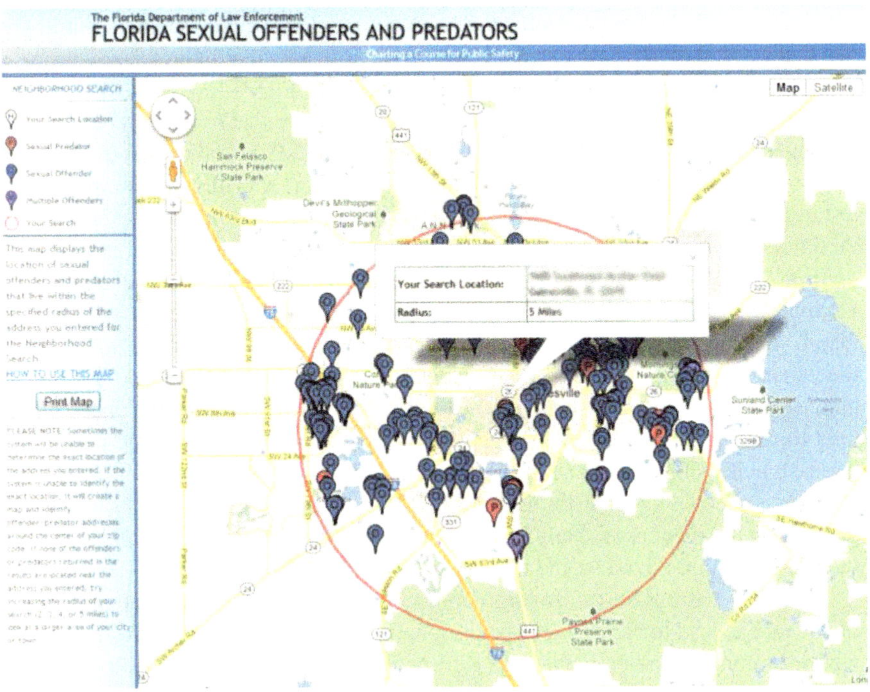

Fig. 6.8 Advanced map generated by Florida's RSO web site of Gainesville, Florida area. Florida Department of Law Enforcement, 2013. http://offender.fdle.state.fl.us/offender/homepage.do

Many states have an approach where if there is a concentration of offenders, the map will be symbolized with a symbol indicating that there are "multiple offenders at this location." This is the case in California and Colorado, for example, and it is a good idea given the high density of offenders in the urban areas in these two states. It is particularly important that multiple offenders at the same address be differentiated in states which do not exclude incarcerated offenders from the map.

At the state level there is typically only one official RSO web site, but many cities and counties have their own sites which link to statewide sites. San Jose municipal police in California is an example of a city with its own web site; likewise Broward County, Florida, and Rockland and Saint Lawrence County, New York, have their own sites. A unique variation is used by Ohio and Indiana which uses the same base map as all the states using the Offender Watch approach, but they have a statewide gateway page that then appears to users to link to the web page of each county sheriff's department.

Alternatively, a user can go directly to the county sheriff's office RSO web site which has the same design for every county in the state, except that it has the name of the county, a picture of each county sheriff and specific contact information for that county. One can search for offenders in any county in Indiana starting from any of these seemingly independent county sheriff's office web sites as long as a valid address is supplied (see Fig. 6.4 for the Porter County Indiana RSO web map gateway page). For users it seems the information is locally maintained, but in fact the data is statewide and the application is built and maintained on servers in Covington Louisiana by Watch Systems, Inc.

Most states also have some information about the crime that resulted in registration. However, some states like Minnesota have only the most general details (i.e., "rape," "sexual assault"), while some other states like New York and Texas have specifics as to the nature of the offense(s), the date it occurred, and the age and gender of the victim(s) as well as details on punishment(s) and restrictions.

In no state is the name of the victim or their location revealed, but potentially victims may discover that information on these web sites could be used to partially identify them. For example, victims of incest might discover that the last name of the parent or sibling involved could be determined along with the age of the victim and in a small town someone could make the connection, although the data is likely only to reveal that there was a sexual assault of a child of a certain age rather than the case being one of incest. The revelation of the last names of some victims and any associated emotional damage has to be balanced against the potential for preventing another crime. Only in Iowa is the age range of the victim and gender of the offender symbolized on the map however.

Providing the public information about the locations and characteristics of sex offenders can lead to harassment and criminal actions directed against them. All RSO web sites contain warnings that it is a crime to misuse the information contained in them. There is good evidence that RSOs have been the victims of harassment, and assaults and a number of cases of arson mostly in rural areas are associated with sex offenders and attributed to the attempt to drive them out of the area.

In 2006 there occurred the murder of two RSOs living 25 miles apart in Maine. The confessed killer was Stephen Marshall, a 20-year-old mentally disturbed Canadian "survivor" of child sex abuse who crossed the border and armed himself with two guns and a list of 29 names and addresses of RSOs in Maine he obtained from the state police web site. He used web maps and his personal laptop to find and stake out the homes of six RSOs, and he went on to kill two of them. It is instructive that both of the victims were guilty of relatively minor offenses. One was a 24-year-old named William Elliot who was registered for "statutory rape" of his teenage girlfriend. The other had committed a single nonviolent offense against a child many years previously; he was 57 years old, married, and working. In 2008 in Washington State, two RSOs were murdered by Michael Mullen seeking revenge for wrongs committed by an unrelated sex offender (Blanco, 2012).

The inclusion of information about the IP addresses of computers used by RSOs and their membership in social networks (some states prohibit certain RSOs from joining or using social networks like Facebook) can be a very valuable tool. However, it is quite easy for knowledgeable persons with ill intentions to create and use an assumed IP address or in some cases get near a home with an unsecured wireless network and use an assumed IP address that is then "hijacked." A 2013 case in Montgomery County, Texas, illustrates the downside of providing the IP address information of RSOs to the public. In this case, messages, sexually explicit photos, and later sexual contact were made with a young girl who was a member of a family that a man had a grudge against. He used a stolen IP address obtained from the state web site of an RSO residing near his target family to make a series of illicit online contacts. After the criminal conduct was discovered, the RSO was initially suspected, but the actual criminal was eventually apprehended and held on multiple charges including falsely incriminating the RSO (Glenn 2013). This case illustrates the potential for information about IP addresses to be misused. There is also the high potential for harassment of RSOs over the Internet, and as more and more jobs depend on web access, there may also be a tendency to preclude offenders from using an essential part of modern life, which may harm their successful rehabilitation.

6.3.8 States Without Web Maps

Eight other states have no link to any web map whatsoever. But that means that 42 of 50 states do use web mapping to disseminate information about RSO locations and details. These details generally include one or more photographs of the offender, personal information about the offender, and other salient information like details of their crimes. However, because the provisions of Megan's Law only require registration and public notification and do not specify if all offenders must have their information released, nine states have opted to not use a GIS as a means of doing so, while other states only release information on some offenders (such as tier II and tier III) and most do not release information about juvenile offenders although Iowa does for the most serious offenders.

Figure 6.2 shows the states in the conterminous United States that use web maps and those that only have searchable tabular lists of RSOs. Also one should note that Both Alaska and Hawaii have web maps, but Alaska only initiated the map in late 2013. Also Washington DC and Guam do have web maps of RSO information, but Puerto Rico, the US Virgin Islands, American Samoa, and the Northern Mariana Islands lack them. At present none of the 109 Native American nations, tribes, and bands that have RSO web sites have web maps on those sites. There is a map on the National Sex Offender web site that shows those states that have Native American entities within them and alphabetical list of all 165 states, territories, and tribes that maintain RSO web sites with a link to the selected site, but not a centralized nationwide web map maintained by the US Department of Justice (NSOPW 2013).

Based on an analysis of the spatial distribution of states that use of web maps for disseminating information about RSO locations, one thing stands out and that is that New England has been resistant to adopting web mapping of RSOs. In New England, only Vermont and Connecticut have such a capability and those were recent actions. Another thing that states that lack such web maps have in common is that most are small in size; this could be a coincidence since New England has much smaller than average sized states. One can argue that rather than the limited spatial extent of New England states, what is at work is a politically induced hesitation at revealing information about RSOs. The American Civil Liberties Union (ACLU) is influential in New England, and they have opposed registration and particularly effective public notification in legislative hearings and in court challenges. Web mapping is likely to be a highly visible means of public notification and therefore might be considered particularly objectionable (ACLU 2001).

The lack of RSO web maps in Alaska until 2013 was arguably largely attributable to Alaska's large size and the sparse population and the consequent difficulty in mapping offenders in the wilderness who do not live or work at valid street addresses. Imagine the difficulty in geocoding a cabin hundreds of miles from a valid street address or the work locations of a trapper or fisherman. While GPS could be used to do this, no state has a requirement that GPS be used to determine valid addresses of RSOs. Some of the mapped locations in Alaska of RSO's are in harbors reflecting that they "reside" on fishing boats, many other offenders in Alaska cannot be mapped at all. The growing use of GPS, along with radio frequency (RF) technology, in electronic monitoring of many RSOs raises this as a likely future development (Vollmann 2009). Not all states that are largely wilderness lack web maps (Montana and Wyoming have good ones), but Nevada just barely has such a capability. Both Alaska and Nevada share the situation where most of the state is uninhabited wilderness except for one major urban area. Both the Anchorage and Las Vegas police departments actively use GIS and keep information about RSOs in their community in their internal GIS, although that information is not part of the publicly available crime maps generated on the Las Vegas Crime View Community web site that otherwise contains a great deal of data, including locations of various sex crimes, if not the criminals involved (Omega Group 2012).

Georgia is a state whose failure to use web maps is puzzling. The difficulty in parsing RSO information in a large state with a major urban area is well illustrated by using the Georgia State Bureau of Investigation's RSO site to explore the issue in Fulton County. For any county in Georgia, one can rapidly obtain an alphabetical list sorted by the last name of all RSOs in that county and by going through the list get information about their address and view their pictures.

For many rural counties, this approach works quite well. But it works poorly in a county with a population over one million. How can users extract useful information from a list that starts with "Aaron" and runs through "Zeigler" and that requires users to scroll through 155 pages with details on 10 RSOs displayed per page? Answering the question "which of these 1,550 offenders is living, working, or going to school in close proximity to my family among the more than 1 million residents of Fulton County?" is a challenge to potential users. Thus in the county where the most people and RSOs in Georgia live, the state web site is least effective.

6.4 Spatial Issues

The spatial analysis of RSO information can be used to understand the proximity of offenders to one's family, to vulnerable populations, and to crime incidents. Spatial analysis could also be used to help site outreach and treatment facilities, optimize home visits and other supervision efforts, and even determine optimal locations for offenders to live and work. There are a variety of spatial accuracy and precision issues that crop up with respect to RSOs. One important set of issues involves the geocoding of RSO locations. There are a whole range of issues involving precision and accuracy as well as privacy concerns related to geocoding. Then there are spatially based strategies for determining proximity of RSOs to locations of interest which might be accomplished through buffer zone generation or by spatially based view and query and the issue of how the information in RSO web maps is interpreted and used by the public (as well as law enforcement professionals that may also use the web sites).

6.4.1 Geocoding Issues

Geocoding of sex offender residential locations is generally accomplished using street centerline data such as that created initially by the US Census Bureau for the TIGER product (Harries 1999). This data has been enhanced by firms like TomTom (since 2012 owned by Apple and the basis of Apple Maps) and Navteq by using GPS-equipped cars to drive many roads, but in any case has an address range for each side of each street as well as street name, street type, street direction, and other locational data such as city, zip code, county, and state. This approach used first in US Census data structures like TIGER is termed "dual address range." An RSO is required to submit an address to the state government agency in charge, and this address is entered in an RSO list along with many other details mentioned above. However, often it is a local agency which actually collects and monitors addresses associated with RSOs. These agencies then report the addresses, especially if they are places of work or the offender has moved to a new address recently. Some offenders move frequently, and depending on the state, community, and nature of their offense, they may have restrictions or none placed on their movements.

In many cases the address may not match a valid address, but that by itself is not a violation of restrictions on sex offenders. This is a very common problem. Rural areas are likely to have geocoding issues because a home (often a trailer home) may not have a physical mailing address. In that case they might have a rural delivery route address, a PO box, or only a zip code as their address. Efforts to provide physical addresses for all residents in rural areas in order to help emergency 911 dispatching will also improve this situation. If a rural address that lacks a physical location is provided to a state that maintains a web map of sex offenders, then it may be noted as "unknown address" or "address cannot be mapped" or "rural address" and not mapped at all. In some other cases, an attempt is made to geocode the inadequately specific address, and the address may end up being geocoded but appear at the wrong location.

An analysis of all 396 RSOs residing in Saint Lawrence County, New York, in 2013 indicated fully 155 did not have addresses that the state of New York could geocode accurately. This is a rural county with many sex offenders due to the presence of two prisons and a state psychiatric hospital, and most of the incarcerated offenders were mapped. However, many of those were mapped based on a post office (PO) box address for a state prison and ended up being located on Main Street in the county seat which is physically where the post office is located and not mapped to the state prison which was located several miles away. In the same county, several unrelated RSOs who only had zip codes for locational information were mapped as residing at the center of a large fallow field, and others were shown as residing in the middle of various forested tracts, and these locations appear to be zip code centroids. The availability of aerial imagery on the New York RSO web site helps users see that some locations are inappropriate, but many states only use street map data which would less clearly indicate incorrect geocoding of offender locations.

It is important to understand that the addresses will not necessarily geocode directly on top of a single home or apartment building. The intention of the geocoding of RSO locations is to give an approximate location. Usually, only the block the offender lives at will be identified. Often this location will be in the middle of the street. In other cases it is on the wrong side of the street. It may often happen that an RSO location is closer to a residence where he or she does not reside than to one where they in fact reside. Alternatively, the point representing the offender location may be close to a large highly visible residence when the offender in fact resides in a small residence hidden in various ways such as under tree canopy, at the back of a farm building, or in a small travel trailer. In other cases, mostly in rural areas, the RSO location is not in question because there is only a single residence on the entire block (or even the entire street).

6.4.2 Multiple Location Issues

In densely populated urban areas, the geocoded location may identify a large high-rise apartment block with hundreds of separate apartments, or it may fall into a space between two large apartment buildings (see Figs. 6.9 and 6.10 for map and "bird's-eye view" from Brooklyn, New York). The issue of dealing with offenders residing in multistory buildings in which perhaps more than one offender might be located spatially on top of one another in a two-dimensional map is one of the several issues that are not well handled by the geocoding and visualization approach used.

There are several state web sites that have a special marker symbol for "multiple offenders" at a single address however, and the detailed attribute data that may be associated with each of several of multiple offenders may contain exact apartment number data although it may just contain the address of the entire building. This and many other forms of spatial ambiguity and inaccuracy are defended by web site developers as preserving the privacy of RSOs, but are mostly a function of the lack of precision in geocoding of addresses using street centerline data. In many cases the exact street address (and apartment number in urban areas) of RSOs is anyway provided in the detailed attribute information on each RSO (Wartell and McEwen 2001).

Although many sex offenders that are supposed to be registered are missing from state web maps because they have failed to register, have absconded, have moved out

Fig. 6.9 Map of Brooklyn, New York, showing apartment footprints and interpolated RSO locations. New York State Division of Criminal Justice Services, 2013. http://www.criminaljustice.ny.gov/nsor/

of state, or have been deported (or died, since dead offenders remain in the system for some time in many states), there are usually more locations on these maps than there are RSOs in a given state. This is because most states also put the place of work and/or place of school. In Virginia two very different 3D symbols are used to portray these two types of locations, and in Kansas the letters "R" and "S" are used to differentiate residence and school locations on the map. Some states have secondary residence information in their maps. Montana is a state where secondary residence information is provided for all offenders mapped. Perhaps this relates to the number of seasonally occupied cabins in Montana or the number of RSOs who live in a camper or tent and continually move about; in any case having information available about more than one residence per offender is not common. Ohio lists a primary residence and secondary residence and multiple places of work as well as places that the offender goes to school (if any). Several states also add features in the community such as schools, parks, day care centers, and even hospitals to their maps. Missouri does this, while Arizona uses the approach where park and school locations can be selected to be used to generate buffer zones around. This relates to concerns about proximity of offenders to these places where children may congregate. Some states list and map offenders that are residing or working out of state, and this includes Arkansas, Tennessee, and Pennsylvania. This cuts down on the temptation of offenders to live near the edge of a state to make their supervision more problematic.

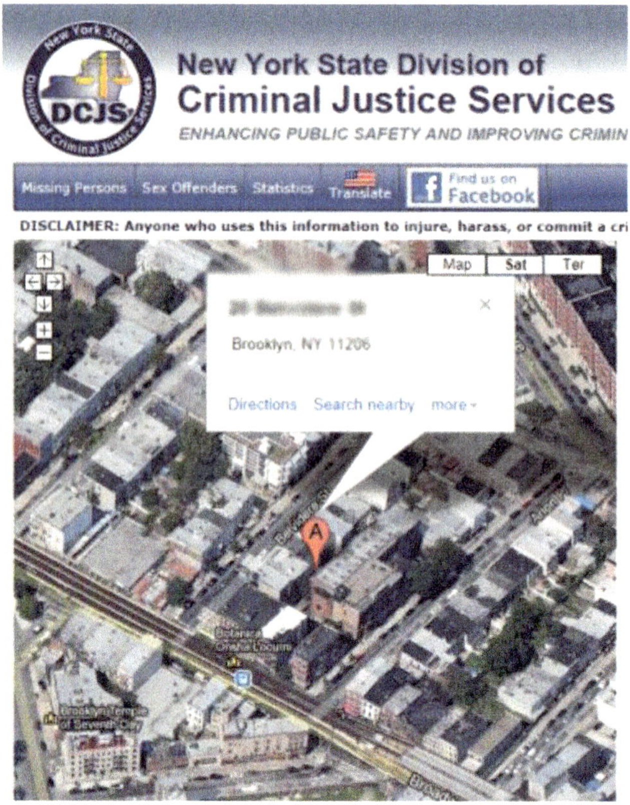

Fig. 6.10 "Bird's-eye" view of Brooklyn, New York, RSO locations. New York State Division of Criminal Justice Services, 2013. http://www.criminaljustice.ny.gov/nsor/

In some cases, an RSO may appear in two (or more) state databases. However, the data is not always consistent. Thomas Leggs is a convicted rapist who was released in 2006 and was classified as a tier III high-risk offender in Delaware where he lived, but he also worked in Maryland. He was also registered in Maryland but listed only as "compliant" in that state's RSO web site. Maryland did not then maintain risk-related information. When he went on to rape and murder an 11-year-old child in Maryland in 2009, the discrepancy between the two databases was brought to light. Another example of a somewhat similar issue is that of an offender serving in the military. He is listed in North Carolina as an RSO residing somewhere at an unknown location on the huge Fort Bragg military base but cannot be mapped, and he is also listed in a federal RSO web site as residing at a federal prison in Kansas.

Another sex offender who cannot be easily mapped is the *Billionaire Pedophile* Jeffery Epstein. This hedge fund manager was convicted of hiring underage prostitutes. He owns a waterside mansion in Palm Beach, Florida, a 10,000-acre cattle ranch in New Mexico, and his own private island in the British Virgin Islands. The Florida State RSO web site does not have a map showing any of these locations and lists his private island (which lacks a street address) as his primary residence. Incidentally, Forbes Magazine has disputed his status as an actual billionaire since his immense wealth is open to question, as apparently is his location (Forbes Magazine 2011).

Other examples of spatial issues are presented by those RSOs who are homeless and therefore tend to move between multiple locations many of which are difficult or impossible to geocode. Homeless sex offenders have been recognized as a particular problem in California, Florida, and Washington. These states, along with 17 other states, had by 2008 enacted residential restrictions, and many communities have used their local land use planning and zoning authorities for a similar purpose. This has been blamed for homelessness among RSOs (Strutin 2008). California has identified over 3,000 offenders as being "homeless," while in contrast Texas does not recognize the category of "homeless sex offender" and thus officially has no homeless sex offenders. Most states officially acknowledge having less than 50 homeless RSOs, but do have some of them (California Sex Offender Management Board 2008).

Some offenders while not exactly "homeless" have avoided scrutiny by living in unconventional locations and/or moving frequently. A case in point is that of the pedophile defrocked priest Gilbert Gauthe who moved from priesthood and then prison in Louisiana to a series of campgrounds in coastal areas of Texas. He generally stayed in a given campground no more than 2 weeks at a time. He also allegedly tried to reoffend, and he served 2 years in Texas from 2008 to 2010 specifically for violating the terms of the Texas sex offender registration law. Thus there is sometimes a "one-to-many" relationship between offenders and their locations in a web map when an offender moves between multiple temporary locations. There is also the more common situation of a "many-to-one" relationship that exists where multiple offenders are housed in the same location, usually an institutional setting like a prison, mental hospital, or "halfway house."

6.4.3 Multiple Names

There are also "many-to-many" relationships in quite a few cases. This comes about since offenders may have multiple names, generally aliases they have employed as well as multiple locations associated with some or all of those names. In the state registry of some states like New York, all these names may be included and listed separately but cross-referenced, and each can be associated with one or more locations on a map. Analysis of 396 sex offender details for Saint Lawrence County in upstate New York and 1,706 sex offender details for Kings County (Brooklyn), New York, includes a listing for the name "Bill" (just *Bill*, no other first or last name) and also listings for "Fred Flintstone" and "Chris Hitler" along with the usual "John Smith" and "Bob Jones" being used as aliases. These names are identified as aliases in the New York RSO web site and map to the location also associated with the full correct name.

The situation where more than one sex offender has the same name also crops up. In Brooklyn there are over 27 men named "Rivera" that are RSOs; of these, three men are named "Angel Rivera." In Bronx County there are also 27 men named "Rivera," and seven of them are named "Angel Rivera." Some of these multiple men named Rivera who are RSOs live on the same block in Brooklyn, although they are in adjacent high-rise public housing project buildings. The same building in Brooklyn does have multiple RSOs in it however, so there is a possibility that two unrelated offenders with the same first and last name could live at the same mapped location. One can gain some measure of anonymity despite being featured on a public RSO web site by having a common name and living in a densely populated urban area. Rural areas likewise provide anonymity because RSO locations cannot be adequately geocoded using current approaches.

6.4.4 Proximity Issues

Those seeking information about their own spatial proximity to sex offenders using web maps usually do so by entering their own street address and other locational information and then seeing all the sex offenders that reside within a specified buffer zone distance of that geocoded "home" address location as portrayed by a point symbol on the map (it is a different symbol than those used for sex offender locations on all web sites, usually a different color). Because population density varies and users may be concerned about a larger or smaller "neighborhood," variable buffer distances are options used in many sites. The "Offender Watch" web design used by multiple states has intervals from .25 miles to .5 miles, 1 mile, 2 miles, and 4 miles. The default buffer option in New Mexico is two miles, while in more densely populated Ohio, it is one mile. Other states have only one possible buffer zone of a mile in radius, while some states have as many as 5 optional buffer distances with that range as high as 10 miles in the case of the more rural state of Iowa and up to 15 miles in Missouri, as portrayed in Fig. 6.11. The buffer zones are usually highlighted on the web map and may even all be displayed simultaneously and concentrically as in Idaho whose sample map is shown in Fig. 6.12, but other states only show offenders within a specified distance of a valid address and do not generate a visible buffer zone "ring" on the map at all.

6.4.5 View, Query, and Use

The anticipated use of RSO web maps is that the public will use the information to take proactive precautions. Also there are a number of cases where information on web maps of RSO locations has been used to help solve crimes or determine that RSOs are in violation of restrictions. In 1997 in Knoxville, Tennessee, the police solved a series of sexual assaults associated with the *Greenway Rapist* by mapping three attacks that occulted over a 2-week period in relation to RSO residence locations. All the attacks occurred along a jogging path and were within easy walking distance

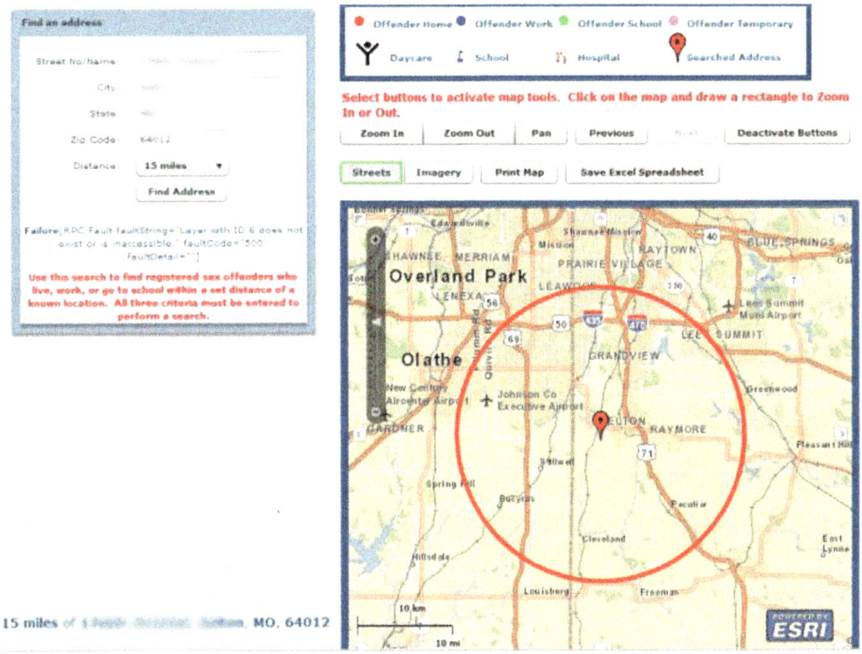

Fig. 6.11 Fifteen-mile buffer zone generated by the Missouri web site extends into Kansas. Missouri State Highway Patrol, 2013. http://www.mshp.dps.missouri.gov/MSHPWeb/PatrolDivisions/CRID/SOR/SORPage.html

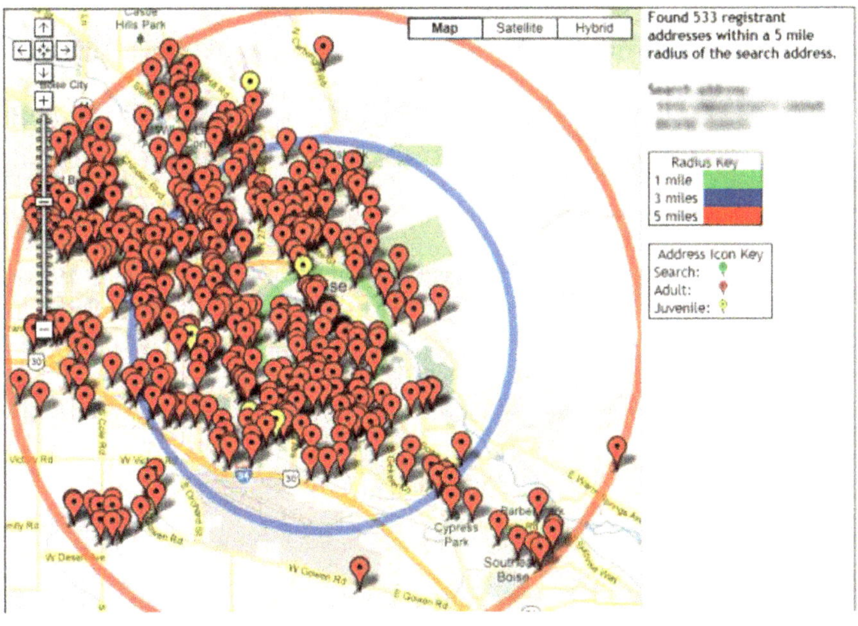

Fig. 6.12 Multiple color-coded buffer zones generated for Boise, Idaho, area by the Idaho state RSO web site. Idaho State Police, 2013. http://www.isp.idaho.gov/sor_id//nij/mapping/ch4_3.html

of the closest RSO's home. He was a recently paroled rapist who was identified from photographs by each victim and is currently serving life without parole (Hubbs 1998). In this case a public RSO web site did not yet exist, although a GIS with statewide RSO and crime incident locations was used by police. This web site has since served as the nucleus for Tennessee's really outstanding RSO mapping site. In another Tennessee case, Holly Bobo was abducted by a man wearing camouflage on April 13, 2011, and has not been seen again. Her abduction location was linked by the public to that of an RSO who absconded soon after her disappearance. Both locations were within 5 miles of each other. The missing RSO was considered a suspect and was being sought by the Tennessee Bureau of Investigation and the FBI. This could be a false accusation, and there has probably been an increase in false reports about sex offenders in relation to crimes simply as a result of their being the main category of offenders in the community the public has locational information and personal details about. In another case in 2012, FBI agents located another missing young woman, this one from North Carolina, who was fortunately still alive and was identified by a member of the public as being in the company of an absconding North Carolina RSO who was subsequently arrested in Florida; information on the RSO web site was used in the identification of the suspect. In a New York State case, also in 2012, an RSO was recognized from a picture posted on a state RSO web site as having been in the company of a young "exotic dancer" at a nightclub. This woman had disappeared. This situation resulted in the RSO being labeled a "person of interest." It was learned that he had bleached the interior of his car and apartment. His presence in a nightclub was a violation of the terms of his parole for prior assaults on young women. Therefore while the investigation into the missing person case is ongoing, he has been arrested for violation of parole restrictions.

In a more typical 2012 case in Indiana, a member of the public was anonymously using the "tip" feature of the Offender Watch web site design and reported an RSO as being in violation of registration rules based on information on the Indiana RSO web site. The RSO had failed to report a new employment location which was in another county from that listed as his place of employment on the web site. Since the offender was a tier III "predator," this seemingly minor infraction resulted in his being re-incarcerated for violation of the terms of his parole. Another 2013 Indiana case was based on a tip submitted to the county RSO site. It was found that an RSO had failed to register as a secondary residence a relative's home where he stayed a few days a week; this secondary residence was within 1,000 ft of an elementary school. The combination of violating a residential restriction on spatial proximity to a school and failure to register at a secondary residence resulted in his arrest.

6.5 Conclusion

GIS and other geospatial technologies have been used in a variety of ways to deal with issues related to sex offenders. Web maps are a very important part of the use of this technology. Unlike other countries in the world, the United States has chosen to make RSO information and, in particular, their locations public. Four fifths of the states in the United States have chosen to use web mapping to do this. While the approaches used are effective and often very informative, they are highly variable and there remain issues with missing data, imprecise (or impossible) geocoding, and poor cartographic and web site design. Nevertheless, the popularity of this approach with states and the public is unquestionable. What is far more questionable has been the results of "outing" RSOs, which may have tended to increase issues of failure to register, homelessness and absconding, and perhaps harassment and false accusations (SMART 2013). Also the issue of whether RSO web maps are part of a proactive family and community defense or a source of unjustified fears is not easily resolved. Determining on balance which is a more common outcome would require more systematic research. Such research might also improve the design and content of RSO web maps and also help make them part of effective public participation strategies.

References

Ackerman A, Harris A, Levenson J, Zgoba K (2011) Who are the people in your neighborhood? A descriptive analysis of the individuals on the public sex offender registries. Int J Law Psychiatry 34(3):149–159

ACLU (2001) Connecticut's sex offender website registry violates rights. Court Rules April 2, 2001. http://www.aclu.org/racial-justice_drug-law-reform_immigrants-rights_womens-rights/connecticuts-sex-offender-website-reg. Accessed 01 May 2013

Bardens D (1990) Life of Landru; The French Bluebeard. Chronicles of Crime, London

Blanco J (2012) "Stephen Marshall" murderpedia, the Internet encyclopedia of murderers. http://www.murderpedia.org/male.M/m/marshall-stephen.htm. Accessed 01 May 2013

Broers M (1999) The Napoleonic police and their legacy. History Today, May

Buckley A (2011) Cartographic design of web-maps. http://blogs.esri.com/esri/arcgis/2011/08/04/cartographic-design-for-web-maps/. Accessed 01 May 2013

Bureau of Justice Statistics (1994) Recidivism of sex offenders released from prison in 1994 http://www.bjs.gov/content/pub/press/rsorp94pr.cfm. Accessed 01 May 2013

Bureau of Justice Statistics (2003) Re-arrest of sex offenders within five years of release. Office of Justice Programs USDOJ, Washington, DC

Bureau of Justice Statistics (2007) Recidivism rates. http://www.bjs.gov/index.cfm?ty=tp&tid=17#pubs. Accessed 01 May 2013

California Sex Offender Management Board (2008) Homelessness among registered sex offenders in California: the numbers, the risks and the response. December 2008. www.cdcr.ca.gov/Parole/Sex_Offender_Facts/docs. Accessed 01 May 2013

Cheney S, Ratcliff J (2005) GIS and crime mapping. Wiley, Hoboken, New Jersey, USA

Craig L, Browne K, Beech A (2008) Assessing risk in sex offenders. Wiley, Chichester, U.K.

Finkelhor D (2008) Childhood victimization. Oxford University Press, New York

Forbes Magazine (2011) Jeffrey Epstein: sex offender, yes. Billionaire, no. http://www.forbes.com/sites/clareoconnor/2011/02/28/jeffrey-epstein-sex-offender-yes-billionaire-no/. Accessed 01 May 2013

Foucault M (1976) Society must be defended. Lectures at the College of France, 1975–1976 (trans: Macey D). Picador Press, New York, USA

Glenn M (2013) Vendetta alleged in sex assault of teen, February 26, 2013. Houston Chronicle. http://www.chron.com/news/houston-texas/houston/article/Vendetta-alleged-in-sex-assault-of-teen-4310940.php. Accessed 01 May 2013

Harries K (1999) Geocoding. In: Mapping of crime, principles and practice. Crime Mapping Research Center, NIJ. https://www.ncjrs.gov/html. Accessed 01 May 01 2013

Hubbs, R (1998) The Greenway rapist case. In: Wartell J (ed) Crime mapping case studies. Police Executive Research Forum. www.ncjrs.gov/pdffiles1/Digitization/178051NCJRS.pdf. Accessed 01 May 2013

Leiphon M (2008) State budget woes. New Geogr 12(29). http://www.newgeography.com/content/00497-state-budget-woes. Accessed 01 May 2013

Leipnik M, Albert D (2003) GIS in law enforcement. Taylor & Francis, London

Leipnik M, Ye X, Wu L (2013) Jurisdictional boundaries and crime analysis policy and practice. Reg Sci Policy Pract 5(1):45–65

Louisiana State Police (2012) History of sex offender registration laws. http://www.lsp.org/socpr/history.html. Accessed 01 May 2013

Mawby R (2011) World policing models. In: Mangai N (ed) International crime and justice. Cambridge University Press, New York, USA

National Center for Missing and Exploited Children (NCMEC) (2013) http://www.missingkids.com/SOTT. Accessed 01 May 2013

National Conference of State Legislatures (2013) Status of adoption of Adam Walsh Act. http://www.ncsl.org/issues-research/justice/adam-walsh-child-protection-and-safety-act.aspx. Accessed 01 May 2013

NSOPW (2013) National sex offender public website. http://www.nsopw.gov/en-US. Accessed 01 May 2013

Omega Group (2012) Crime view community overview. http://www.theomegagroup.com/police/crimeview_desktop.html. Accessed 01 May 2013

Roberts J (2013) Personal communication to author from J. Roberts, Director of Operations for Watch Systems, Inc. Covington, LA

Sieber J (2006) Public participation GIS. Ann Am Assoc Geogr 96(3):491–507

Slocum T, McMaster R, Kessler F, Howard H (2008) Thematic cartography and geovisualization, 3rd edn. Prentice Hall, Upper Saddle River, New Jersey, USA

SMART (Office of Sex Offender Sentencing, Monitoring, Apprehending, Registration and Tracking) (2013) Sex offender registration and failure to register. Office of Justice Programs, Washington, DC

Stimson C, Noronha M (2011) Complying with sex offender registration standards. Heritage foundation reports. http://www.heritage.org/research/reports/2011/07/get-smart-complying-with-federal-sex-offender-registration-standards. Accessed 01 May 2013

Strutin K (2008) Sex offender residency restrictions. Criminal justice resources. http://www.llrx.com/features/sexoffenderresidency.htm. Accessed 01 May 2013

Thomas T (2011) The registration and monitoring of sex offenders. Routledge, New York, USA

U.S. Senate (2009) Sex Offender Notification and Registration Act (SORNA): barriers to timely compliance by states. Sub-Committee on Crime. U.S. Government Printing Office

Vollmann J (2009) Electronic monitoring of dangerous sexual offenders. Corrections Magazine: 5(2). http://www.corrections.com/articles/21373-electronic-monitoring-of-dangerous-sexual-offenders. Accessed 01 May 2013

Wartell J, McEwen T (2001) Privacy in the information age: a guide for sharing crime maps and spatial data. Crime Mapping Research Center, Institute for Law and Justice, Washington, D.C.

Washington State (1990) Department of Social and Health Services. Community protection act of 1990. http://www.dshs.wa.gov/scc/Laws.shtml. Accessed 1 May 2013

Watch Systems, Inc (2013) Offender watch. http://watchsystems.com/offenderwatch.php. Accessed 01 May 2013

Winick B (2005) Civil commitment: a therapeutic jurisprudence model. Carolina Academic Press, Durham, North Carolina, USA

Chapter 7
The SDIK Police Model: How to Make the Invisible Visible

Manuel Rodríguez Herrera and Daniel Salafranca Barreda

Abstract The current public policy for international security is fundamentally different to that prevailing at the end of the last century. A radical change in context means that traditional strategies and urban policing models have to be revised and adapted to a new reality. Police science, new policing strategies, and approaches have evolved to give answers to new challenges in security. The police model "Science, Data, Intelligence, Knowledge" (SDIK) incorporates technical-scientific and geospatial innovations to understand emerging communities and uncover evidence of criminal activity. SDIK launched a project to study new religious movements in communities and neighborhoods of the city of Castellón de la Plana (Spain). The project combines two main objectives: first, to achieve knowledge of the impacts and perceptions of new religious organizations and, second, to establish connections between new religious organizations and other activities, some of which may be patterns of illegal or disorderly behavior. To ensure safety and peaceful coexistence in cities, it has become imperative to make visible the apparently invisible.

Keywords New religious movements • Community policing • SDIK police model • GIS

7.1 Introduction

There is no doubt that, independent of the scale of jurisdiction be it state, region, or city, the management of public security has become a fundamental strategy in the policies implemented by different governments following increases in terrorism in

---

M. Rodríguez Herrera (✉) • D. Salafranca Barreda
Castellón Police Department, City Council of Castellón,
Castellón de la Plana 12001, Spain
e-mail: manuel.rodriguez@castello.es; daniel.salafranca@castello.es

G.A. Elmes et al. (eds.), *Forensic GIS: The Role of Geospatial Technologies for Investigating Crime and Providing Evidence*, Geotechnologies and the Environment 11,
DOI 10.1007/978-94-017-8757-4_7,

recent decades. There is no question about the necessity for the increase in the importance of security, which has been translated into an enormous increase in investments in resources by different administrations. These investments have accompanied an important set of social changes that have driven models and strategies being used by police forces in many countries, with the aim of giving answers to the many different security demands and problems prompted by the new circumstances.

Worldwide events as the terrorist attacks that took place in September 2001 in New York or in 2004 in Madrid, and the recent assault in the city of Boston, 2013, have uncovered substantial deficiencies in public security systems. At the same time, these events have driven important changes in strategy to improve security management practices in the majority of western countries, with the final aim of being able to offer efficient answers to these new risks. These strategies not only pursue the demands, which can appear unexpectedly, but also consider the specific conditions in which the new risks are arising. Such new security strategies should be based on overall lines which allow the combination of productivity and prediction as key elements and strategic planning as a basic element of territorial vectors and resources management.

Everybody has to account for new strategic courses in which the "new public security" is based. Consequently, police management of the social and religious diversity and new religious movements has been positioned as a central element, which allows for the better understanding of the reality of our cities, as well as offering better answers to their risks and demands (United Nations 2010). In the context of new religious movements, we should recall the events of March 1997, when 39 people of the Heaven's Gate sect committed suicide in San Diego, California. We should also remember the FBI's assault on the Branch Davidian community at Waco, USA, with the death of 80 people, and the terrorist attack at the Tokyo metro using sarin gas executed by the Shoko Asahara's Aum Shinrikyo sect in March 1995. But the event with the greatest social impact was, with no doubt, the collective suicide of 913 members of the People's Temple sect, in Guyana, November 1978. This is the social image that public opinion associates with new religious organizations, a new reality that has converted our cities to environments where these developing situations of conflict happen, environments in which new measures and tactics have to be articulated and orientated to reduce risk and increase real safety.

Before this new scene, there is no doubt that police organizations and public administrations in general have to give answers and adapt to the kind of society in which they are integrated. This new society is in a state of steady change, a mutability being a constant and inescapable condition when planning answers to public security problems. Equally, the social legitimacy of these answers will be conditioned by the capacity to answer citizens' expectations and demands. As an example of this adaptation necessity to the new social reality of our cities, and by way of an innovative strategy in the public security frame, the city of Castellón, Spain, has implemented a community-policing project based on the SDIK police model (Castellón de la Plana 2013). The project has taken as a key element for analysis and study the "new religious movements" which have developed in the city since the first years of the twenty-first century.

The aim of the project has been twofold. On the one hand, a global perception of security (and insecurity) has been taken, which has its roots in the establishment of religious organizations in different neighborhoods of the city. On the other hand, these organizations are an emerging social reality, making it important to understand and establish the connections between the stated goal of the organization (the public and known objective) and the parallel activities in which the practices of such organizations are based within the sphere of a community, which on many occasions are neither made public nor are widely known. It is not at all the objective of the SDIK policing model to establish regulatory guidelines which every community has to adopt in or practice as a definition of religion. SDIK starts from an accurate definition of the urban reality, the analysis of the distribution patterns, which show the police facts of interest and their relationship with other variables, such as urban or sociodemographic variables. Study of the features and dynamics of the different groups and collectives compose a micro-space with high diversity, which is enriched by citizen requests received through emergency phone calls or other information obtained directly from police officers.

All these data are a base for the application of new technologies such as Geographic Information Systems (GIS), to manage police activity, which has allowed us to achieve the last objective, as the title of this chapter states, make visible the invisible. Geospatial technology generates appropriate solutions to problems that may arise alongside the development of new religious communities. For example, reduction of the stigmatization of recognizable or distinctive groups, insecurity, perception, or fear of the potential commission of serious crimes, as well as knowing accurately the different activities and relations established among the members of such groups. Due to the scope and complexity of the desired goals, it is necessary to remind the reader that the tools and techniques utilized to achieve the stated goal of this project are within the confines of the laws and regulations of both Spain and the European Union.

The authors wish to make it clear that it is not the intention of this project to promote or provide any type of special information or service to advance the powers of the police force at the expense of the individual or collective liberties of the public. At the same time it cannot be forgotten that this project is attempting to provide the most accurate portrayal possible of the city of Castellon de la Plana and consequently to understand those surrounding circumstances that are, or could become, a risk to the security of the citizenry. The circumstances discussed in this project should be viewed as circumstances that require special attention and analysis in order to ensure general welfare and security in the city, but not at the expense of invading the privacy or individual liberties of the citizens of Castellon.

7.2 Methodology

7.2.1 *Towards a New Risk Society*

Security means different things at different times to different people. In life, security and risk coexist in a permanent way. Concepts of risk and security are subjective, associated with other problems related to life, for example, property, health, power, and control. People's daily lives contain invisible risks and this reality produces different perspectives of risk, which tends to reinforce cultural differences. The progression of society towards a risk society, combined with the new reality of globalization, has produced a change in the sources of citizens' insecurity, so much so that redefining such insecurity has become obligatory. This revised concept of insecurity emphasizes uncertainty and presupposes big challenges with respect to the effectiveness and legitimacy of the institutions. It is not only a question of disorder, offenders, and enforcement officers but is a complex problem with implications in other areas of well-being, quality of life, and the individual rights of citizens. The step from a restricted concept of disorder to another wider one of a more modern sense of security affects the way of understanding the nature of public security. Three dimensions become important: objective risk, subjective risk, and tolerance. In the analysis of security, it is necessary to know better the social distribution of insecurity and its connections with other social problems.

The new context of objective and subjective risks highlights enormously the new social image that public opinion ascribes to new religious organizations. A new reality has transformed cities to new environments in which new situations of conflict take place, environments in which new measures and tactics have to be articulated, all of them with the aim of reducing risk and increasing real security for citizens. When speaking about risk, we are starting from a culturally constructed vision. Different cultures offer different meanings to risk depending on the situations, events, objects, and personal relationships involved. Social construction of risk in micro-spaces (community, neighborhood, or district) is partially determined by three aspects: social relationships, behavior strategy, and culture.

Culture is based on beliefs and values that justify behaviors and provide guidelines for social relations. At the same time, these behaviors and guidelines influence representations of what does or does not constitute risks and dangers. People "look at the world under a particular perspective legitimized by the social relations of the world to which they belong" (Dake 1992). According to Douglas (1996), the perception of risk is built in terms of its contribution to the maintenance of a particular way of life. It is possible to say that individuals choose what they fear (and in what way) to support their way of life. Thus somehow the risk and the danger are determined by a given socioeconomic structure as well as the culture in which those individuals are immersed. People perceive and accept risks from a cultural reference of the society to which they belong, and we should take into account that the security of some could be the insecurity of others. The fact that risk is socially constructed does not mean that all individuals of a given society perceive such risk the same way, e.g., the perception that a neighborhood with average rates of Muslim population should not have the same risk ascribed to a cultural Islamic center, as the community does not see risks in the same way as experts, who frequently have a technical vision of such risks. Today, governments are forced to adapt to the needs generated by the growing demand of the society, due to continuous development with regard to two issues of great importance for the survival and life in society: quality of life and public safety. It is entirely necessary to respond directly to these demands by providing services to citizens.

7.2.2 New Religious Movements

Prior to the start of the SDIK project, it was necessary to address the complexity and ambiguity of the concept of new religious movements (NRMs), without forgetting that this concept becomes more complex the moment we discover the existence of a new social and plural reality. Two realities should be considered while examining the term "sect" in the twenty-first-century society. The first consists in that it is impossible to continue with the divergence between church and sect as defined by the classic historic and social context, as the church is no longer the supreme entity in Western society. The church no longer monopolizes power in a lawful way to access the supernatural, nor is it the only organization that represents symbolic universes in our society. The second reality consists in that many new religious movements have not emerged from dissidence from inside Christianity, Hinduism, Buddhism, or Islam among other religions or main cults, but may be determined by local character groups or even international ones, whose only authority is based on their founders or leaders.

What is more, we can assign the label NRMs to religions which are not religions in a strict way, although they may be religions in the way they are settling in the European Union with respect to legal freedoms. For example, Islamic or Protestant churches are implementing their religious manifestations in ways they previously have not. Although there is no doubt that we should not and we cannot say that these religions are new, they are new in the way they are developing inside our communities. Despite the decline of mainstream religion in modern times, and the abandonment of traditional churches, emerging groups of a religious nature (undercover as spiritual associations) and new meditation cultures or techniques where people find a better way of satisfying spiritual needs show that religion, at least in its spiritual manifestation, continues to be a powerful force (Ramos 2006).

This complex situation is not unusual for the city of Castellón de la Plana, a medium-sized city of about 180,000 inhabitants on the Mediterranean coast of western Spain. Castellón is experiencing a fast and deep social transformation, involving changing compositions and social dynamics in many neighborhoods, which are becoming ever more complex and diverse. In the last 10 years, the city has seen different religious groups or cults established. Directly or indirectly, these groups have modified the daily structure of cultural centers, sometimes resulting in conflict situations, which have affected public order and the peaceful coexistence of the community.

7.2.3 The SDIK Police Model

Before this new diverse and complex scene, the local police[1] of Castellón de la Plana planned different questions to be answered through the project reflected in this chapter: is the police the proper avenue to provide an answer to this new situation of conflict? Do we really know the reality of these new movements and religious groups? Could there be hidden interests and motives behind these organizations?

With the aim of giving an answer to these and other questions, the first objective raised by the local police of Castellón was to determine which strategy was correct to address the questions that had been raised by the development of NRMs. After reviewing scientific literature related to police science, it was determined that the best strategy was to implement a new work model which gathered all of the particular necessities of the local police of Castellón, a new police model of proximity that has been named as the "SDIK model." The SDIK police model may be characterized as "intelligence-led policing" based, among other aspects, on not giving purely reflex answers to new problems and conflicts of security but on a proactive and preventive service supplemented through the implementation of other work methodologies which allow more satisfactory results. The answer was crystal clear: it was necessary to adapt the activities of the local police of Castellón to the new reality.

The SDIK police model is a management model built on four essential pillars: science as a global element (Science), ability to document the results derived from the implementation of police strategies (Data), implementation of models based on police intelligence as a basic tool (Intelligence), and finally management of knowledge applied to security (Knowledge). The SDIK model is based on the analysis of all types of information, in order to have the necessary data needed to know what has happened, is happening, and can happen. The police cannot nor should we treat the information as isolated and independent events.

7.3 New Religious Movements in the City of Castellón (Spain)

For there to be a preventative activity, the police are required to be proactive, and proactivity presumes that unlawful acts have a predictable component (pattern recognition). The main focus of this work is looking for patterns of illegal or disorderly behavior by religious organizations in the community. In order to provide a specialized response to the very specific condition of the presence of new religious movements in the city of Castellón, the first step was a redefinition of the social framework that would provide a better understanding. One of the fundamental characteristics that differentiate our contemporary society from the past is the new

[1] Spain is a country of approximately 47 million people. Its public security forces are organized at three distinct, nonoverlapping levels.

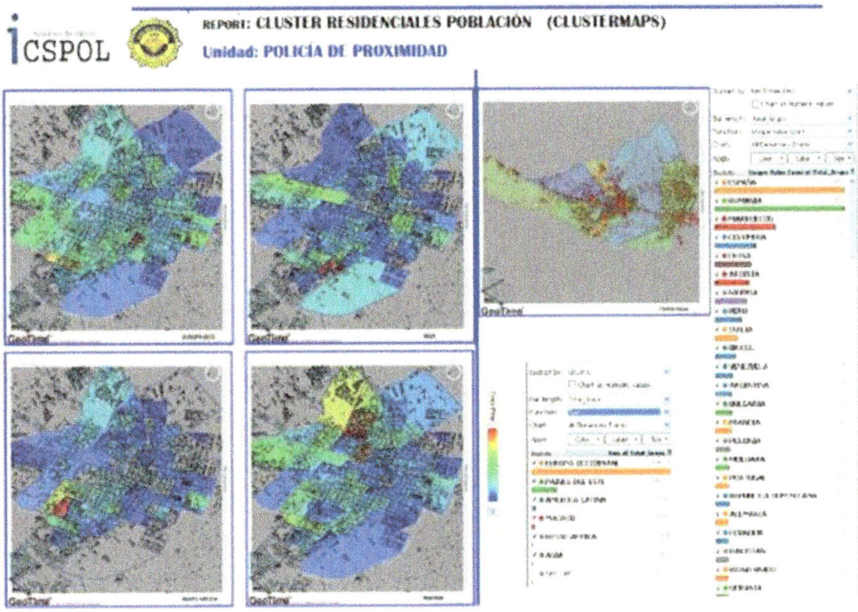

Fig. 7.1 Nationality clustermaps

nature and scale of urban phenomena and processes, a concern, both theoretical and normative, about how to properly define the urban reality (Fig. 7.1). This change in nature and scale represents a starting point in the definition of new urban-spatial realities, allowing analysis of the distribution patterns that present the facts and relationships with other urban variables.

Having made this first step, which allows us to obtain a more precise knowledge of the social reality of the city (knowledge not offered by the traditional tools that were being used), the second step was to locate the various religious organizations and religious centers by using a GIS (Geographic Information System) to create a specific layer that placed religious organizations in the city, making way for a better understanding both from the purpose of study, as well as the whole country (Fig. 7.2).

The potential of an integrated GIS tool in a police organization is to improve understanding of socio-policing, but the exploitation of this potential must be through qualitative and quantitative data integration. Georeferencing is not just a computer application; it is a tool integrated into the SDIK police model that allows us to understand and analyze complex spatially related problems in geographical areas defined as a neighborhood. Spatial analysis allows us to manage information that, when properly processed, facilitates obtaining direct knowledge about how NRMs are established, how they develop, and what roles religious centers belonging to various faiths play in society. The incorporation of such new technologies of information and communication as GIS, and new methodologies in the field of social research, specifically those of qualitative research, is an important thematic and methodological potential for the study of socio-police issues.

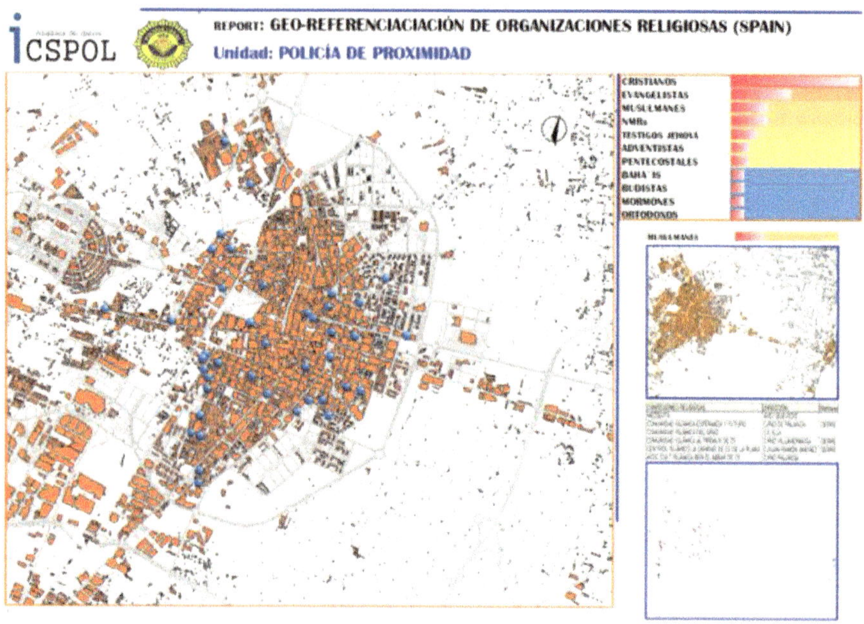

Fig. 7.2 Georeferencing religious movements

Today, police investigation should not be restricted to a pencil, notebook, tape recorder, or a simple computer. Technological development has provided various tools and business applications, and with them new forms of flexible research environments, new data types, and new ways to collect, store, analyze, and transform them into descriptions, explanations, and conclusions aimed at intelligence communities to reduce uncertainty in decision-making. However, we must not forget that new technologies cannot replace the deductive powers of the police analyst and that the quantitative phase of data analysis represents the complementary side of qualitative research.

Two disadvantages that have been attributed to police investigation are the lack of scientific methodological rigor and the apparent lack of validity that social research can make when approaching a micro-space (neighborhood, community, or district). For instance, we refer to the collection of information that may be relevant to the subject of the investigation according to a preliminary theoretical orientation, which subsequently undergoes police social adaptations as you progress in fieldwork. It is advisable to collect everything considered relevant, regardless of the situation or informant, even when the record becomes changeable.

In order to better understand this project, it is necessary to clarify that the services provided by the local police can be classified into two distinct categories. On the one hand, there are those services that are "reactive" in nature, or in other words services that have been directly requested by the citizen. Examples of a reactive service would be an emergency call to the police due to a fight in a public area or the robbery of a motor vehicle. On the other hand there are those services that are considered to be "proactive" in nature. These proactive services are designed to prevent the aforementioned reactive services and promote a greater sensation of security, insofar as these services are designed to reduce disorder. Examples of this could include, but not be limited to, traffic control and stops to enforce drink driving laws, beat officers that patrol specific areas of the city on foot or bicycle, or other campaigns designed to meet the security needs of a particular community. The rapid social changes we are witnessing as spectators and actors, materialized in an unstoppable proliferation and diversification of styles and ways of life, encourage the use of qualitative research in a twenty-first-century policing environment. From the time the police changed their perception of NRMs as being alien to one of NRMs being citizens, rather than foreign bodies entrenched in a diverse society, interest was generated in the incorporation of community-policing attempts to resolve conflicts. The relationship between police and society becomes completely different, and therefore, new methodologies are being adapted to a reality in which the minimal use of force is necessary and social efficiency is emphasized amid social diversity (Gandarillas 2011).

The main feature of policing today is integration into the social tissue and orientation of police work towards solving concrete problems of citizens. All this foundational theory and methodology has been applied by the local police in Castellón City within the policing project *Project ProxPol* (Proxpol 2012). The methodology, based on the SDIK police model, is divided into five main phases: selecting locations; getting information; capturing, transcribing, and ordering information; encoding and integrating information; and reporting.

7.3.1 Phase I: Selecting Locations

The research aims to describe and understand social phenomena that are transforming neighborhoods with new religious centers from the point of view of each citizen and collectively constructed perspectives in the community. It is based on the analysis of citizens that have made calls to the communication room or who have reported incidents in a nearby area to a religious center (Fig. 7.2) and in pursuit of their possible meanings. The first question we must ask in this methodology is: How many cases do we need to look at in order to make a solid scientific approach? Although it seems difficult to answer this question, since the selection of the informants in this first phase of research, sample selections are based in relation to space-time incidents near a religious organization.

The representativeness of the observations and places lies in a careful choice based on two mathematical algorithms in a Geographic Information System (Cluster Finder and Nearest Neighbor) having the actor or social space selected for the study objectives (Fig. 7.3). The relevance of this methodology has a direct connection with the need to adopt the heterogeneity that typifies the socio-political approach.

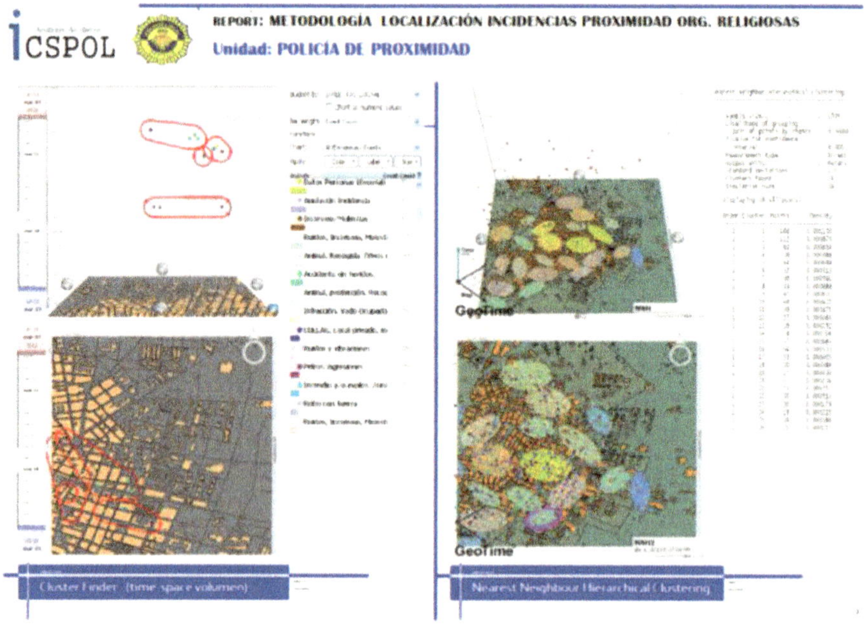

Fig. 7.3 Methodology (Geotime 5.4 OCULUS)

Territory has different neighborhoods that vary in their relevance in understanding the sociopolitical phenomenon of the religious organizations established in a community, so judicious selection of the scenes is particularly important to guarantee success of the investigation of the community police. We define an environment as the space within which the police can find demonstrations of sociopolitical reality that gain their attention. The suitability of those individuals chosen to collect accessibility conditions among the variety of potential neighborhoods facilitates empathy between the police and citizens. The experience and training of the police and their ethical responsibility place the only limits on achieving this ideal framework.

Once we have defined the study sample, community police are integrated into the social fabric of the neighborhood, obtaining complete, deep, and current information about the assigned territory (cluster), and identifying safety problems, such as formal and informal community resources deployed in the community (human intelligence or humint sensors).

A similar technique to determine access points is a clustering test based in the nearest neighbor hierarchical clustering of special clusters[2] (Fig. 7.4).

[2] To display and process the information generated by http://www.icpsr.umich.edu/CrimeStat (CrimeStat III), we used the commercial software http://bairanalytics.com/atac/ (ATAC Workstation).

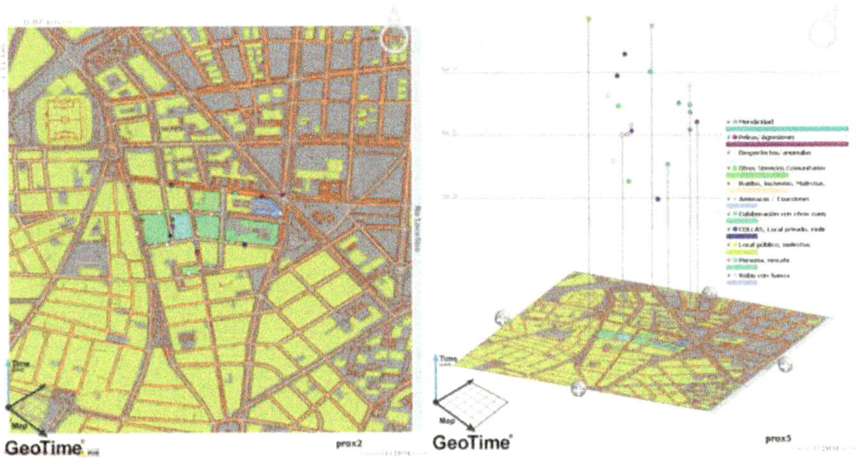

Fig. 7.4 Cluster finder methodology

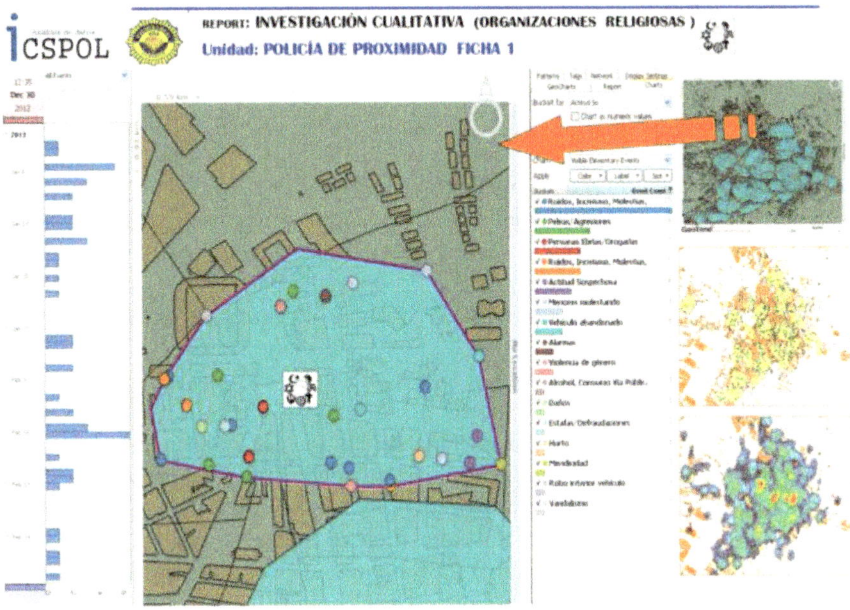

Fig. 7.5 Location selection

7.3.2 Phase II: Getting Information

Selecting a scene must meet the theoretical assumptions established by the police in the design of this research, an option that should search for balance between what is desirable and what is possible when there is some, but not all, knowledge of the field (Fig. 7.5).

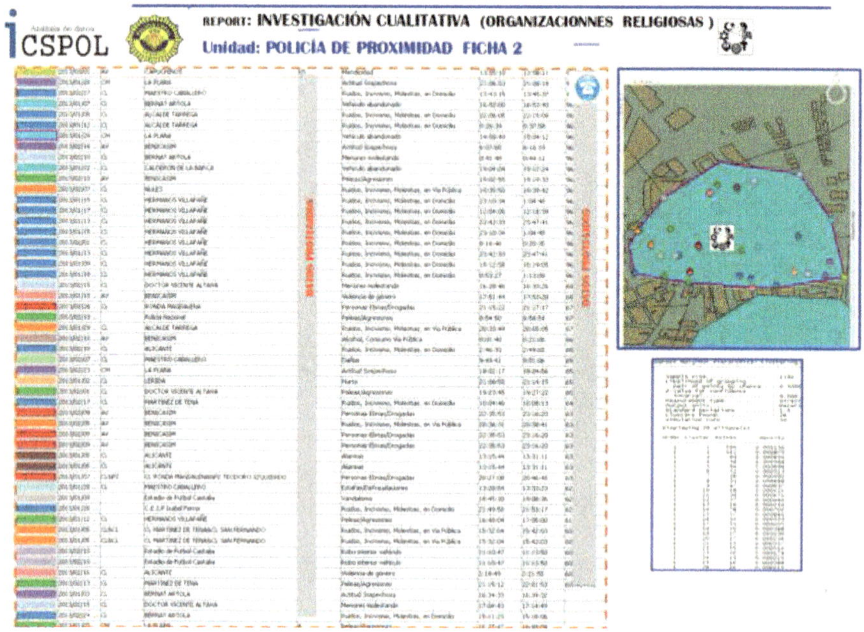

Fig. 7.6 Report 112 emergency telephone (911)

The most desirable approach would be to cover all assigned clusters over a period of time with a distribution of specialized police (two groups) for six clusters for a period of 30 days. The access and the possibilities of penetration in them and the roles needed to perform community policing will determine the scene. Our approach to selecting clusters is by obtaining the feedback from citizens who required police service during the period.

In the framework of a rollout of human intelligence sensors throughout the city, community agents previously trained in social research (e.g., semi-structured interviews) have the duty of establishing initial contact with citizens who have applied to the police service for the incidents reported in religious centers (Fig. 7.6). The investigator must verify the meeting day to avoid interference with other citizen activities, collecting all the information about open advocacy, and also all information that can be extracted from the interviews (Rubin and Rubin 1995), contextualized in terms of its temporality, space, corporeality (individuals who lived in the area), and the relational context (bonds that were generated during the experiments) (Moraima and Auxiliadora 2008).

7.3.3 Phase III: Capturing, Transcribing, and Ordering Information (Fernández 2006)

The police cannot cover the entire society, but they can try to access those stimuli that, through their experience and knowledge, are indicated as significant for research. To collect the information, the police apply an instrument that meets SDIK methodology (interview technique in depth) far from the typical police interrogation. The interview and interrogation are situations where the subject and object are in a space and one party asks and the other party answers, examining the answers with other information in its possession, allowing additional questions, the final task being obtaining any physical evidence. In an in-depth interview, in contrast to the interrogation, the police must refrain from any confrontation with the citizen. Its function is to hear the experience and gain the knowledge of the citizen, the objective is unique: get a good rapport with the interviewee (del Val Cid and Gutierrez 2010). Reaching this rapport enables the interview to flow through the appropriate channels, so the police must act to encourage the citizen to speak, avoiding channeling speech, trying to break the distance in the classic police-citizen relationship. Training police beforehand in interview technique helps them to understand the measure on which citizens are willing to cooperate depends on the ability of the preparation of an interview guide, containing themes and subthemes to be followed. This guide provides a scheme with points to be addressed, taking into account that it is not a closed model and whose order is not followed mechanically.

Facts expressed by citizens record a narrative of the features and episodes of daily life and experience. The usefulness of an in-depth or open-ended interview is aimed at obtaining information in a pragmatic way, i.e., how the subjects act and reconstruct the system of social representation in their individual practices. One of the pillars of the SDIK model is the creation of a network of key informants (people who have extensive knowledge about the environment or the problem under study), an array of humint sensors over the territory. The key moment of the investigation is that it starts the registration of the data, pending development in the police agencies that consists of laborious and essential tasks to continue the collection of information.

Data collection methods have been adapted to new technologies of communication. We should add to our field journal and memory the technical resources such as digital voice recorders and digital cameras, according to the laws of each country (Ryan and Bernard 2003). The police field journals (Fig. 7.7) are built with the intention to describe, interpret, and explain certain situations of socio-policing reality, not to share a global perspective of detected problems. Only the sum of the field journals from other police officers and the research time spent in a given territory, along with the appropriate selection of informants, will finally give sufficient light towards unraveling the puzzle of conflicts in a community.

7.3.4 *Phase IV: Encoding and Integrating Information*

To simplify and to make sense of all the interviews is one of the most challenging tasks of the police intelligence analysts. The challenge of simplifying and giving meaning to the complexity contained in the field notes and verbatim transcriptions is a phase that entails some difficulties (making value judgments about the meaning of contiguous blocks of text), which implies relying on software to encode the data, finding hidden patterns, labeling themes, and developing a system of categorization (Fig. 7.8).

In this study we have chosen the QDA Miner software + DEMO Provalis WordStat. Currently police do not have an international code book (organized lists

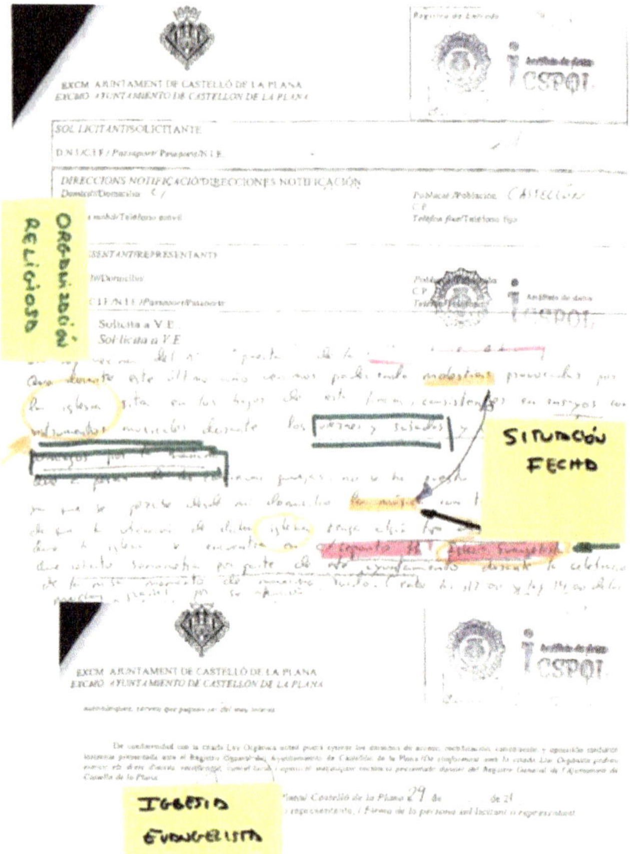

Fig. 7.7 Field journal

that include a detailed description of each code, exclusion criteria, and exclusion) by which common texts can be analyzed in a work team. The coding is not a rigid process, but having a basic shared codebook facilitates collaborative work among others. Each police organization is able to develop and refine its code book as the investigation proceeds.

Once the analysts have identified a number of elements (themes, beliefs, concepts, etc.) the following step is to identify how these elements are interrelated in a theoretical model (Fig. 7.9).

7.3.5 *Phase V: Reporting*

The analysis and interpretation of materials for community police in fieldwork starts from the initial collection. The police continue to collect information day by day in various clusters assigned (feedback) or from the network of informants

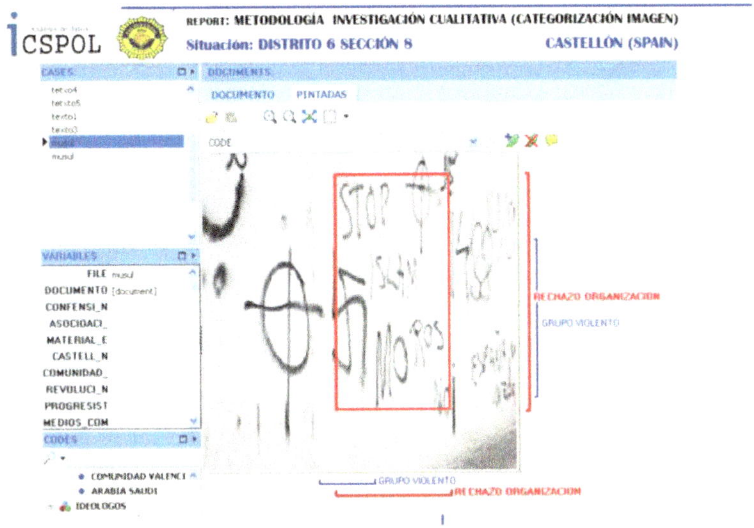

Fig. 7.8 Image categorization

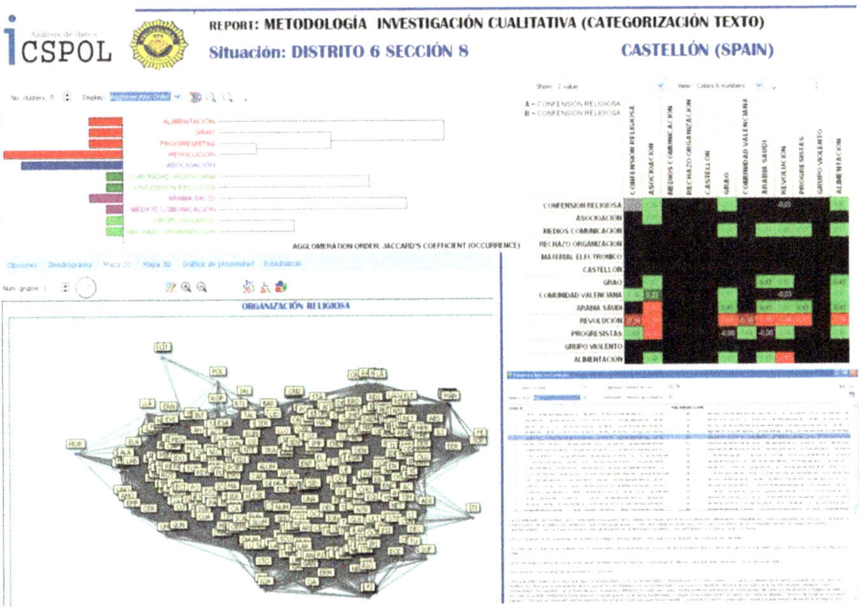

Fig. 7.9 Text categorization

generated in the first stage of contact with citizens, extended and interpreted systematically and continuously. The final report is presented by this analysis, but we do not like to talk about a final intelligence cycle, because in doing so we would not account for the beginning of other cycles and lines of research (local, national, international). Therefore, we leave this as an open-ended process.

7.4 Results

The results obtained so far show that the SDIK methodology accompanied by a GIS captured the social fabric of the community and may be useful for the extraction of information. Knowing the experiences and viewpoints of citizens has a qualitative value in developing the police intelligence cycle and in making decisions in the field of coexistence in and security of territory. The implementation of the SDIK methodology has provided the local police of Castellón a more exhaustive knowledge of the social fabric of the city, especially and particularly the set of circumstances related to the growth of new religious movements, a phenomenon of emerging importance in recent years that may affect public safety. This knowledge and the application of different methodological phases of SDIK have facilitated an understanding of the various religious centers and an assessment of their effect of each on the micro-space in the city, as in each community or neighborhood. This knowledge has facilitated the implementation of specific measures to minimize risks and problems, as well as the anticipation of certain conflicts that were previously not addressed with sufficient intensity.

Similarly, the work system has allowed police to better locate all public safety services provided by the city, allowing them to be placed in an improved position in relation to the community, thus allowing a closer relationship between police and the community and an increased capability to obtain quantitative and qualitative information. But certainly the most important factor for improvement has been the implementation of georeferencing systems and the technical and scientific treatment of the spatial information obtained. This step has greatly improved the cycle of information management and knowledge applied to police activity, which in turn has allowed improvement of the initially proposed objective, to learn more about what the city makes visible to us, and to discover what, sometimes, has been invisible to us.

Since beginning this project, there have been a variety of results which have helped support its goals. Also, there has needed to be some clarification and reorganization of the initial goals of this project that were oriented to see a reduction in the number of "administrative" infractions of the law, e.g., common delinquency and public disturbances. However, for the purposes of this chapter, the authors have decided to include only those results which are specifically relevant to the general purposes of this project. The results have also been divided into two specific categories which will be referred to as primary results and secondary results.

One of the primary objectives of this project has been to achieve a more complete understanding of those religious communities that have been established within the previous 10 years. This time period was considered relevant as it corresponds with a period of intense growth in immigration for the city. Consequently, this objective provided a basis for the utilization of georeferencing as a tool to gather information and facts for use by the local police. This first objective has provided for a more thorough understanding of the leaders, as well as those responsible for these newly established groups, as well as the activities in which they have been engaged. The process has revealed to the local police that many of these activities, in fact, have been illicit in nature. Furthermore, these activities have often been to the benefit of some of the religious leaders thus blurring the distinction between the religious and the illegal.

The improvement in data gathering and subsequently the better understanding of the community has allowed for the establishment of improved relations between the local police and leaders of the newly established groups in Castellon, which in fact was the secondary goal of the project. The result has been to begin an improvement of the ability of the local police to meet the demands of these communities as their needs are identified. Some of the results of these secondary results include the following: the sale of unlicensed foods was found to be commonplace near and around mosques. Legislation exists regarding the handling and sale of food in public in Spain for a variety of health and other reasons. Through the use of better data, and better understanding between cultures, this practice of unlicensed selling of food in public has been eliminated.

Another practice which has been eliminated has been the repair of motor vehicles on public streets without previous authorization. This practice was commonly associated with the proximity of mosques and Orthodox churches commonly frequented by Romanian immigrants. It is believed that reducing the illegal repair of cars on public streets has led to a severe reduction, to the point of near elimination, of the sale of stolen car parts. While this finding still has yet to be confirmed, the downward trend of stolen car parts seems to be related. Also of importance, the distribution points of illicit drugs has been better refined and highlighted. Location of distribution points and the sale of illicit drugs has been the result of interviews with neighbors in specific parts of the city. These areas have, in the main, coincided with the recent establishment of certain religious establishments and were discovered through the use of the SDIK method. Finally, the illegal sale of cell phones and cell phone SIM cards linked to certain religious institutions was discovered in Castellon. The detection of these sales led to a marked decrease in these sales. It should be noted that the illegal sales of these items was directly connected to congregants of some newly established religious organizations.

Consequently, four out of the six areas that were used for reference by the local police within the city of Castellon for the purposes of this project showed a decrease of 25–30 % in the number of complaints made to the police, as well as complaints made to the police about the religious institution located within these areas. The complaints typically dealt with noise, vehicular complaints, and other suspicious activities.

Another point of interest was the fact that due in large part to the cooperation of local religious centers of Evangelistic groups and the local police, two previously commonplace infractions have been detected and reduced. Firstly, in subsidized housing projects, certain housing units had numbers of people far in excess of what was legally acceptable. In certain instances, it was found that these units were being used for purposes beyond living space. Often these other activities were for economic gain. A second discovery due to the cooperation of both groups was that children between the ages of 13 and 16 years were not attending school. Because of this situation, the authorities were able to take proper measures to ensure the education of these children, as well as any appropriate measures with the parents. Finally, due to the cooperation of the local police, in conjunction with citizens of the study areas, they were able to successfully close down three unauthorized religious establishments. These religious establishments were in clear violation due to not having the proper accreditation, nor licensing. In all three cases the "religious" establishments were engaged in the illegal sale of food and music entertainment.

7.5 Discussion

One cannot pass up the opportunity to reflect on the results that have been obtained after the implementation of the SDIK model to a problem as specific as the development of NMRs. First, it has achieved a more detailed approach to these types of organizations, and at present the police and the community have a much greater knowledge compared to the information available prior to the commissioning of SDIK. This methodology has proven itself necessary to the implementation of certain measures to provide information about NMRs, such as the number of members of these groups, their individual interests and aspirations, and the important aspect of their funding sources and expenses. Second, it is remarkable that the activity directed towards obtaining information on each one of the micro-spaces has made visible much information that has no direct relationship with the NMRs. However, this information has been useful in taking specific measures to respond to the general security problems of the community otherwise unrelated to these organizations. This fact reinforces the effectiveness of the chosen strategy, as it has favored knowledge management and intelligence applied to all police activity, not just activity related to new religious movements.

The authors are in agreement that there are two principal areas of the SDIK model that could be improved. The first area for improvement is the training at the local level of police and the second being the sharing of the model and its purpose with other police bodies operating in local and neighboring police jurisdictions. The authors both agree that having a more formal approach to educating the local police forces could have had a significant effect on the results. Areas of improvement that should be specifically addressed are in interviewing and interrogation. These two areas could lead to significantly improved data quality and quantity and a better understanding of the areas and groups being observed. Furthermore, through better training the local police would have been able to perform better during times of higher volumes of calls for support and service. In times of higher than average number of phone calls, the local police were put into a reactive position trying to stop and solve criminal activity, rather than being in a proactive position preventing those crimes from starting.

Also, it would have been of interest to share and coordinate the SDIK model with other police entities that provide different services, yet operate within the same jurisdiction. One of the reasons is that it would provide even more complete information regarding the habits and behaviors of the inhabitants. Also, it would have been more efficient to accumulate information about other similar or related activities, which were not the purview of the local police. In not having open lines of communication regarding all information, it is nearly impossible to have a complete picture of Castellon, nor is it possible to formulate other conclusions about future illicit activity or preventative measures. Finally, the authors were pleased with the overall reception by the public who were represented in this project. However, the initial reception was one of being wary of the local police's motives; eventually the initial suspicions of the public were overcome and participation became voluntary and successful.

The authors would like to explain how information of the project was shared. In the case of police entities, information about the results was shared directly without reservation. As to the affected groups within the community, information was shared indirectly through three different agencies. The information shared with the public could be defined as limited within the confines of that which was pertinent. One of such agencies is the Agencia de Mediacion para la Integracion y la Convivencia social de Castellon (AMICS) (Agency of Mediation for Integration and Social Coexistence). This agency is dedicated to collecting, analyzing, and dispersing of all information as relates to the local community. A second agency is the Mesa Interreligiosa de Castellon (Interreligious Table (Council) of Castellon). The Mesa Interreligiosa de Castellon is dedicated to sharing all information pertinent to local religious entities. Lastly is Unidad para la Gestión Policial de la Diversidad (GESDIPOL) (Police diversity management) whose objective is to share matters that relate to social, religious, and cultural diversity. This last organization was created within the organization of the local police.

In evaluating the initial operation of the SDIK model, the authors feel that it has undoubtedly been a success. The local police have been able to provide better local security and law enforcement, as well as to have improved information as related to new religious and immigrant groups in the city of Castellon. While not directly measured, it is the hope that the citizenry noticed a reduced sense of insecurity in certain parts of the city. More importantly, the exhaustive nature of improved understanding and information relating to the leaders of these groups new to Castellon has also led to a decline in conflict and problems between the local police and the aforementioned new religious groups. Consequently, the local police have been better able to meet the demands of the affected groups as well as work towards preventative measures in affected areas.

References

Chaney S, Ratcliffe JH (2007) GIS and crime mapping. Wiley, Hoboken

Castellón de la Plana (2013) http://www.castello.es/web30/pages/contenido_web20.php?cod0=2&cod1=9&cod2=30&cod3=13. Accessed 10 Oct 2013

Dake K (1992) Myths of nature: culture and the social construction of risk. J Soc Issues 48(4):21–37

del Val Cid C, Gutierrez J (2010) Prácticas para la Comprensión de la Realidad Social. McGraw Hill, Madrid

Douglas M (1992) Risk and justice. In: Risk and blame. Essays in cultural theory. Routledge, Londres, pp 22–37

Fernandez L (2006) Como analizar datos cualitativos. Butlletí La Recerca, Barcelona, pp 1–13

Galdon G (2011) Crisis Económica y Gestión de la Seguridad Ciudadana: Mapas de Delincuencia. En Revista Catalana de Seguridad Pública nº 24, Generalitat de Catalunya, Barcelona, 221, pp 79–107

Gandarillas MA (2011) Memoria del Congreso Internacional sobre Seguridad Ciudadana en la Sociedad Diversa. Cultivalibros, Madrid

Geotime 5.4 © OCULUS www.geotime.com/. Accessed 10 Oct 2013

Hill B, Painich R (2013) Fundamental of crime mapping. Jones and Bartlett Publishers, Burlington

Johnson CL (2010) Police use of intelligence networks for reducing crime. LFB Scholar Publishing LLC, El Paso

Moraima M, Auxiliadora L (2008) El análisis de contenido: una forma de abordaje metodológico. Laurus, volumen 14, Venezuela, pp 129–144

PROXPOL (2012) Modelo de proximidad basado en el conocimiento. http://www.gigapp.org/es/component/jresearch/?view=publication&task=show&id=500. Accessed 10 Oct 2013

QDA Miner software + DEMO Provalis WordStat. http://provalisresearch.com/downloads/updates/. Accessed 15 May 2013

Ramos M (2006) Nuevos Movimientos Religiosos en España. Thesis. Universidad de Granada

Rubin HJ, Rubin IS (1995) Qualitative interviewing. The art of hearing data. Sage, Thousand Oaks

Ryan GW, Bernard HR (2003) Data management and analysis methods. In: Denzin NK, Lincoln YS (eds) Collecting and interpreting qualitative materials. Sage, Thousand Oaks, pp 259–309

United Nations (2010) Sistemas Policiales de Información e Inteligencia. UN, New York

Vallés MS (2007) Técnicas Cualitativas de Investigación Social. Síntesis Sociología, Madrid

Chapter 8
Spatial Analysis of Fear of Crime and Police Calls for Service: An Example and Implications for Community Policing

Jamison Conley and Rachel Stein

Abstract Individuals' fear of crime exhibits a complex spatial relationship with not just actual crime incidents, but a mix of actual crime, perceptions of crime, neighborhood disorder, and collective efficacy. If people have a high fear of crime, they may be more likely to report suspicious or criminal activities to the police. The fear of crime individuals maintain is most often directly linked to the fear of violent crime; however, a spatially explicit examination of the impact of violent crime calls for service to police officers, neighborhood disorder and collective efficacy on the fear of crime is still needed. In the current study, we examine the relationship among all of these factors using measures of spatial correlation and spatial regression. While the reactive policing strategy of responding to calls for service is more cost-effective than community policing, targeted proactive strategies might be more useful for long-term crime prevention. Our findings illustrate the potential of spatial analysis in informing policing strategies, by highlighting variation in the spatial relationships between fear of crime, violent crime incidents, collective efficacy, and neighborhood disorder. Using the results of this type of analysis can lead to a better use of police resources to avert crime.

Keywords Fear of crime • Violent crime • Community policing • Geographically weighted regression

J. Conley (✉)
Department of Geology and Geography, West Virginia University,
330 Brooks Hall, PO Box 6300, Morgantown, WV 26506, USA
e-mail: jamison.conley@mail.wvu.edu

R. Stein
Department of Sociology and Anthropology, West Virginia University,
307 Knapp Hall, PO Box 6326, Morgantown, WV 26506, USA

G.A. Elmes et al. (eds.), *Forensic GIS: The Role of Geospatial Technologies for Investigating Crime and Providing Evidence*, Geotechnologies and the Environment 11,
DOI 10.1007/978-94-017-8757-4_8,

8.1 Theoretical Framework for Understanding Crime and Fear of Crime

The fear of crime affects people's everyday behavior. Researchers have defined the fear of crime as the emotional response people feel toward the possibility of victimization (Taylor 2001). The level of fear impacts how often people leave their homes or are engaged in the public sphere (Taylor 2001; Taylor et al. 1985). In addition, the possibility and frequency of interaction with neighbors is affected when people are afraid of being victimized. The fear of victimization presents a reciprocal relationship between the resident and the neighborhood. Oftentimes, a resident's fear is brought about by the conditions of the neighborhood. This includes the physical and social environment of the area (Bursik and Grasmick 1993; Doran and Lees 2005; Skogan 1990; Taylor 2001; Warr 1990). As a result of the fear residents feel, they are likely to disengage from the neighborhood. Fearful residents fail to maintain the upkeep of their property and do not pursue relationships with their neighbors (Skogan 1990). These actions, or inactions, further exacerbate the deteriorating conditions in the neighborhood, which foster feelings of fear.

Neighborhood research on crime and the fear of crime is often framed using the broken windows theory and the theory of collective efficacy (Bursik and Grasmick 1993; Doran and Lees 2005; Skogan 1990; St Jean 2007; Taylor 2001). Both of these theories indicate neighborhoods with high levels of disorder will also experience increased rates of crime (Sampson et al. 1997; Wilson and Kelling 1982, 2006). In addition, the theories suggest informal social control plays an important role in limiting the opportunities available for crime. Social control is the central underlying causal factor of crime in the theory of collective efficacy, while broken windows theory proposes a noncausal correlation between the level of disorder and informal social control.

The central concept in broken windows theory is neighborhoods with high levels of physical and social disorder will have increased crime rates (Wilson and Kelling 1982). Physical disorder is defined as the visible evidence of deteriorating conditions in a community. The disruptive activities of people in the neighborhood signify social disorder. Empirical evidence suggests that communities with high levels of physical disorder are also likely to experience frequent occurrences of social disruptions (Braga et al. 1999; Skogan 1990; Wilson and Kelling 1982). Signs of physical disorder suggest a certain lack of social control in the neighborhood, thereby creating situations in which social disruptions are more likely (Skogan 1990; Xu et al. 2005). The presence of some disorder in a neighborhood leads to even more disorders. Wilson and Kelling (1982) propose higher levels of disorder will eventually lead to criminal acts.

The theory of collective efficacy proposes that there are underlying factors causing both crime and disorder (Sampson and Raudenbush 1999; Sampson et al. 1997). These factors include the social cohesiveness of the neighborhood, the level of trust among residents, and the willingness of neighbors to help one another (Sampson et al. 1997). Overall, these elements represent the level of informal social control, also termed collective efficacy, present in a community (Sampson et al. 1997). Residents in neighborhoods with high levels of collective efficacy have the ability to identify and work toward collective goals. These goals are centered on building and maintaining strong ties among residents in a community and ensuring that the neighborhood has relatively low levels of crime and disorder (Janowitz 1975).

Residents in neighborhoods with strong collective efficacy are more likely to intervene when there is a problem and work together to correct the problem (Sampson et al. 1997). Research indicates that high levels of social control directly and positively limit the existence of social disorder, which, in turn, decrease the visible elements of physical disorder (Novak and Seiler 2001; Sampson and Raudenbush 1999; Skogan 1990; Wilson and Kelling 1982). A neighborhood in which residents are joined by strong network ties contains the element of social control necessary to constrain social disruptions. Also, in such a neighborhood the residents care enough about their neighborhood to manage the upkeep of physical properties, thereby reducing physical disorder.

8.1.1 Neighborhoods and the Fear of Crime

Research on the fear of crime indicates the actual crime rates in a neighborhood have a varying impact on the fear of crime held by residents. The role of disorder and neighborhood collective efficacy has a greater impact on how fearful residents are of victimization. Empirical research on the direct link between crime rates and fear of crime finds a range of correlations from relatively weak to strong, depending on the type of crime involved (Rountree and Land 1996; Taylor 2001). Taylor (2001) reports local crime rates are only moderately related to the feelings of fear held by residents. When only violent crime rates are considered, high rates of violence in a neighborhood are consistently related to increased levels of fear (Brunton-Smith and Sturgis 2011; Porter and Purser 2010).

The level of collective efficacy in a neighborhood conditions residents' fear of crime. If residents perceive the people in their neighborhood truly work together as a social unit in opposition to crime, then they are likely to have less fear. In contrast, those who feel isolated from their neighbors and their community will have higher levels of fear. Strong collective efficacy represents the willingness of people to intervene in the social disruptions that erupt in the neighborhood (Sampson and Raudenbush 1999; Skogan 1990). This action gives residents a sense of empowerment that they can do something to lower the crime in their neighborhood. If residents feel they have some control over the disturbances in their neighborhood, they are likely to have decreased levels of fear. In contrast, neighborhoods characterized by frequent social disruptions in which residents do not intervene foster an environment of fear. Fearful residents are not likely to seek out others and maintain only limited networks in the community. Residents who work together to maintain order in the neighborhood build collective efficacy. The concepts of collective efficacy and social disorder are interdependent and both are related to fear.

Residents who are actively engaged in the neighborhood are more likely to care about the appearance of the area than those who are disconnected from the neighborhood social networks (Skogan 1990). An area plagued by dilapidation is often perceived as an unsafe area. While physical disorder is not in itself frightening, the implication is that people who live in these disordered areas do not care about the upkeep or safety of their neighborhood (Bursik and Grasmick 1993; Skogan 1990). The physical environment affects social interactions as well. Skogan (1990) reports that residents in neighborhoods with greater disorder tend to have higher levels of fear because they interact less with their neighbors (see also Doran and Lees 2005). Even so, it is important to recognize that long-term residents in disordered areas can normalize the disorder. In fact, researchers indicate that a dramatic change in the level of disorder must occur for residents to perceive the incivilities as a problem (Perkins et al. 1993; Taylor 2001).

8.1.2 Theoretical Concepts and Policing

Research indicates residents are more concerned about the elements of physical and social disorder in their neighborhood than they are about even more serious crimes that might occur close to their homes (Skogan 1990). Residents are faced with evidence of incivilities on a daily basis and confront these issues as a part of their daily routines. The disorder is more present and more relevant to residents than potential incidents of criminal behavior. Wilson and Kelling (1982) suggest if police in the neighborhood concentrate their efforts on improving deteriorating conditions and imposing controls to curb social disruptions, the perception is that police officers care about neighborhood residents' quality of life (see also Dixon and Maher 2005; Trojanowicz and Bucqueroux 1992; Wilson and Kelling 2006). Braga et al. (1999) found strict adherence to maintaining social order, and officers' involvement in removing the physical signs of disorder helped to reduce criminal incidents in high-crime neighborhoods in Jersey City. The physical improvement efforts included removing trash from the streets and securing vacant lots. Efforts to control social order included officers dispersing groups of loiterers and initiating "stop and frisk" of suspicious persons. These efforts demonstrated the police officers were not only aware of the incivility problem but were willing to be involved in the upkeep and order of the neighborhoods.

Proactive policing strategies focused on eliminating disorder can also be effective to curb more serious criminal behaviors (Braga et al. 1999; Xu et al. 2005). The policy implications associated with broken windows theory are that the levels of physical and social disorder in a community need to be decreased to lower crime rates. Vindevogel (2005), for example, finds employing private security guards and sanitation workers to clear graffiti and trash in the downtown areas leads to increased foot traffic and greater feelings of safety. When police officers help facilitate the feelings of safety among residents, people are likely to be engaged in public spaces. Wilson and Kelling (2006) indicate policing strategies that minimize disorder will also help build collective efficacy among neighborhood residents. The willingness of people to walk in their neighborhood builds trust and cohesion among residents by providing opportunities for people to get to know their neighbors (Sampson and Raudenbush 1999; Skogan 1990).

The theory of collective efficacy indicates that instead of a focus on minimizing disorder, the social networks among neighborhood residents need to be established and strengthened to prevent crime from happening. If residents have a strong bond of trust, they are willing to look out for one another and intervene to stop criminal behaviors (Sampson et al. 1997). Morenoff et al. (2001) indicate that strong collective efficacy represents a level of social control in neighborhoods, which significantly lessens homicide rates. Understanding the relationship between disorder, collective efficacy, fear of crime, and violent crime incidents can help define policing strategies that are most effective in a particular neighborhood. Much research on neighborhood crime concentrates on collective efficacy and disorder at the neighborhood scale, which is often defined according to census tract boundaries. Studies that consider the neighborhood as an aggregate of street blocks across a census tract area often ignore potential pockets of disorder or efficacy within the neighborhoods (St. Jean 2007). That is, there are smaller block or block group areas within census tracts that might be characterized by high levels of disorder, while the rest of the neighborhood experiences minimal disorder. A spatially explicit analysis of neighborhoods can identify how the pockets of disorder or efficacy might relate to fear of crime differently across the geographic area. If these patterns are recognized, proactive policing resources can be utilized in a more effective manner to curb criminal behaviors.

8.2 Spatial Analysis of Fear of Crime and Calls for Service

To assess the relationships between fear of crime and calls for service, we conducted both nonspatial and spatial regression analyses. We conduct a secondary analysis using data from a study conducted in 2009 focused on situational policing in two neighborhoods in a large city in Delaware (Nolan et al. in press). The two neighborhoods are characterized as residential areas, located southeast of the business district, or the downtown area of the city. The data include variables on fear of crime, collective efficacy, residential perceptions of social and physical disorder, and calls for service to the police department.

Census demographic information from 2010 for each of the census tracts is presented in Table 8.1. All of the areas of study encompassed in this research are characterized as low-income, high-crime areas. The total population in the neighborhoods is 5,563. Approximately half of the residents are male (51 %). The population in these neighborhoods is relatively young, with about 30 % of the population comprised of people under 18 years of age. The median age across the census tracts is 32 years. In terms of racial composition, whites represent over half of the population, about 63 %. Almost 18 % of the population reported they are of Hispanic origin, regardless of race.

Table 8.1 Census demographic information

	N	Study area (%)
Total population (two census tracts)	5,563	
Sex		
Male	2,716	48.82
Female	2,847	51.18
Racial composition		
White	3,006	62.65
African American	1,792	37.35
Hispanic (of any race)	1,057	17.50
Age		
Under 18 years old	1,655	29.55
Median age	32.2	
Median household income	34,873	
Employment (population aged 16+)		
In the labor force		
Employed	2,320	56.99
Unemployed	227	5.58
Not in the labor force	1,524	37.44
Individuals below the poverty level	906	16.29

The median household income across the census tracts is approximately $35,000. Low income is somewhat problematic in this area of the city; about 16 % of the individuals live in poverty. The census data indicates 5.6 % of the population is unemployed, while just over half of the people in this area maintain employment (57 %).

8.2.1 Data

The data are in two different formats. Most of the data are measured by surveys, which we have geocoded to the faceblock on which the respondent lives. Faceblocks are one side of a city block. For example, a city block such as Maple Street between the 3rd and 4th Avenues would contain two faceblocks: one for the northern side of the street and another one for the southern side of the street. Of the 306 faceblocks in the study area, we have survey responses from 105 residents living on 76 faceblocks. These surveys capture the residents' fear of crime, collective efficacy, and perceptions of disorder. The last piece of data which we use are the calls to police for the calendar year 2009, which are geocoded to the most exact location possible given the records, whether that is an address or an intersection of two streets. The calls to police included every call, although for this analysis, we extracted only the 533 calls for violent crimes. We are using these calls as a proxy for the neighborhood's crime rate. While imperfect, because some crimes may go unreported and others generate multiple calls, this type of data has been successfully used in previous research (Braga et al. 1999).

The survey questions measuring fear of crime were (1) How safe do you feel being out alone during the day? and (2) How safe do you feel being out alone at night? They are both coded as Likert scales ranging from 1 (very safe) to 4 (very unsafe) with high values representing greater fear. We combined these into a fear index with an unweighted average.

The survey questions measuring collective efficacy were measured through both social cohesion and shared expectations. Social cohesion was measured by the following questions, which are also coded as Likert scales ranging from 1 (very likely) to 4 (very unlikely): (1) people are willing to help their neighbors, (2) this is a close-knit neighborhood, (3) people in this neighborhood can be trusted, (4) people in this neighborhood generally don't get along with each other (reverse coded), and (5) people in this neighborhood do not share the same values (reverse coded). Shared expectations were measured by five questions that asked residents of the likelihood that their neighbors could be counted on to take action if (1) children were skipping school and hanging out on a street corner, (2) children were spray painting graffiti on a local building, (3) children were showing disrespect to an adult, (4) a fight broke out in front of their house, and (5) the fire station closest to home was threatened with budget cuts. The index for collective efficacy is an unweighted average of all ten questions, and the questions marked as reverse coded are structured to ensure that higher values for all questions indicate less collective efficacy.

We also measured the residents' perceptions of physical and social disorder by asking the residents to identify whether each of the following conditions is a "big problem," "somewhat a problem," or "not a problem" in their neighborhood. We code "big problem" as 2, "somewhat a problem" as 1, and "not a problem" as 0. Social disorder was measured by (1) people insulting or bothering others as they walk down the street, (2) groups of teenagers hanging out, (3) high noise levels, (4) bad elements moving in, and (5) people fighting and arguing. We did not define the terms for the residents, letting them identify how much noise constituted a high level or what they considered bad elements. Physical disorder was measured by (1) vandalism, (2) vacant houses, (3) people not keeping up their properties or yards, (4) litter and trash, and (5) vacant lots filled with trash. In both cases, the indices are unweighted averages where greater disorder has higher values in the index.

To convert the crime data to the faceblock spatial resolution, we created a surface of violent crime data using kernel density estimation (KDE) (Fig. 8.1) using a kernel radius of 200 m, chosen to smooth out the surface to eliminate noise resulting from only using 1 year of data, while retaining spatial variation across the study area. Even within this small study area, and after accounting for the lower rates along the boundaries of the study area due to edge effects, there is a strong variability of the density of calls to police. The surface values at each of the points along the center of the faceblocks are then extracted from this surface to represent the density of calls to police for violent crimes in the local area surrounding the faceblock. These were joined to the survey responses in the regression methods.

To summarize, we have five values for each faceblock in our study: residents' fear of crime, residents' perceptions of social disorder, residents' perceptions of physical disorder, collective efficacy, and calls for service. Higher values

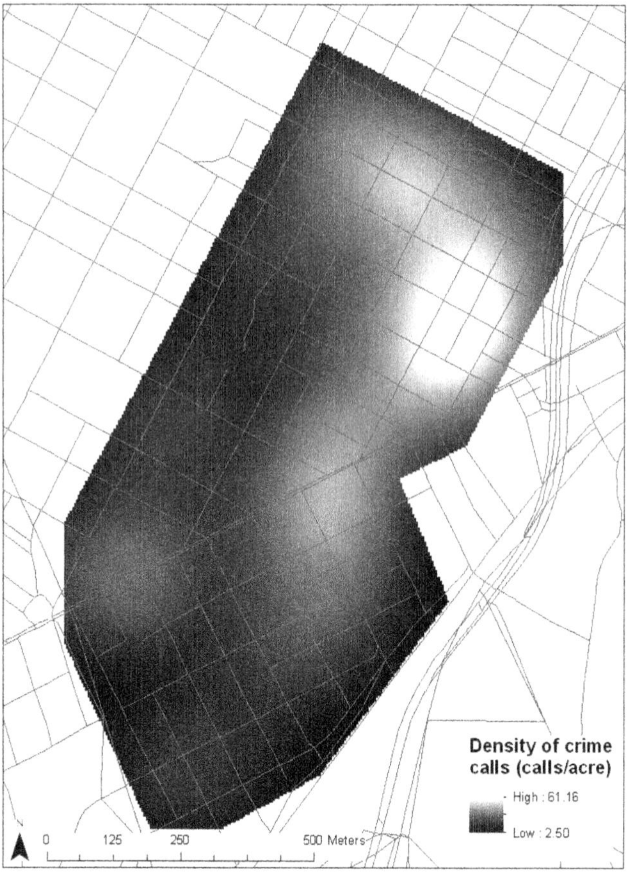

Fig. 8.1 Kernel density estimate surface of calls to police

represent greater fear, more disorder, less efficacy, and more crime calls. Therefore, all expected relationships in the regression analysis are direct or have positive coefficients.

8.2.2 Methods

We applied two forms of regression to model relationship between fear of crime and calls for service. The regression techniques are standard ordinary least squares regression (OLS) and geographically weighted regression (GWR) (Brunsdon et al. 1998). We choose GWR because the relationships between these variables may not be consistent across the entire study area, as is assumed by OLS regression. GWR allows, for example, a strong relationship between fear of crime and calls for service

Table 8.2 Regression results of fear versus calls for service

	Calls for service
Coefficient	135.5
p-value	(0.059)

Table 8.3 Regression results of fear versus calls for service, residential perceptions of disorder, and collective efficacy

	Calls for service	Social disorder	Physical disorder	Collective efficacy
Coefficient	55.5	0.038	0.016	0.459
p-value	(0.432)	(0.412)	(0.703)	(0.0008)

in parts of the neighborhood where the police are viewed by the residents as effective, but a weaker relationship where the residents do not trust the police and therefore do not call them even if there is a high crime rate. Thus, GWR will be able to isolate the pockets of disorder or efficacy as highlighted by St. Jean's (2007) research. The GWR analyses used an adaptive-distance bandwidth set to include the 30 nearest neighbors from the dataset.

The first model in both regression frameworks is a bivariate model relating the fear of crime index from the survey as the dependent variable against faceblock-specific value of the KDE surface for calls for service as the independent variable, which is a proxy for the violent crime rate. Because fear of crime can be mediated by other neighborhood factors, as discussed above, we also tested a multivariate model including some of these neighborhood factors, namely, residential perceptions of social disorder, their perceptions of physical disorder, and collective efficacy.

8.2.3 Results

8.2.3.1 OLS Results

The linear regression results of the first model simply comparing fear and calls for service (Table 8.2) are marginally significant, showing a marginally significant positive relationship between the two. The coefficient is in terms of a unitless index, so is not directly interpretable by itself.

The multivariate linear regression including the mediating factors (Table 8.3) shows a stronger relationship between fear and collective efficacy rather than calls for service. While all of the relationships are direct, as expected, the only significant relationship is that between collective efficacy and fear of crime, in which higher levels of collective efficacy reduce the fear of crime. In this multivariate regression, the relationship with efficacy is significant, even as the relationship between fear of crime and calls for service is not.

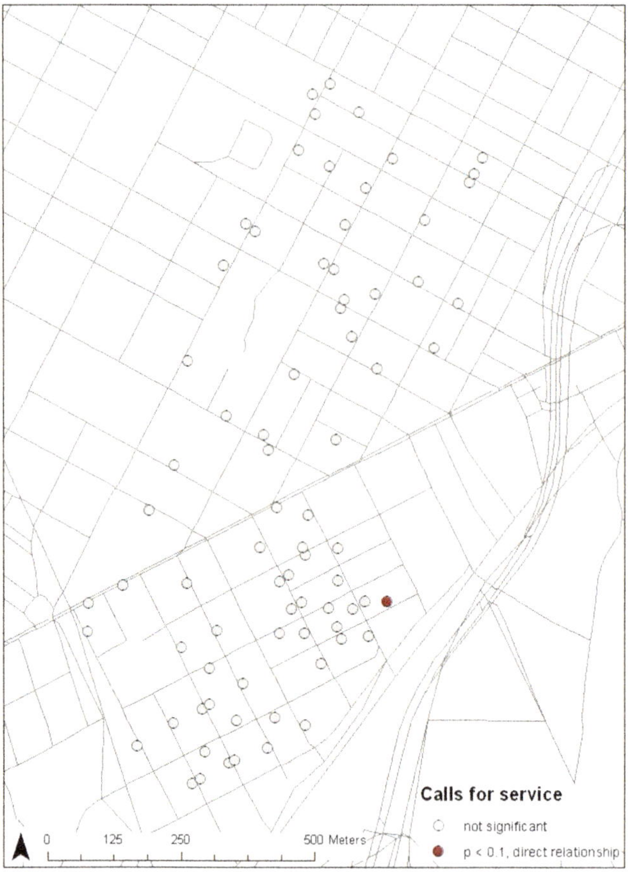

Fig. 8.2 GWR significance map of fear versus calls for service for violent crime

8.2.3.2 GWR Results

The relationship between fear and calls for service in the first statistical model (Fig. 8.2) does not yield a strong relationship; the relationship is not significant in the vast majority of the study area. The single faceblock where the relationship is significant may be an outlier or an artifact of edge effects from its position in a corner of the study area. Alternatively, the explanation might hinge on the fact that a playground is located on the south side of this faceblock. Residents who are fearful in this area might be more willing to call the police in efforts to maintain a safe environment for the children who frequent the public space. This reinforces the OLS results in which the bivariate relationship between these two processes was weak, albeit statistically significant. As with our OLS analysis, our second GWR analysis added in additional terms to see if the relationship between crime and fear

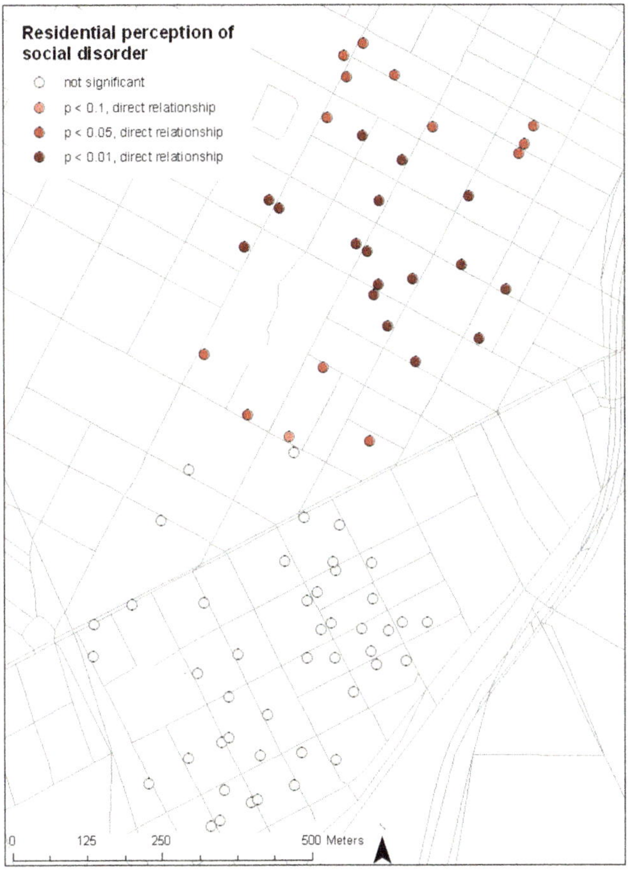

Fig. 8.3 Multivariate regression—significance of residential perceptions of social disorder

of crime is mediated by other factors. Maps for those other factors are given in Figs. 8.3, 8.4, 8.5, and 8.6.

These maps show a complex spatial pattern, which is masked by the nonsignificance of most variables in the linear regression results. While the multivariate OLS only had significance for collective efficacy, the multivariate GWR shows pockets of local significance for all variables. This pattern shows stronger relationships in two regions of the study area. In the areas where there is a direct relationship between perceptions of fear and calls for service, there is also a direct relationship between perceived physical disorder and fear of crime. Likewise, where the relationships between fear and calls for service are inverse ones, so is the relationship between physical disorder and fear. Collective efficacy is significantly and directly related to perceived fear in much of the area, while perceptions of social disorder are only related to fear in the area where the physical disorder and fear of crime relationships are inverse. However, in these areas, the relationships of fear with

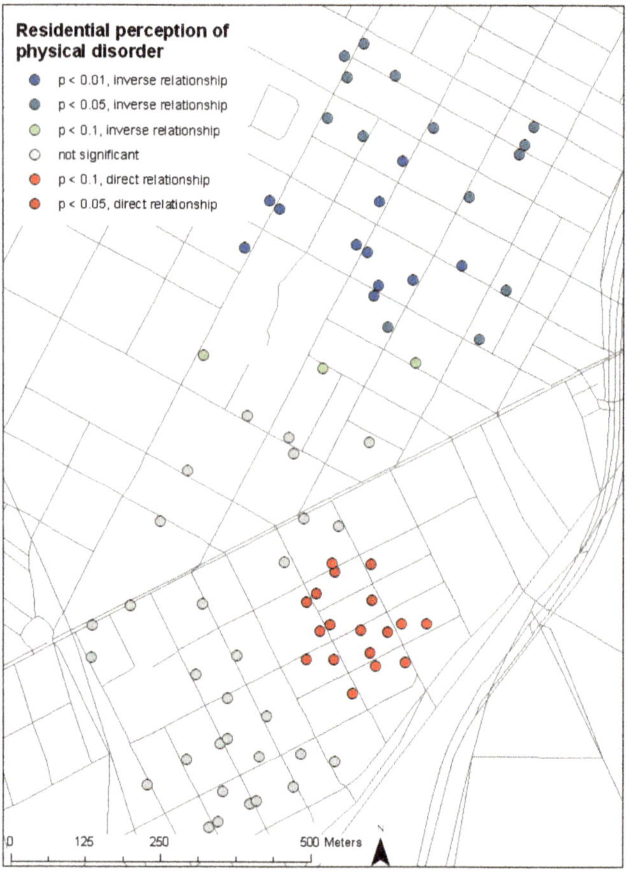

Fig. 8.4 Multivariate regression—significance of residential perceptions of physical disorder

social disorder and collective efficacy (which are in the expected direction) are stronger than the relationship of fear with physical disorder (which is in the opposite direction).

The patterns of relationships that emerge among variables in the spatial analysis support the existence of pockets of disorder and collective efficacy within neighborhoods, as proposed by St. Jean (2007). While we can hypothesize about why these differential explanations of the fear of crime exist, it is beyond the scope of the current research to definitively explain the causal elements. In fact, the physical characteristics of the neighborhoods are relatively similar across the study area. The neighborhoods are primarily residential and are composed mostly of row houses. One potential explanation for the patterns is that certain characteristics of the residents might be influencing the results. Residents who have lived in the neighborhood for a long period of time, for example, might have less fear than residents who are relatively new to the neighborhood. People who have lived in the area for an

Fig. 8.5 Multivariate regression—significance of collective efficacy

extended period are accustomed to their surroundings. These residents experience fear of crime only when there are dramatic changes in the level of disorder or crime in the neighborhood (Taylor 2001; Warr 1990). The impact and direction of the relationships between the variables of fear, calls for service, collective efficacy, and disorder are impacted by the length of residence and changes in crime or disorder in the neighborhood. Because our study is based on one point in time, we cannot accurately gauge how these changes impact fear.

This study highlights the application of GWR at a neighborhood scale. At other scales, different patterns may emerge related to other geographic processes. For example, a study including the entire city rather than these two economically homogeneous neighborhoods could illustrate the impact of higher or lower socioeconomic status neighborhoods on fear of crime, which can then inform city-wide policing strategies and policies. On the other hand, analysis at that wider scale could miss block-level factors like the potential impact of a

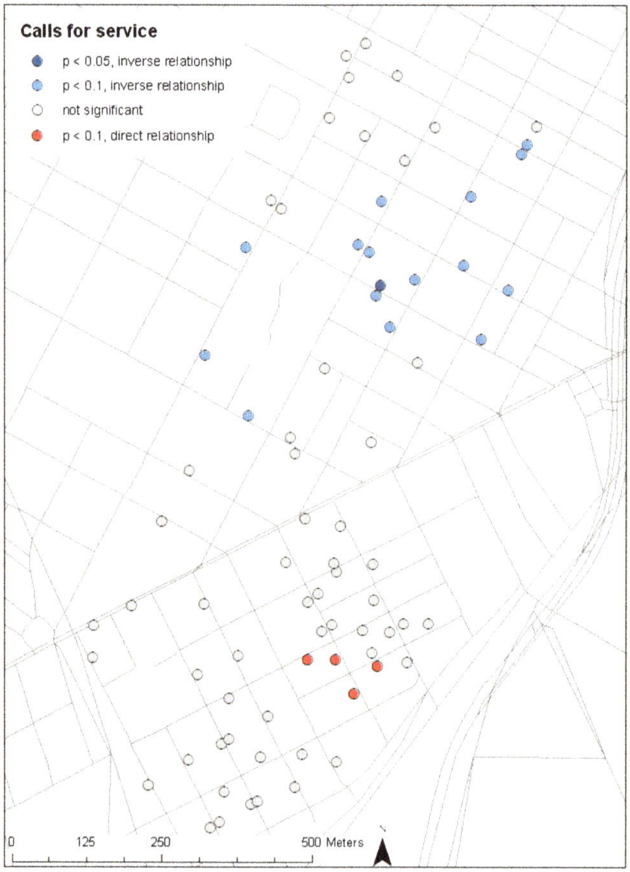

Fig. 8.6 Multivariate regression—significance of violent crime calls for service

playground, as we highlight above. The results of a geographically weighted regression are dependent upon local conditions; thus, replicating this study in the cities of New York, Los Angeles, and Dallas would likely yield three different spatial patterns of the relationship between fear of crime and the other neighborhood characteristics. However, even though local conditions make each individual analysis unique, GWR is useful nonetheless because it recognizes and illustrates the uniqueness of different areas, allowing the police to tailor their strategies to local conditions. Local analyses like what we have presented here can guide the approach of officers to the local community, recognizing spatial variation of attitudes within the neighborhood. These results show the importance of applying GWR within a policing context. Important lessons for the allocation of policing resources may be missed if only OLS is conducted. These implications are expanded upon in the following section.

8.3 Potential Uses by Law Enforcement Agencies

While policing strategies in the United States have historically taken a centralized reactive approach in which the main task is responding to complaints from residents of a community, the proactive community policing model has become more prominent since the 1980s (Goldstein 1990). In this approach, police officers develop more permanent relationships with residents to develop a rapport which, in turn, makes the officers more responsive and accountable when problems arise. In an ideal community policing model, police officers are responsible for a small part of the community, which allows them to develop strong relationships with the neighborhood residents (Weisel and Eck 1994).

The success of community policing programs also stems in large part from building community ties among residents, not just between residents and police officers. This fosters collective efficacy in the area, increasing the level of informal social control (Reisig and Parks 2004). Neighbors who are willing to look out for one another and trust one another are more likely to spend time in public places, which will make them more likely to get to know their neighbors and thus build stronger interpersonal networks (Sampson and Raudenbush 1999; Skogan 1990; Wilson and Kelling 2006). The social cohesion embedded in these networks provides a source for informal social control, making the residents more willing to intervene in problems that arise in their community. When this happens, potential offenders perceive an environment with little opportunity for criminal behaviors, lowering the risk of crime. In this way, greater trust among a community's residents will then make them more likely to informally observe their neighborhood for suspicious behavior and intervene to prevent crimes, thus reducing the crime rate more effectively than the traditional reactive model (Sampson and Raudenbush 1999).

Community policing, however, can be more resource intensive than the traditional reactive model of policing. Not only does community policing require officers be on location in the neighborhoods, it is expected that officers are responsible for smaller patrol areas to build rapport with community members. Additionally, community policing can lack the immediately noticeable results of proactive strategies with their more publicly visible reactive approach, with its sirens and arrests. The combination of expense and lower profile has led to recent changes and reductions in community policing efforts as a cost-cutting measure (Venkatesh 2012; Dumke 2013; Moore 2013).

The results of this chapter illustrate the value of maintaining targeted community policing as a crime reduction strategy in specific areas within a neighborhood, where they are likely to be most cost-effective. The local police force could use this information to guide decisions on how to allocate its resources. For example, a proactive community policing strategy may be appropriate for the study area in the bottom part of the figures, where the relationship between fear of crime and calls for service, as well as the relationship between fear and physical disorder, is inverse or contrary to what is theoretically expected. In this area, the residents' concern for their safety seems to be driven more by an absence of collective efficacy and a great deal of social disorder, both of which exhibit the expected direct relationships and both of which have lower *p*-values (comparing Figs. 8.3 and 8.5 against Figs. 8.4 and 8.6). These multivariate relationships indicate that areas with higher fear have less efficacy and more social disorder, but fewer calls to the police for violent crimes. The apparent contradiction of a greater fear of crime with fewer calls to the police may be explained by the residents not trusting the police to be effective; if they do not feel the police will solve the crime or make them safer, they do not consider it worthwhile to call the police even if they have high levels of fear and perceive a great amount of social disorder.

The police may want to target a more community-oriented strategy to particular segments of the neighborhood. This strategy should build trusting relationships with the residents of the community, which will reduce the opportunities for potential offenders in these specific areas of the neighborhood (St. Jean 2007). While community policing efforts can make residents feel safer (Pate 1986; Reisig and Parks 2004; Skogan 1990), these efforts ideally require officers to walk small areas several times during each shift. These efforts can be undertaken in conjunction with other outreach strategies such as storefront substations and placing officers in public housing projects. Local community policing efforts improve residents' perceptions because the visibility of police officers can decrease perceptions of disorder (Pate 1986), at least inasmuch as the residents trust the police force. Therefore, for these areas of the neighborhood, a community policing strategy aimed at improving trust between the residents and police officers can be more effective than the reactive strategy suggested by the broken windows theory in which the police react to physical disorder by cleaning it up. While Wilson and Kelling (2006) suggest a strategy of reducing physical disorder can also foster collective efficacy, we find that since physical disorder is, in these pockets, inversely correlated with fear of crime, a reactive strategy of reducing the physical disorder may not be as effective at ensuring the residents feel safer in their community. The spatial analysis approach can help the local forces identify places where the strategy of reducing disorder is ineffective and thus where a community policing effort may cause the greatest reductions in crime incidence.

By decentralizing the policing organization and engaging the police in community efforts to proactively reduce all of the potential contributors to crime, including disorder and efficacy, policing resources can be used more efficiently (Weisel and Eck 1994; Goldstein 1990; Sampson et al. 1997). This spatial analysis demonstrates the utility of GIS and spatial statistics in effectively analyzing the underlying contributors to crime rates and utilizing policing resources to meet the individual needs of each community. The value of GIS is reinforced because these patterns were all obscured by the nonsignificant coefficients in the OLS regression, but revealed by the GWR analysis.

References

Braga AA, Weisburd DL, Waring EJ, Mazerolle LG, Spelman W, Gajewski F (1999) Problem-oriented policing in violent crime places: a randomized controlled experiment. Criminology 37:541–580

Brunsdon C, Fotheringham S, Charlton M (1998) Geographically weighted regression–modeling spatial non-stationarity. The Statistician 47:431–443

Brunton-Smith I, Sturgis P (2011) Do neighborhoods generate fear of crime? An empirical test using the British Crime Survey. Criminology 49:331–369

Bursik RJ, Grasmick HG (1993) Neighborhoods and crime: the dimensions of effective community control. Lexington Books, Lanham

Dixon D, Maher L (2005) Policing, crime and public health: lessons for Australia from the 'New York miracle'. Crim Justice 5:115–143

Doran B, Lees B (2005) Investigating the spatiotemporal links between disorder, crime, and the fear of crime. Prof Geogr 57:1–12

Dumke M (2013) City officials say they're committed to community policing, just not to funding it. Chicago Reader. http://www.chicagoreader.com/Bleader/archives/2013/01/10/city-officials-say-theyre-committed-to-community-policing-just-not-to-funding-it. Accessed 31 May 2013

Goldstein H (1990) Problem-oriented policing. Temple University Press, Philadelphia

Janowitz M (1975) Sociological theory and social control. Am J Sociol 81:82–108

Moore N (2013) Chicago revamps community policing program. WBEZ. http://www.wbez.org/news/chicago-revamps-community-policing-program-104778. Accessed 31 May 2013

Morenoff J, Sampson R, Raudenbush S (2001) Neighborhood inequality, collective efficacy, and the spatial dynamics of urban violence. Criminology 39:517–559

Nolan J, Althouse R, Stein RE, Bennett S (in press) Situational policing. U.S. Department of Justice, Office of Community Oriented Policing Services, Washington DC

Novak K, Seiler C (2001) Zoning practices and neighborhood physical disorder. Crim Justice Policy Rev 12:140–163

Pate AM (1986) Experimenting with foot patrol: the Newark experience. In: Rosenbaum D (ed) Community crime prevention. Sage, Newbury Park, pp 137–156

Perkins D, Wandersman A, Rich R, Taylor R (1993) The physical environment of street crime: defensible space, territoriality and incivilities. J Environ Psychol 13:29–49

Porter JR, Purser CW (2010) Social disorganization, marriage, and reported crime: a spatial econometrics examination of family formation and criminal offending. J Crim Justice 38:942–950

Reisig MD, Parks RB (2004) Community policing and the quality of life. In: Skogan WG (ed) Community policing: can it work? Wadsworth, Belmont, pp 207–227

Rountree PW, Land KC (1996) Perceived risk versus fear of crime: empirical evidence of conceptually distinct reactions in survey data. Soc Forces 74:1353–1376

Sampson RJ, Raudenbush SW (1999) Systematic social observation of public spaces: a new look at disorder in urban neighborhoods. Am J Sociol 105:603–651

Sampson RJ, Raudenbush SW, Earls F (1997) Neighborhoods and violent crime: a multilevel study of collective efficacy. Science 277:918–924

Skogan W (1990) Disorder and decline: crime and the spiral of decay in American neighborhoods. University of California Press, Berkeley

St Jean PK (2007) Pockets of crime: broken windows, collective efficacy, and the criminal point of view. The University of Chicago Press, Chicago

Taylor RB (2001) Breaking away from broken windows: Baltimore neighborhoods and the nationwide fight against crime, grime, fear, and decline. Westview Press, Boulder

Taylor RB, Shumaker S, Gottfredson S (1985) Neighborhood level links between physical features and local sentiments: deterioration, fear of crime, and confidence. J Archit Plan Res 21:261–75

Trojanowicz R, Bucqueroux B (1992) Toward development of meaningful and effective performance evaluations. Michigan State University, National Center for Community Policing, East Lansing

Venkatesh S (2012) How the federal government is killing community policing. New Republic. http://www.newrepublic.com/article/politics/107675/how-big-brother-killing-community-policing#. Accessed 31 May 2013

Vindevogel F (2005) Private security and urban crime mitigation: a bid for BIDs. Crim Justice 5:233–255

Warr M (1990) Dangerous situations: social context and fear of victimization. Soc Forces 68:891–907

Weisel DL, Eck JE (1994) Toward a practical approach to organizational change: community policing initiatives in six cities. In: Rosenbaum DP (ed) The challenge of community policing. Sage, Thousand Oaks, pp 53–72

Wilson JQ, Kelling GL (1982) The police and neighborhood safety: broken windows. Atlantic 127:29–38

Wilson JQ, Kelling GL (2006) A quarter century of broken windows. Am Interest 2(1):168–172

Xu Y, Fiedler ML, Flaming KH (2005) Discovering the impact of community policing: the broken windows thesis, collective efficacy, and citizens' judgment. J Res Crime Delinq 42:147–186

Chapter 9
Using GIS to Monitor and Investigate Police Use of Force: The Spatial Distribution of Force Factors

Matthew J. Hickman, Loren T. Atherley, and Geoffrey P. Alpert

Abstract This chapter describes the potential utility of geospatial analysis as applied to monitoring, understanding, and responding to police use-of-force incidents. Data are drawn from 1,240 official use-of-force reports in the City of Seattle, representing a two and a quarter year period. Each report includes officer identifier information, suspect demographic information, a categorization of the type of suspect resistance and how force was applied, as well as information about location, booking, injuries, and evidence of impairment or suspected impairment. In addition to coding these administrative data elements, the authors coded the content of the officer narratives with regard to the nature of officer-suspect interactions as well as relevant Graham factors. Force and resistance levels were classified based on complementary scales drawing from previous research. All use-of-force data was attributed to geocoded incident locations. A series of maps and basic spatial statistics demonstrate how GIS can assist in understanding the nature and scope of police use of force and potentially improve the quality of police monitoring and accountability mechanisms.

Keywords Police • GIS • Use of force • Internal investigations • Social justice

M.J. Hickman (✉)
Department of Criminal Justice, Seattle University, Seattle, WA, USA
e-mail: hickmanm@seattleu.edu

L.T. Atherley
Seattle Police Department, Compliance and Professional Standards Bureau, Seattle, WA, USA

G.P. Alpert
Department of Criminology and Criminal Justice, University of South Carolina, Columbia, SC, USA

G.A. Elmes et al. (eds.), *Forensic GIS: The Role of Geospatial Technologies for Investigating Crime and Providing Evidence*, Geotechnologies and the Environment 11, DOI 10.1007/978-94-017-8757-4_9,

9.1 Introduction

There are three broad themes in the study of policing that suggest a strong utility in using Geographic Information Systems (GIS) to monitor and investigate police use of force: (1) the importance of police use of force to understanding democratic policing; (2) the importance of place in criminology/criminal justice generally, and in policing specifically; and (3) the importance of understanding the spatial context of police use of force for effective monitoring and response, particularly as it pertains to issues of social justice and questions about disparate impact.

First, the use of force by the police is a topic of central concern for democratic policing. We entrust our police with tremendous powers of coercion and therefore demand oversight to ensure that democratic ideals are being upheld. The commonly embraced themes of *accountability* to the public, *transparency* and *openness* to external scrutiny, and *fairness* in policing suggest the principal concern of democratic policing is not the behavior of citizens, but rather the actions of the police (Hickman 2010). Yet, the second largest democracy in the world (and arguably the leading global advocate of democracy) actually has very little to offer in terms of knowledge about the nature and extent of police use of force. The late police practitioner and scholar James Fyfe (2002: 99) remarked that "… we still live in a society in which the best data on police use of force come to us not from the government or from scholars, but from the *Washington Post*." Robert Kane (2007: 776), a student of Fyfe's, likewise has argued that "… it is both ironic and unacceptable that in American democratic society, the police, which function as the most visible representatives of the crime control bureaucracy, collect data on members of the public in the form of arrest and complaint reports without systematically distributing comprehensive data on their own activities that produced those crime statistics."

A recent literature review (Hickman et al. 2008) identified 36 studies reporting an incident-based rate of police use of force and found that the majority of studies were based on data from a single jurisdiction and the methods were quite diverse (including arrest reports, household surveys, independent observations, police surveys, suspect surveys, and official use of force forms), as were the units of analysis (arrests, contacts, citizen encounters, disputes, police stops, potentially violent mobilizations, suspect encounters, and calls for service). Somewhat unsurprisingly, across 36 studies reporting on the amount of nonlethal force used by the police, rates varied from about a tenth of one percent up to almost 32 %. The bottom line is we simply do not know how often the police in the United States use force upon its citizens, to what degree, and with what consequences (Kane 2007).

Second, the field of criminology/criminal justice broadly, and policing specifically, has widely embraced the importance of place (Sampson 2013). The currently popular mode of operation in policing is even called "place-based" policing (Weisburd 2008; Weisburd et al. 2009), owing to its clear transition away from criminology's traditional focus on high rate offenders (Wolfgang et al. 1972) to high rate locations (Sherman et al. 1989). The use of GIS to analyze police crime data, combined with a microlevel focus (down to the level of "street segment"), has resulted in potentially new or greater efficiencies for the police, along with renewed interest in the effects of targeted foot patrol (e.g., Ratcliffe et al. 2011) and dosage models aimed at maximizing deterrence effects at hot spots (Koper 1995) which have been increasingly employed in practice.

Third, the literature suggests that neighborhood context is important in explaining variation in coercive and analogous police behaviors. Prior to the development of the literature on the neighborhood context of police behavior, much of the research focus had been on individual officer characteristics and group socialization. Yet classic observational studies of the 1960s (Black and Reiss 1967; Skolnick 1966; Wilson 1968; Rubenstein 1973) demonstrated differential police behavior across neighborhoods and provided much of the impetus for later quantitative study. For example, Smith's (1986) study of data drawn from the Police Services Study (PSS) found that the racial composition of neighborhoods (percent non-white and racial heterogeneity) predicted police coercion including physical force and verbal threats. Alpert and Dunham (1988) used neighborhood as the unit of analysis to explore the public's perception of the police and police practices. They found that "... residence in a specific neighborhood is a more influential factor than gender or ethnicity in explaining variance in attitudes toward policing" (1988: 119). Theories regarding ecological variation in police behavior also grew out of this genre of empirical research. More recently, drawing data from the Project on Policing Neighborhoods (POPN), Mastrofski et al. (2002) found that police disrespect toward the public (racial slurs, cursing, derogatory statements, name calling, etc.) was more likely in neighborhoods characterized by concentrated disadvantage. Terrill and Reisig (2003), using the same data, found that concentrated disadvantage and homicide rates were predictive of the level of force used by police. As another example, Kane (2002) demonstrated that structural disadvantage, population mobility, and change in Latino population predicted police misconduct across police precincts and districts of New York City.

Given evidence of ecological variation in police behavior, including the use of force, it is clearly important to understand the spatial distribution of force in order to have effective monitoring and response. The initial development of early warning systems (and later early intervention systems) tended to be based on simplistic "three strikes" types of approaches in which individual officers who exhibited a predefined pattern of behavior (such as three use-of-force incidents in a fixed period of time) were flagged for intervention. These models can be considered inefficient to the extent that they ignore variation in police behavior by location (and other factors, such as officer assignment and specific training). For example, it may be that a certain level of use of force is considered "normal" (i.e., average) in some locations but very abnormal in other locations. Should we be worried about officers who exceed a particular fixed threshold, or officers whose behavior is substantially different from their peers in the same work context or the department as a whole? This also has potential implications for the allocation of investigative resources, which may be more efficient under a context-based approach particularly in larger jurisdictions.

Ignoring underlying theoretical processes may also contribute to inefficiency, insofar as officer characteristics and work environments interact. For example, subsequent personnel actions, such as mandatory training, counseling, and transferring officers to other districts or units, could result in negative or otherwise unanticipated outcomes (e.g., placing an officer in another work context that actually increases his/her risk of use of force).

Finally, it is important to move away from understanding force in terms of its raw incidence and toward consideration of the dynamics and context of force events. Alpert and Dunham's force factor model (Alpert and Dunham 1997, 2004; Alpert et al. 2001) has been well received as an efficient means of understanding the level of police force relative to suspect resistance, using both the highest level of force in the event and a sequential analysis of the use-of-force event (see Terrill et al. 2003). This approach challenges traditional use-of-force management systems and allows for an analysis of the highest levels of force and resistance as well as those at different stages of the event. There is to date no research on the spatial distribution of force factors, but the literature strongly suggests there should be important spatial variation in these force differentials.

In sum, there is ample evidence in the literature that suggests there is strong utility to using GIS in monitoring and investigating police use of force and doing so with rich information about the specific event dynamics. In the next section, we describe the data and methods we used to develop information about police use of force in Seattle. In subsequent sections, we provide a basic descriptive overview of the use-of-force data and then move into the spatial distribution of force incidents with an eye toward highlighting the potential applications in monitoring, understanding, and responding to police use-of-force incidents.

9.2 Data and Methods

The data were drawn from official Seattle Police Department (SPD) Use of Force reports that were produced to the US Department of Justice as part of their recent (2010–2011) "pattern or practice" investigation in that city. The authors requested and were provided with the same material, a series of PDFs containing scanned images of SPD use-of-force reporting forms and related supporting documents during the period January 2009 through March 2011, amounting to 1,240 use-of-force records in all.

Data for this study were coded from the content of the official SPD Use of Force reports.[1] Each report included officer identifier information, suspect demographic information, a categorization of the type of suspect resistance and how force was applied, as well as information about location, booking, injuries, and evidence of impairment or suspected impairment. The first four pages of these reports (more if supplemental forms are required) are comprised of text entries and check boxes which were coded into a computer database using a numeric coding system. The remainder of the report packet varied depending on the presence of supplemental documentation, report narratives, photographs, Washington Crime Information Center (WACIC)

[1] The researchers recognize the potential bias of officer accounts of use of force. These reports include statements from the officer, the witnesses, the investigating supervisor, and the chain of command involved in the review process, but not the subject of the action. Further, these reports are written for the purpose of documenting justification for the application of force and as such may contain an inherent bias toward police. This is not to say that these reports are inaccurate; rather, these are official statements given by sworn officers whom we presume are giving the most accurate statement possible.

Table 9.1 Suspect resistance and officer force levels

Level	Description	Level	Description
Resist 1	No resistance. The subject is offering no resistance or threat	Force 1	Officer presence in uniform or marked police vehicle
Resist 2	Verbal resistance to complying with lawful orders. Subject may challenge authority or standing and may present as "dead weight"	Force 2	Issuance of lawful orders and light physical contact to include guiding, leading, and/or handcuffing. No intentional infliction of pain for the purpose of compliance
Resist 3	Use of posture and verbal threats of physical violence. Subject may attempt to intimidate or otherwise pose a physical threat to officers	Force 3	Chemical agents for the purpose of crowd dispersal or distraction. Tactic is often reserved for large gatherings, civil disobedience, and fight disturbances
Resist 4	Physical noncompliance including refusal to give up hands for cuffing and attempts to flee	Force 4	Physical control tactics such as pain compliance holds, joint manipulation, and open handed strikes
Resist 5	Active physical resistance to compliance. Subject may attempt to strike officers, kick, and struggle free from holds and compliance positions	Force 5	Advanced physical control tactics including closed fisted strikes, knee and elbow strikes to the body and the extremities
Resist 6	Use of nonlethal weapons to injure or otherwise actively assault officers. Drug paraphernalia, beverage containers, and rocks may be employed as cutting and impact weapons	Force 6	Intermediate weapon use, deployment of electronic control weapons and impact weapons for pain compliance, and strikes to the body and extremities
Resist 7	Use of lethal force as presented by whatever means are available: firearms, knives, and motor vehicles	Force 7	Use of lethal force including carotid artery holds, head strikes, and intentional discharge of firearms

reports, Labor and Industry (L&I) claim forms, and computer-aided dispatch (CAD) call logs as well as document routing information and other administrative documentation. These data fields were entered as they appeared in the record.

9.2.1 Force Factors

In addition to the variables contained in the form, the narrative reports were read and officer-suspect interactions were coded with officer force and suspect resistance classified based on complementary scales drawing from previous research (Alpert and Dunham 2004). For each record in the dataset, a *static force factor* was calculated, comparing the maximum force applied by the officer to the maximum level of resistance (i.e., officer force level minus suspect resistance level). The static force factor takes on numeric values ranging from −6 to 6 (see Table 9.1).

For example, if the maximum level of suspect resistance was "Resistance 4," and that was met with a comparable level of officer force ("Force 4"), this would result in a static force factor equal to 0, indicating a proportional response. If the officer used a higher level of force, say Force 5, then the corresponding force factor would be +1, indicating the officer used one level of force higher than the degree of suspect resistance. If the officer used a lower level of force, say Force 3, the force factor would be −1.

In addition to the static indicator of maximum force and resistance, we also coded *dynamic force factors* throughout the force incident. Narratives from all 1,240 cases were read, and we coded up to ten iterations of dyadic action/reaction using the same coding scheme as was used to code the static force factor. This analytic scheme helps assess how force incidents evolve from the perspective of the officer and provides a dynamic overview of the incident from start to finish. Where multiple officers or suspects are involved, their actions are represented as a composite.

Weapon draws are a problematic area in use-of-force research. Some parties contend that a weapon draw is a use of force, while others contend that it is at most a threat of force, and still others argue either side depending on the level of the draw, other actions, and the motives of the officer with regard to safety.[2] Lethal force, for the purposes of this study, was only coded when the officer used an impact weapon on an area of the body with potential to cause lethal injury (such as a head strike), used a carotid artery hold/restraint, or discharged a firearm.

All incident addresses were geocoded using an address locator generated from Census street files. Of the 1,240 use-of-force reports, 12 incidents occurred outside the greater Seattle metropolitan area and were not geocoded for spatial analyses. The projection and datum for all geospatial data is the State Plane for Washington North and North American Datum 1983 HARN, and the units of measurement are US feet. All statistical analyses reported here were performed in SPSS 19 (IBM), and all spatial analyses were performed in ArcGIS 10 (ESRI).

9.3 Analysis

We begin with a brief descriptive review of the 1,240 force incidents, and then move into the spatial analyses with a focus on understanding the spatial distribution of force factors in Seattle. First, it is important to recognize that the use of force in Seattle, like other cities, is a relatively rare event. Standardized by arrest activity, the

[2] From a legal perspective, courts have upheld the right of officers to protect themselves against unknown or potentially threatening situations by drawing and pointing a firearm when it is reasonable to do so. During the study period, SPD officers did not report weapon draws consistently. In fact, the lead author was on scene at one of the use of force incidents that occurred during the study period, as a ride-along observer on unrelated field research (Hickman et al. 2011). In that incident, there was a weapon drawn and pointed at the suspect that went unreported in the subsequent use-of-force report. Sometimes there are references in reports to other officers providing "lethal cover," or references are made to "felony stop" or "high risk stop" procedures. Insofar as weapon draws were not consistently reported, we were faced with a substantial measurement problem.

police in Seattle use force at the rate of about 2.4 incidents for every 100 arrests.[3] On average, Seattle police made about 1,880 arrests per month during the study period, and there were an average of 46 use-of-force incidents per month. This rate is fairly consistent over the study period, with monthly rates varying from 1.9 to 3.1 per 100 over the two and a quarter year period. This rate has also been consistent over the past decade in Seattle, as reported in SPD annual reports. Use of force is more frequent on weekends and during the late evening/early morning hours.

9.3.1 *Type of Offenses*

Officers recorded the type of incident on their use-of-force reporting forms. This indicates the general nature of the dispatch or on-view incident and associated incident characteristics. These characteristics are not mutually exclusive, so several type characteristics may be indicated for a particular incident. Forty percent of the incidents were reported as involving some type of felony, and 38 % involved non-felonies. Fights or disturbances were indicated in 30 % of incidents. About 1 in 5 use-of-force incidents involved violent crimes and/or were drug related. Eighteen percent were characterized as dealing with a mental/suicidal suspect. Domestic violence was indicated in 15 % of use-of-force incidents.

9.3.2 *Suspect Characteristics*

The vast majority of use-of-force incidents (94 %) involved a single suspect on whom force was used. The median age of suspects was 29, and ages ranged from 12 to 78 years. Eighty-seven percent of suspects were male. Regarding race, 45 % were white, 40 % black, 5 % Asian, 3 % American Indian, and 6 % "others." These racial proportions are roughly comparable to the race distribution for all arrestees during the study period. For example, whites comprised 50 % of all arrestees and 45 % of those upon whom force was used during the study period, blacks comprised 38 % of all arrestees and 40 % of those upon whom force was used, Asians comprised 6 % of all arrestees and 5 % of those upon whom force was used, and American Indians comprised 3 % of both the arrestee and force populations.

Some suspects exhibited impairments that affect judgment and behavioral decision-making. There was no evidence of impairment noted for 26 % of suspects

[3] Finding truly comparable use of force data remains a challenge for researchers. The earlier noted review of 36 studies found great variability in methods, definitions, and units of analysis, leading to widely varying rates. However, among those studies using official use-of-force reports for a given time period and calculating rates based upon arrests during the same time period, Seattle's rate is fairly comparable. For example, Terrill (2003) reported a rate of 1.8 per 100 arrests in San Antonio, TX; Williams and Hester (2003) reported a rate of 0.9 per 100 in Polk County, FL; the Seattle Police Department previously reported a rate of 2.1 per 100 arrests (Seattle Police Department 2001); and Croft (1985) reported a rate of 1.9 per 100 arrests in Rochester, NY.

in use-of-force incidents. About half of suspects exhibited signs of intoxication (either drugs or alcohol). In about a third of cases, suspects specifically indicated to the officer(s) that they had been using drugs and/or alcohol (32 %), exhibited rambling or incoherent speech (33 %), or smelled of alcohol (36 %). Slurred speech (24 %) and poor balance (17 %) were also noted impairments. A possible mental illness, suicidal intentions, or delusional behavior was noted in 28 % of incidents.

Suspects fled from officers in about a quarter of incidents, by either foot (26.5 %) or motor vehicle (3.1 %). These figures also include a small subset of cases in which suspects fled by both motor vehicle and foot. Suspects complained of injury in 51 % of incidents (although this refers only to the first suspect reported and is thus conservative), and suspects were reported by officers as visibly injured in 69 % of incidents, but the injuries were sometimes a result of something that happened prior to police involvement. Among those suspects complaining of injury, 20 % specifically refused treatment. The majority were treated by EMS personnel at the scene or at a police precinct. In 14 cases (or about 1 % of all use-of-force incidents), suspects were treated and admitted to a hospital.

9.3.3 Officers Involved

Eighty-one percent of all incidents involved force used by either a single officer (48 %) or by two officers (33 %). Another 12 % involved force used by three officers. There were 650 unique officers involved either as first responding officers or as backing officers in the 1,240 use-of-force incidents during the study period. It is important to note that not all officers were accounted for; we recorded up to six officers involved, but there were a few incidents that exceeded that range, including one incident that involved 29 officers. About a third (34 %) of the officers in the use-of-force data were involved in a single incident during the study period, about half (53 %) in one or two incidents, two-thirds (66 %) in one to three incidents, and about three quarters (76 %) in one to four incidents. Ninety-five percent of officers in the use-of-force data were involved in less than ten incidents during the study period. Thirty-one officers (about 5 %) were involved in 10 or more incidents during the study period, and there were three officers (about 0.5 %) who were involved in 20 or more use-of-force incidents during the study period.

Officer injuries were reported in 17 % of incidents, although this figure relates only to the first officer reported and is thus conservative. Officers typically self-treated or were treated by EMS personnel.

9.3.4 Force Factors

In nearly half (48.5 %) of all incidents, the highest level of suspect resistance was "defensive resistance" (see Table 9.1). Next most frequent was "active resistance," with one quarter (25 %) of incidents falling in that category. In about 4 out of 10

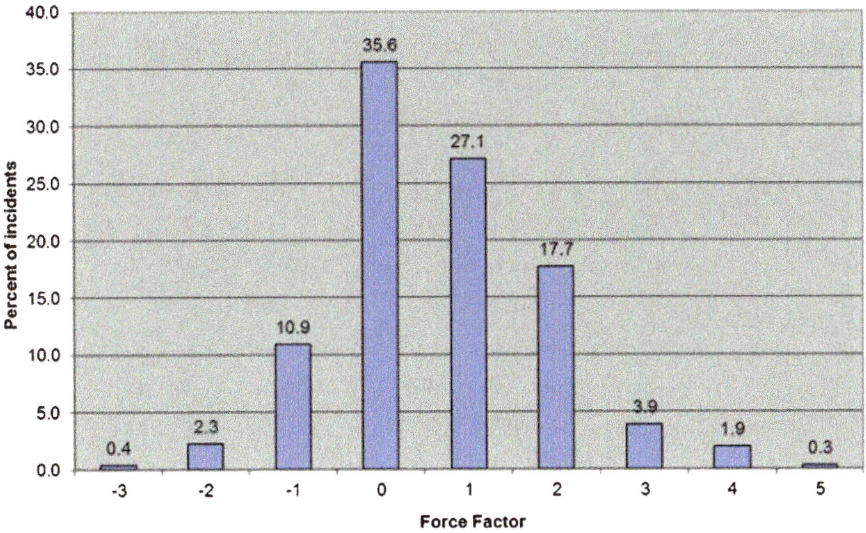

Fig. 9.1 Percentage of use of force incidents by force factor (force–resistance)

incidents, the highest level of officer force was "defensive physical force." The next most frequent was the "use of an intermediate weapon," with almost 3 out of 10 incidents falling in that category. About one quarter (24 %) of incidents involved officers using "offensive physical force."

The distribution of force factors (i.e., level of officer force minus level of suspect resistance) is displayed in Fig. 9.1. A value of "0" indicates that police force was proportional to suspect resistance, positive values indicate officers using a higher level of force compared to suspect resistance, and negative values indicate officers using a lower level of force compared to suspect resistance. In general, officers following a force continuum model will use a higher level of force in order to overcome a given level of suspect resistance, thus +1 events (and to some extent, +2 events, some of which may be an artifact of skipping over OC [Oleoresin Capsicum] spray,[4] or going to an intermediate weapon such as the Taser rather than physically fighting an individual) are to be expected. Overall, about 80 % of use-of-force incidents fell in the 0–2 range, and about 14 % fell in negative categories. The remaining 6 % were in the 3–5 range.

[4]Oleoresin Capsicum spray or "OC spray" as it is referred to here is a chemical irritant utilized in policing. A lachrymatory agent (from the Latin for "tear," *lacrima*), pepper spray, as it is more commonly known, is a derivative of a chemical common to the genus Capsicum, capsaicin. Used in a variety of delivery methods, OC spray, as referenced here, is an emulsified spray at concentrations between 5 and 10 % and is usually directed at the face of a single subject, causing pain, temporary blindness, and respiratory symptoms, or a crowd (usually for respiratory effect) and is an effective pain compliance/crowd dispersal method accepted by most civilian law enforcement agencies.

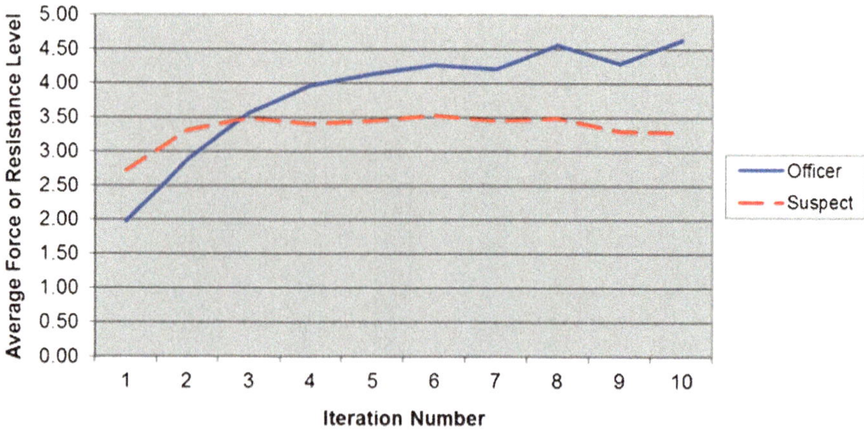

Fig. 9.2 Dynamic force and resistance levels

Since the static force factor only captures the maximum levels of officer force and suspect resistance, an examination of dynamic force factors may help to understand the evolution. As previously mentioned, we coded up to ten dyadic iterations of officer-suspect actions. More than half of all cases end by the fourth iteration. Figure 9.2 shows the average levels of officer force and suspect resistance at each iteration. Overall, officers on average entered situations at a force deficit, and transitioned to a force surplus by the third iteration, reflecting initial attempts to reason with noncompliant suspects either with voice commands or limited physical contact.

9.3.5 Understanding the Spatial Distribution of Force Incidents and Force Factors

The West Precinct, which serves the downtown core of the city, has the largest proportion of use-of-force incidents (34 %), followed by the South Precinct (21 %). The North and East precincts each account for about 17 % of incidents. The Southwest Precinct had the lowest proportion of use-of-force incidents (10 %). Each of these precinct areas serves distinct commercial and/or residential communities and has differing social services and criminal justice infrastructure.

9.3.5.1 West Precinct

Serving the Magnolia, Queen Anne, Downtown, and a portion of the East census neighborhood districts, the West Precinct is a diverse mix of demographics (see Table 9.2). The precinct covers the geographic area south of Lake Union and the

Table 9.2 Precinct characteristics

	West (Magnolia/Queen Anne, Downtown, and part of Lake Union)	South (Greater Duwamish and Southeast)	North (Northwest, North, Northeast, Ballard, and part of Lake Union)		East (Central, East, and part of Lake Union)	Southwest (Southwest and Delridge neighborhoods)
Population	>75,600	91,600	>233,400		>80,000	82,900
Age	Median ages are 36 and 42 in Magnolia/Queen Anne and Downtown, respectively	Median ages are 37 and 38 in Duwamish and Southeast, respectively. Highly localized concentrations of young persons (5–17 years)	Median age is lower in Northeast (30) as compared to Northwest (36), North (37), and Ballard (38)		Median ages are 36 and 33 in Central and East, respectively	Median ages are 42 and 34 in Southwest and Delridge, respectively. Delridge has a higher concentration of persons between the ages of 5 and 17
Race/ethnicity	Magnolia/QA White: 84.8 % Black: 2.2 Asian: 6.6 Hispanic: 4.7 Downtown White: 60.1 % Black: 10.7 Asian: 21.2 Hispanic: 5.9	Duwamish White: 27.9 % Black: 16.9 Asian: 41.7 Hispanic: 12.1 Southeast White: 37.6 % Black: 25.0 Asian: 26.3 Hispanic: 7.8	Northwest White: 77.3 % Black: 4.2 Asian: 10.0 Hispanic: 6.2 North White: 68.5 % Black: 7.1 Asian: 14.2 Hispanic: 7.8	Northeast White: 77.8 % Black: 2.0 Asian: 13.3 Hispanic: 4.2 Ballard White: 86.8 % Black: 1.9 Asian: 4.8 Hispanic: 4.9	Central White: 59.6 % Black: 21.4 Asian: 9.1 Hispanic: 7.3 East White: 75.0 % Black: 7.8 Asian: 9.6 Hispanic: 6.3	Southwest White: 86.2 % Black: 2.0 Asian: 4.9 Hispanic: 4.8 Delridge White: 56.5 % Black: 13.3 Asian: 15.5 Hispanic: 12.7

Note: All census figures are taken from the US Census Bureau Decennial Census 100 % Count data 2010, as processed by the City of Seattle, Department of Planning and Development, and do not necessarily reflect actual neighborhood communities. Source material can be accessed on the web at http://www.seattle.gov/dpd/cityplanning/populationdemographics/decennialcensus/2010/default.htm

Ship Canal, north of the West Seattle Bridge, and west of Interstate 5 to the Seattle Waterfront. In addition to high rise, high density apartment, and condo complexes, the West Precinct is home to the social center of Seattle, the seat of government for King County and the City of Seattle, regional headquarters for many federal agencies and the financial center of the Pacific Northwest.

During the 1999 World Trade Organization Riots (often referred to as the "Battle in Seattle"), the West Precinct, along with the East Precinct, saw the bulk of police control action. Today many social demonstrations take place in this area. Additionally, the West Precinct is home to one of the largest open air drug markets on the west coast and hosts a large population of homeless and disadvantaged persons seeking social services, which are also located in the area. The area is also characterized by two nightlife districts, Belltown and Pioneer Square, which are associated with a significant number of use-of-force incidents during the evening and weekend hours.

9.3.5.2 South Precinct

The South Precinct, including the Greater Duwamish and Southeast neighborhood districts, is the second largest precinct area in the city (see Table 9.2). The South Precinct follows Martin Luther King Boulevard and Rainier Avenue south through the Rainier Valley. Located between Interstate 90 to the north and the Seattle City limits to the south, Lake Washington to the east, and the Duwamish industrial district to the west, the South Precinct serves a mixed population of residential and industrial areas amid gentrifying neighborhoods.

The relatively high concentration of young persons of color in the South Precinct may contribute to a higher association with gang violence and activity than anywhere else in Seattle. A cursory analysis of crime data from the first quarter of 2013 suggests the area may have a slightly higher rate of "Criminal Homicide."[5] However, compared to the West and North Precinct areas, the incidence of assault is nearly a third lower. Similarly, burglary and larceny occur at a lower rate. Of the instances of "Forcible Rape," none occurred in the South Precinct areas.

9.3.5.3 North and East Precincts

The North and East Precinct areas are characterized primarily by residential neighborhoods and academic institutions with localized concentrations of divergent demographics (see Table 9.2). The North Precinct, Seattle's largest precinct area, encompasses all areas north of the Lake Union Ship Canal/Montlake Cut waterway and includes the neighborhood districts of Ballard, Lake Union, Northeast, North,

[5] Crime statistics are maintained on a monthly basis and are published for public viewing on the Seattle Police Department's website. Statistics were accessed on the web at http://www.seattle.gov/police/crime/stats.htm

and the Northwest areas. The East Precinct is bounded by the Lake Union Ship Canal/Montlake Cut waterway to the north, Lake Washington to the east, Interstate 90 to the south, and Interstate 5 to the west, and serves the East and Central neighborhood districts.

While primarily residential, characterized by localized micro-downtown communities such as Greenwood, Laurelhurst, Madison Valley, Capitol Hill, and the Central District, these precinct areas also host two major universities (the University of Washington and Seattle University). Rates of "Criminal Homicide" are higher in the East Precinct, though caution should be used in generalizing such limited data. Similarly, rates of burglary and larceny are higher in the North Precinct; however, given the area and population contained therein (277,000 people) living in a largely residential setting, a higher incidence of these events is to be expected.

9.3.5.4 Southwest Precinct

Finally, the Southwest Precinct, containing the Southwest and Delridge neighborhood districts, has the lowest share of force incidents in this study. The Southwest Precinct is bound to the west and north by Puget Sound and Elliot Bay, the east by the Duwamish River, and the south by the city limits. The Southwest Precinct serves a primarily residential population with a relatively low occurrence of police use of force.

9.3.6 *Within Precinct Variability and Hot Spots*

Precinct-level figures on use of force mask important local spatial variation within Precincts. Figure 9.3 is a point map of the greater Seattle area, where each point indicates a use-of-force incident. As can be seen, use-of-force incidents are distributed across the entire city. A cursory visual inspection suggests some clustering and other patterning in the distribution of incidents. Statistical tests for clustering verify the visual impression: the Average Nearest Neighbor test is significant with an observed mean distance between nearest neighbor incidents of 461 ft, compared to an expected mean distance of 690 ft if the incidents were spatially random.[6]

In Fig. 9.4, we present a kernel density estimation (KDE) using the incident data. The cell size was set to 50 ft, and the search radius set to 1,000 ft (search radii greater than 1,000 ft resulted in over-smoothing and less than 1,000 ft in under-smoothing). Thus, in estimating the density of use-of-force incidents at a particular cell in the raster, incidents within a 1,000 ft radius are considered (with incidents located closer to the center of the cell weighted more than those near the edge of the search radius). Following Gorr and Kurland's (2012) suggestions, the density surface was

[6]The resulting nearest neighbor ratio is .67, and the corresponding z-score is −22.21, indicating significant clustering of incidents at $p < .001$. The study area was specified as 83.9 square miles, the official land area of Seattle as determined by the Census Bureau.

Fig. 9.3 Use of force incidents in Seattle, January 2009 through March 2011

symbolized using 1/3 standard deviations as the classification method. Rather than display the entire density surface, the boundaries of the resulting use-of-force "hot spots" (using the two highest density classes) are depicted in the map as contours.

There are a number of locations where higher densities of police use of force are evident (see Fig. 9.4). Starting in the North Precinct, four small clusters were identified.[7] In the downtown core of the West Precinct, there is a large clustering that is difficult to disentangle due to the substantial concentration of activity in that area.[8]

[7] The northernmost cluster bridges the Olympic Hills and Cedar Park neighborhoods of Lake City, in the vicinity of 125th NE and Lake City Way. The next two northernmost clusters are located in the North College Park neighborhood. The first of these clusters is actually centered on the SPD North Precinct Headquarters (an artifact of uses of force within the police facility), while the second cluster consists of incidents along Aurora Ave. between N 85th and N 90th Streets. The last cluster in the North Precinct is in the University District (the area around the University of Washington) in an area along University Way NE between NE 45th and NE 50th Streets.

[8] The cluster includes the Belltown, Pike Place Market, Central Business District, Pioneer square, and International District neighborhoods and smaller clusters around Seattle Center (the fairgrounds site of the 1962 World's Fair, and location of the iconic Space Needle) including lower Queen Anne.

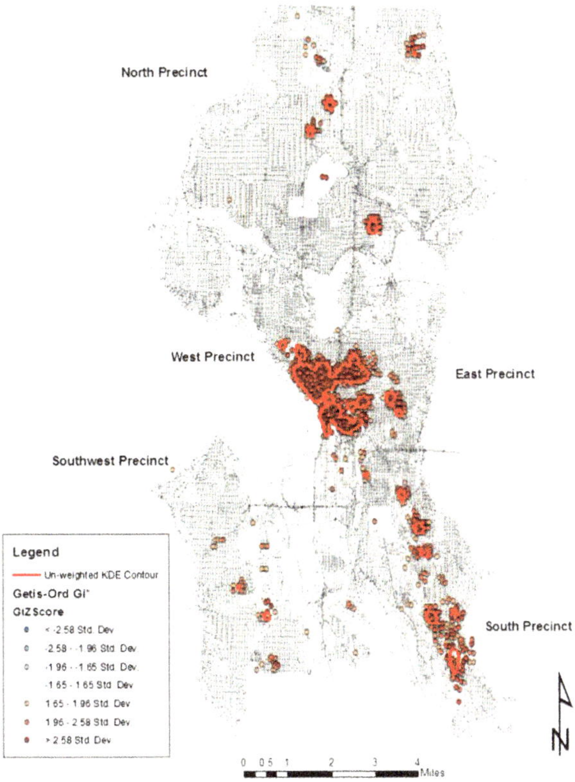

Fig. 9.4 Kernel density estimate contours and Getis-Ord Gi* statistics

In the East Precinct, there are three general areas of clustering.[9] In the South Precinct, seven clusters were identified.[10] Finally, there were two small clusters identified in the Southwest Precinct.[11]

The empirical validity of the hot spots identified via KDE was assessed by calculating Getis-Ord Gi* statistics for the incident data (aggregated to census blocks). Following Gorr and Kurland (2012), we created a census block centroid layer and attributed aggregate incident data to the centroids. The standardized Gi* statistics are symbolized as point markers at the census block centroids (with orange

[9] These are associated with the Capitol Hill neighborhood (particularly the Pike-Pine Corridor and Broadway Ave), the Yesler Terrace and First Hill neighborhoods, and the Minor and Mann neighborhoods of the Central District (particularly around Garfield High School).

[10] These include one located within the North Beacon Hill neighborhood and the rest in neighborhoods of the Rainier Valley, including Mount Baker, Columbia City, Brighton, Dunlap, and Rainier Beach.

[11] These include one within the High Point neighborhood of Delridge and the other centered on the SPD Southwest Precinct Headquarters (an artifact of uses of force within the police facility).

indicating significant clustering at $p < .05$ and red indicating significant clustering at $p < .01$). For purposes of clarity, centroids with nonsignificant Gi* values (i.e., less than 1.65 or greater than −1.65) are not symbolized. As can be seen, the KDE hot spots are supported by statistical evidence of clustering at the underlying census block level. However, there is also some evidence of clustering at other locations not highlighted by the KDE technique.

9.3.7 Neighborhood Characteristics at Hot Spot Locations

9.3.7.1 North Precinct Hot Spots

As indicated above, despite its geographic size and representative population, the North Precinct area accounts for only 17 % of use-of-force incidents. Generally, the area north of the Lake Union Ship Canal and Montlake Cut waterways is primarily residential. Characterized by medium and small neighborhoods ranging from middle class to wealthy demographics, the size and intensity of hot spot activity reflects the highly localized nature of these interactions suggesting an intensive qualitative analysis of the surrounding area may be probative.

The intersection of NE 125th St. and Lake City Way NE (Highway 522) is the junction of two major Seattle arterials. Situated between Interstate 5 and Lake Washington, the route is popular with commuters transiting between Seattle and the outlying suburbs to the northeast. Additionally, two major public transit lines converge here. Drawing from Environmental Criminology and Routine Activities Theory, there is limited research in the area (Poister 1996), but some reason to indicate that areas associated with public transit infrastructure and junctions have a higher instance of low level crime (Buckley 1996).

Aurora Avenue (Highway 99) between N 85th and N 90th streets, the Crown Hill, Greenwood and north Green Lake neighborhoods, has historically been known for drug use and prostitution. Aurora Avenue is a significant north/south thoroughfare lined with a number of low cost motels catering to hourly customers and weekly rate customers. Prostitutes, pimps, and drug users often frequent these low cost motels as temporary housing options. Many of the use-of-force reports reviewed from this area were associated with drug activity and domestic violence type disturbances located at the motels in the area.

The third and final hot spot in the North Precinct, centering on the University Way NE area between NE 45th and NE 50th streets, is immediately adjacent to the University of Washington. The overall demographic of the surrounding area is primarily transient and attributable to students at the University of Washington. Businesses along University Way cater primarily to this population with eateries, alcohol establishments, and support businesses. Among the population is a small demographic referred to by locals as “Ave Rats” who are young, mostly homeless persons, who survive by panhandling.

A review of crime aggregator services[12] indicates increased rates of criminal behavior understood by environmental criminology as criminogenic risk factors. In the area of NE 125th Street and Lake City Way, crime statistics indicate assaults, burglary incidents, and vehicle prowls in the immediate residential area. Along Aurora Avenue between N 85th and N 90th streets, crime statistics indicate a number of warrant services and disturbances. Crimes reported in and around the University Way hot spot indicate a high number of property crimes (vehicle theft and auto prowl), shoplifting, and a significant number of assault incidents.

9.3.7.2 West Precinct Hot Spots

West Precinct hot spot activity is characterized both by the population and the businesses that make up the area. Seattle has a thriving nightlife and a substantial population of young professionals who frequent three major nightclub districts. Two of those districts fall in the West Precinct: Belltown and Pioneer Square. Additionally, the Pike-Pine Corridor (also associated with East Precinct hot spot activity), Pike Place Market, Downtown business district, and International District are strongly associated with active open air drug markets and homeless services.

Intoxicated subjects provide an unpredictable hazard to law enforcement. Producing a wide range of results, the effects of alcohol intoxication impair judgment as well physical coordination and narrow the effectiveness of alternatives to force, such as verbal reasoning. Large congregations of people and close proximity to street traffic, such as has been observed by the researchers around the time when alcohol establishments are legally mandated to close (0200 hours), present a public safety management condition which places police in highly volatile control situations where force is often the only viable option.

A notable characteristic of the Belltown, Pioneer Square, and International District neighborhoods is homelessness. On any given day the streets along the major transit corridor, 3rd Avenue, and the adjoining streets, homeless persons are a prominent feature. Due in part to the concentration of social services (Washington State Department of Social and Health Services, Social Security Administration), emergency services (shelters, needle exchanges, mental health services), and community custody corrections offices (Washington State Department of Corrections, Daily Reporting Center, WA DOC Outstations) and the availability of drugs, the homeless have little reason to venture out of the downtown area.

Some of the underlying causes of homelessness, mental illness and drug addiction, present a hazard to law enforcement for reasons similar to alcohol. Those suffering from Axis I mental disorders, characterized by a lack of reality testing, can be highly volatile when untreated or while decompensating from a more stable condition; similarly, those engaged in drug use can be difficult to reason with. Erratic

[12] Aggregator services such as www.spotcrime.com and www.crimereports.com are linked to computer-aided dispatch (CAD) and records management systems (RMS) and report crime statistics for public use.

and potentially violent behavior forces law enforcement into an aggressive stance which often leads to use of force for reasons of officer safety and to prevent more severe injury to the subject of an enforcement action. This condition is observed in the data and manifests a high number of use-of-force incidents involving drug use and potential illness for relatively minor public disorder offenses.

9.3.7.3 East Precinct Hot Spots

East Precinct hot spot activity is characterized by alcohol establishments and gang territory. The nightclub district of Capitol Hill is also home to two higher education institutions. The Yesler Terrace hot spot encompasses one of the largest public housing facilities in the city. The Central District hot spot is located near the somewhat notorious intersection of 23rd and Union streets and Garfield High School, known for gang, drug, and even terrorist activity (discussed below). Each of these locations manifest unique police citizen encounters.

Located in close proximity to Seattle University and Seattle Central Community College, the Capitol Hill nightclub district caters to students and the overwhelmingly young demographic that lives in the area. Also unique to this neighborhood is the center of the Lesbian Gay Bisexual Transgender and Queer (LGBTQ) community for the Pacific Northwest, considered to be the second largest such community in the United States. A recent article in the *Seattle Post Intelligencer* called Capitol Hill Seattle's "gayest neighborhood" (Cohen 2012). One of Seattle's oldest neighborhoods, the area in the immediate vicinity of the hot spot is characterized by a mix of small to medium apartment and condo complexes as well as single family residences. In addition to the service industry businesses and LGBTQ organizations, a handful of charity social service agencies are located in this area to serve the homeless.

9.3.7.4 South Precinct Hot Spots

With the exception of the North Beacon Hill hot spot, all of the hot spot activity in the South Precinct area centers along the Rainier Avenue and Martin Luther King Way arterials. The North Beacon Hill hot spot activity coincides with a small business district serving the Beacon Hill neighborhood. Historically, Rainier Avenue South and Martin Luther King Junior (MLK) Way have been characterized by low income housing, depressed neighborhoods, small single family homes, and gang and drug activity.

Geographically, the Rainier Valley is situated between Lake Washington and Beacon Hill and runs southeast toward the City of Renton. Extending south from the East Precinct, Rainier Valley neighborhoods are organized along Rainier Avenue South running south from First Hill and the International District and MLK Way south from the Central District. Support businesses are located all along this corridor (drugstores, grocery stores, restaurants, home improvement stores, etc.). Additionally, the north/south corridor serves public transit operations by King County Metro.

The drug trade is prominent in the Rainier Valley, both legal and illegal. A recent article in the *Seattle Times* noted that marijuana dispensaries outnumbered Starbucks 145 to 139, according to data published by the City of Seattle (Martin 2012). Marked by a green cross, similar to the European symbol for pharmacy, legal medical marijuana dispensaries are prominent features along Rainier Avenue South and MLK Way. Also prominent are the hallmarks of the illegal drug trade. With easy access in north and south to more affluent parts of the city, on any given day vehicles can be seen engaged in street level drug transactions with groups of young men in a variety of neighborhoods throughout the Rainier Valley.

The final hot spot area, moving south through the Rainier Valley toward the south city limits, is centered in the Rainier Beach neighborhood, in close proximity to Rainier Beach High School. Student ethnicity is reported as 57 % black, 18 % Asian/Pacific Islander, 13 % Hispanic, 9 % white, and 2 % Native American. Performance on state standardized exams is reported as ranging from 8 % (10th grade Algebra I) to 50 % (9th grade Geometry) across all four high school grades. According to the website Northwest Gangs, the areas associated with hot spot activity in the South Precinct coincide with known gang territory.[13]

9.3.7.5 Southwest Precinct Hot Spots

The Southwest Precinct is a relatively "cool" location for use of force. With the exception of activity in and around the Southwest Precinct facility, the High Point neighborhood remains the only significant hot spot area. Historically a depressed area, property values have been steadily increasing for the last decade. Median home prices are reported at the time of writing as $263,000. The High Point Neighborhood is one of Seattle's oldest gang territories, according to Northwest Gangs.

Driving through the High Point neighborhood, it appears, visually, to be one of the least developed neighborhoods in Seattle. Comprised primarily of small single family homes, the High Point public housing project is also a prominent feature with 425 family housing units replacing the original 716 units which were torn down in 2003.[14] Still, some dilapidated structures and vacant lots punctuate the neighborhood, in stark contrast to the public housing development.

[13] In the North Beacon Hill neighborhood, from Beacon Avenue South to South Atlantic Street between 12th and 5th Avenues, the "South Side Locos 13" control territory. In the Columbia City neighborhood the "Genesee Blocc Crips" control the area between South Dakota Street and South Oregon Street (north and south), Rainier Avenue South, and Cascade Avenue South, to the east. The area between Columbia City and Rainier Beach is a transient mix of six different gangs in large overlapping turf areas and small, self-contained territories. In the area of Holly Park, three gangs, "Holly Park," "74 Hoover Criminals," and "Vatos Locos 13," are shown in competing territories coinciding with hot spot activity.

[14] Northwest Gangs indicates that the renovation of the High Point "Projects" may be responsible for the redistribution of High Point gang affiliates throughout King County as families were forced to move to alternative housing during redevelopment.

Fig. 9.5 Force factor-weighted and unweighted KDE contours

9.3.8 *Force Factor-Weighted Hot Spots*

Although the density maps are valuable for understanding locations within the city where officers may use force more often than others, this only speaks to the incidence of use of force without regard to the level of force or, more importantly, the differential in officer force relative to suspect resistance. In Fig. 9.5, we add a second KDE using the same incident data and search parameters as in the previous map, but the incidents were weighted by incident force factors. Thus, an incident with a force factor equal to three would count as three incidents in the density estimation. Incidents having a zero or negative force factor are essentially weighted to zero. This results in a density surface that reflects hot spot areas where officers not only use force but tend to use a higher level of force relative to suspect resistance levels. As can be seen by the comparison of the two KDE layers, there are three categories of hot spots. First, there are some hot spots where there is a high density of use of force as indicated by the first KDE layer (red contour), but the use of force at those locations tends to be more proportional or lower than suspect resistance, and thus, there is no corresponding force factor-weighted KDE hot spot (green contour). Second, there are some hot spots where there is a only a weighted KDE hot

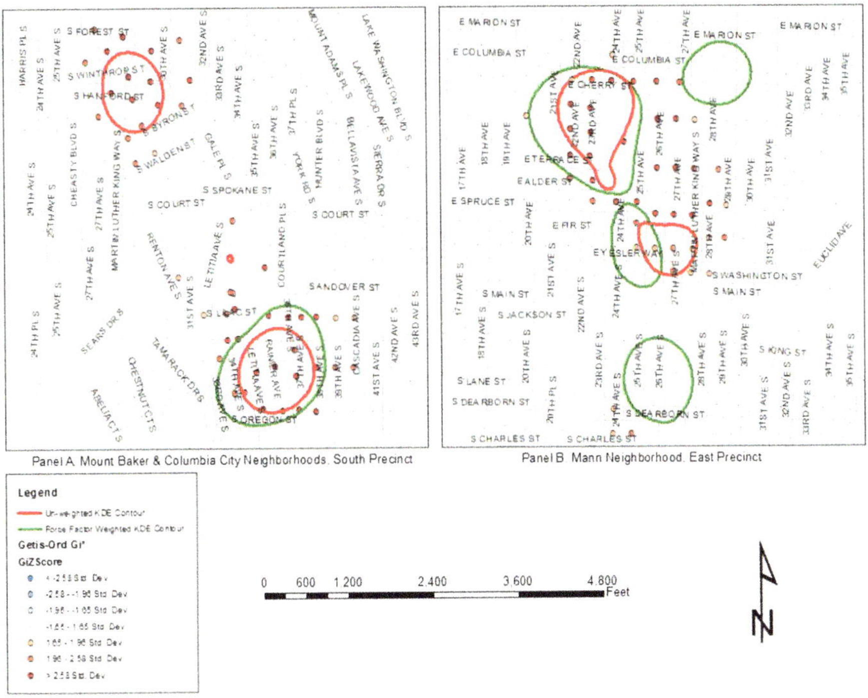

Fig. 9.6 Comparison of force factor-weighted and unweighted KDE hot spots

spot (green contour) and no corresponding unweighted hot spot (red contour), indicating that use of force is not generally concentrated at that location, but the uses of force tend to be less proportional. Care must be exercised with this second category as the weighted KDE may not be supported by statistical evidence of clustering; for example, just one or two nearby incidents with very high force factors could result in a high density value. Third, there are some hot spots where there is correspondence between both unweighted and weighted hot spots (i.e., overlapping red and green contours), indicating that force is generally concentrated at that location and also tends to be less proportional.

A more detailed view is provided in the two panels of Fig. 9.6. Here, we have focused on a section spanning the Mount Baker and Columbia City neighborhoods in the South Precinct (panel "a") and a section of the Mann neighborhood in the East Precinct (panel "b"). In panel "a" there are two hot spots identified based upon the density of incidents (red contour), but only the southern location (around S. Genesee St. and Rainier Ave.) has a corresponding force factor-weighted hot spot (green contour). Thus, police use of force is concentrated at both locations, but force in the northern hot spot tends to be more proportional whereas force in the southern hot spot tends to be less proportional.

In panel "b" there are four areas of interest. Two of the locations are identified with overlapping weighted and unweighted hot spots, indicating that use of force is concentrated at both locations and tends to be less proportional. The other two locations are identified only by the force factor-weighted technique, indicating areas where police use of force may not be generally concentrated, but where uses of force tend to be less proportional. While the Getis-Ord Gi* statistics are not significant at these locations, there are four to five use-of-force incidents underlying each of these weighted hot spot locations.

Extending the analysis into the specific cases at these locations, there are important differences in the nature of the incidents. In panel "a," the incidents in the northern hot spot have a mean maximum officer force level of 4.4, and the mean suspect resistance level is also 4.4. The mean force factor is equal to zero. The underlying seven incidents resulted in charges including VUCSA (Violation of Uniform Controlled Substances Act), warrants, trespass, assault, and one count of obstructing. In contrast, incidents in the southern hot spot had a mean maximum officer force level of 4.9, and mean suspect resistance level of 3.7. The mean force factor is equal to 1.2. Here, out of 10 incidents, five resulted in charges of obstructing, and three resulted in no charges at all.

In panel "b," there are two areas of overlapping weighted and unweighted hot spot contours, and two additional weighted hot spots. The larger overlapping group, to the West of Garfield High School, includes 12 incidents with a mean maximum officer force level of 5.1, and mean suspect resistance level of 3.8. The mean force factor is equal to 1.3. Charges included assault, VUCSA, VUFA (Violation of the Uniform Firearms Act), property damage, warrants, harassment, resisting, and obstructing. Two incidents resulted in no charges.

The area of the East Precinct known as the Mann Neighborhood, surrounding Garfield High School, has a long history of violent crime. Similarly, the Columbia City and Mount Baker neighborhoods have long been the focus of youth violence prevention and anti-gang measures. In addition to calls for service, on-viewed incidents and social contacts initiated by officers assigned to community policing beats account for activity here.

As discussed, the Columbia City and Mount Baker neighborhoods have undergone significant development in past years. Driving south on Rainier Avenue South, the road is lined with businesses. In addition to the usual franchises (e.g., fast food restaurants, grocery stores, and the like), small businesses such as nail salons, ethnic specialty restaurants, and boutiques line the road. As you approach the center of the hot spot (South Genesee Street and Rainier Avenue), there is a dramatic shift in the landscape.

Viewing this area through a Crime Prevention Through Environmental Design (CPTED) framework, the approach along Rainier Avenue South signals several hazards (excess vegetation, inadequate lighting). An excursion into the surrounding neighborhood reveals the homes in the immediate vicinity (contained within the geography of the hot spot), three to four blocks east and west, are in poorer condition (siding and paint are in need of repair, lawns are not well kept, external lighting may not be functional) compared to those homes extending away from the borders of the hot spot. Additionally, within this area, vehicles parked along the side streets are older and show more damage than those parked in the neighborhood immediately outside the area.[15] Finally, the commercial buildings and to a lesser extent some of the residences and residential out buildings (sheds and garages) show evidence of "tagging" or graffiti corresponding to known gangs.

As you approach the Mann neighborhood in the East Precinct, a similar shift can be noted. Though the immediate area surrounding the center of the hot spot (Garfield High School) is residential, several small businesses are prominent in a de facto "downtown." In a state of gentrification similar to the Columbia City/Mount Baker neighborhood area, several of the businesses are new or have recently undergone renovation. A few of the businesses proudly display signs stating they are under new management.

Though both neighborhoods are predominantly minority (primarily black or African American and Asian), the Mann neighborhood prominently indicates the East African origin of the ethnic population. Several of the restaurants and businesses advertise in Arabic script, and the neighborhood is known for strong ties to the Ethiopian and Somali immigrant population. In recent years an FBI investigation focused on a Mosque in the neighborhood of 23rd St. and E. Union with ties to a "terrorist training camp" in southern Oregon.

While the area within the hot spot shows some signs of physical and social disorder, it is not as prominent as the Columbia City/Mount Baker hot spot. Some gang tagging is evident but several new housing projects have been constructed in the past 5–8 years. Townhouses and small apartment complexes are now interspersed in older residential neighborhoods primarily composed of single family residential structures. These structures appear well maintained but show some signs of disrepair, decreasing in frequency with distance from the center of the hot spot. As you approach the older neighborhoods of Capitol Hill to the north, Madison Valley and Madrona to the northeast and east and First Hill to the west the area becomes dramatically more affluent within six blocks of the hot spot.

In both hot spots police use-of-force incidents are characterized by calls for service and on-viewed and/or community policing operations. Common to these interactions are the presence of large groups of youths, gang activity, and drug and property crimes such as residential burglary and robbery. Many of the actions taken by the officers in this area are more aggressive for officer safety reasons. The use of traps, takedowns, and strikes are generally accompanied by a suspicion that the subject of the use of force may be concealing a weapon. Additionally, with regard to residential burglary investigations where a known subject is at large, canine officers are commonly called to track and apprehend the subject.

[15] An informal visual survey of the neighborhood indicates vehicles are 7–10 years old, on average, as compared to 1–5-year average of the surrounding neighborhood.

9.4 Discussion

9.4.1 Utility of Geospatial Analyses for Internal Use of Force Investigations

The ability to match physical locations with trends in use-of-force incidents (and potentially inform supervisory practices) is a critical means of assessing police performance. Each of the hot spots identified in this study yields valuable information about how policing is administered in those areas. Some hot spot locations are artifacts of the criminal justice system (e.g., police precinct facilities), while others are reflective of a predominant social activity (alcohol establishments), and other locations are remnants of social injustices and tend to confirm prevailing criminological theories (such as General Strain and Social Disorganization). The usefulness of this information is in how it is applied by police managers to improve service to the community.

Geospatial analysis of use of force provides a more accurate view of how force is distributed across a community. Typical statistical reports issued by police departments simply tabulate the number of incidents over a particular period of time and do so at gross levels of aggregation. These reports do not speak to variability in the proportionality of police responses nor to issues of disparate impact at the neighborhood level. In spatial context, use-of-force data can be used by command staff and Internal Affairs, or in the case of the City of Seattle, civilian oversight agencies such as the Office of Professional Accountability (OPA), to support investigations of claims of misconduct, mismanagement, and disparate impact.

The force factor-weighted hot spots differ in nature from the unweighted hot spots in important ways. Taking the example from the Rainier Valley (panel "a" of Fig. 9.6), the southern hot spot is probably of more concern than the northern hot spot. Uses of force in the southern hot spot are characterized by a large proportion of "obstructing" offenses as well as no charges, both of which can be associated with "contempt of cop" or otherwise disrespectful suspect behavior perhaps unnecessarily leading to the use of force.

9.4.2 Utility for External Monitoring

External monitoring of the police, both by organized entities and, with increasing technological developments, concerned members of the general public, can be greatly enhanced by exploring the spatial distribution of force events. Reformers should understand that maps are a powerful way to communicate the nature of a problem and may include important behavioral patterns that are largely "invisible" to typical statistical analyses of administrative data.

Government accountability and transparency are primary concerns for civil liberties groups and the public alike. For this reason, political concerns have taken an active interest in promoting transparency, especially in law enforcement. Recent action by the US Department of Justice to compel police reform has led to consent decrees[16] in several jurisdictions, including the Seattle Police Department. When millions of dollars in public money are at stake, compliance, based in objective fact, is essential to maintaining confidence and support for these critical accountability programs.

9.4.3 *Utility for Addressing Issues of Social Justice*

Finally, there are important social justice implications. From this perspective, one might judge agencies in terms of disparate impact on at-risk populations. A key question is whether the force factor-weighted hot spots generally correspond with the locations of historically disadvantaged communities and/or involve at-risk suspects. We believe that our qualitative analyses suggest there is strong correspondence, but we reserve judgment for future research to make the quantitative connection in this and other jurisdictions.

There are a variety of potential explanations for violent interactions between the police and the public. Some of the information in this and other studies point to community composition. The location of alcohol establishments appear to be correlated with police use of force. Similarly, artifacts of the criminal justice system such as police precinct locations naturally tend to experience more use-of-force incidents. From a social justice perspective, it is important to separate these kinds of incidental occurrences from systemic factors. A rich and deep literature has established the correlation between concentrated disadvantage and criminality, particularly in populations already predisposed to marginalized social statuses, racial and ethnic minorities chief among them.

Enforcement priorities and institutional history bring law enforcement officers and groups of young minorities into conflict. With an emphasis on drug and property crimes, likely for their obvious and highly visible impact on the community and public safety, law enforcement officers are inclined to resort to aggressive officer safety tactics. Given the often violent nature of these offenses and offenders and an intense desire by the perpetrators to evade detection and detention, a significant potential for harm exists both to the officer and offender.

The effect of chronic negative interactions with law enforcement is erosion in citizen attitudes toward the police, as well as less likely compliance and deference to authority. Evidence of this historical erosion can also be found in pop culture. Shirts that proclaim "Stop Snitchin" and songs like "Fuck da Police" by NWA are indicative of a fractured trust between citizens and their appointed protectors. Certainly, a significant amount of this animosity can be traced to a tumultuous

[16] A "consent decree" is a legally binding agreement carrying the enforcement authority of a court order. Within the US system of multijurisdictional civilian law enforcement, the consent decree has been used since the mid-1990s by the US Department of Justice to enforce police reform compelled by a finding in violation of one or more civil rights statutes (Violent Crime Control and Law Enforcement Act of 1994, 42 U.S.C. § 14141, the Omnibus Crime Control and Safe Streets Act of 1968, 42 U.S.C. § 3789d, and Title VI of the Civil Rights Act of 1964, 42 U.S.C. § 2000d).

history fraught with overt racism, corruption, and abuse of power. In the modern, more transparent era of policing, one would hope these kinds of interactions are becoming increasingly rare; however, nearly 150 years of organized, state-sponsored policing leaves a lasting collective memory.

Our analysis suggests a strong correlation between disadvantaged populations and force-factor hot spot activity. Gentrification has contributed to a concentrating effect, perhaps emphasizing use of force in highly localized areas. Still, these neighborhoods have historically had a long history of violent crime. In the comparatively affluent surrounding neighborhoods, Seattle enjoys an extremely low rate of violent crime. That force factor hot spots occur in such close proximity to other crime hot spots would suggest that violent interactions with the police are an inevitable outcome of concentrated criminality. If only 3 % of all arrests result in reportable use of force, then areas with concentrated arrest activity will tend to appear as hot spots in geospatial analysis of this type. Naturally, the inclination is to associate other factors in an effort to explain this condition.

9.5 Conclusion

Police use of force cannot be fully understood apart from its spatial context. Any approach to monitoring and responding to officer behavior, absent the spatial context in which that behavior occurs, will be less efficient and could even be potentially misleading. Our descriptive overview of use of force in Seattle painted a fairly benign picture of infrequent and generally proportional police-citizen force interactions. It is only after considering the data in spatial context that important stories emerge about where use of force is concentrated, to what degree, and with what potential consequences for the relationship between the police and the communities served. Hopefully, other researchers will replicate this place-based analysis in Seattle and other jurisdictions and assist police managers and community members to understand the nature and extent of police-citizen encounters that result in force. The findings from this research and replications will advance our knowledge and guide our responses with the ultimate goal of reducing the number and level of force incidents.

References

Alpert G, Dunham R (1988) Policing multi-ethnic neighborhoods. Greenwood Press, Westport

Alpert G, Dunham R (1997) The force factor: measuring police use of force relative to suspect resistance. Police Executive Research Forum, Washington, DC

Alpert G, Dunham R (2004) Understanding police use of force: officers, suspects, and reciprocity. Cambridge University Press, Cambridge

Alpert G, Dunham R, Smith M, Kenney D, Madden T (2001) The force factor: measuring police use of force relative to suspect resistance. U.S. Department of Justice, Washington, DC

Black D, Reiss A (1967) Patterns of behavior in police and citizen transactions. In: Studies of crime and law enforcement in major metropolitan areas. Government Printing Office, Washington, DC, pp 25+

Buckley JB (1996) Public transit and crime: a routine activities/ecological approach. Ph.D. dissertation, Simon Fraser University

Cohen A (2012) Which neighborhoods are Seattle's gayest? And which are more-affordable alternatives? *Seattle Post Intelligencer*. http://www.seattlepi.com/realestate/article/Which-neighborhoods-are-Seattle-s-gayest-3637442.php. Accessed 9 Feb 2013

Croft E (1985) Police use of force: an empirical analysis. Ph.D. dissertation, University at Albany, State University of New York

Fyfe J (2002) Too many missing cases: holes in our knowledge about police use of force. Justice Res Policy 4(special issue):87–102.

Gorr W, Kurland K (2012) GIS tutorial for crime analysis. ESRI Press, Redlands

Hickman M (2010) Democratic policing: how would we know it if we saw it? (Ch.21). In: Rice S, White M (eds) Race, ethnicity, and policing: new and essential readings. NYU Press, New York, pp 489–504

Hickman M, Piquero A, Garner J (2008) Toward a national estimate of police use of nonlethal force. Criminol Public Policy 7(4):563–604

Hickman M, Fricas J, Strom K, Pope M (2011) Mapping police stress. Police Q 14(3):227–250

Kane R (2002) The social ecology of police misconduct. Criminology 40:867–897

Kane R (2007) Collect and release data on coercive police actions. Criminol Public Policy 6(4):563–604

Koper C (1995) Just enough police presence: reducing crime and disorderly behavior by optimizing patrol time in crime hotspots. Justice Q 12(4):649–672

Martin J (2012) Seattle pot dispensaries finding business climate no longer sunny. The Seattle Times. http://seattletimes.com/html/localnews/2019077879_dispensaries06m.htm. Last accessed 9 Feb 2013

Mastrofski S, Reisig M, McClusky J (2002) Police disrespect toward the public: an encounter-based analysis. Criminology 40:519–551

Poister TH (1996) Transit-related crime in suburban areas. J Urban Aff 18(1):3–75

Ratcliffe J, Taniguchi T, Groff E, Wood J (2011) The Philadelphia foot patrol experiment: a randomized controlled trial. Criminology 49(3):795–831

Rubenstein J (1973) City police. Noonday Press, New York

Sampson R (2013) The place of context: a theory and strategy for Criminology's hard problems. Criminology 51:1–31

Seattle Police Department (2001) Use of force by Seattle Police Department officers. Seattle Police Department, Seattle

Sherman L, Gartin P, Buerger M (1989) Hot spots of predatory crime: routine activities and the criminology of place. Criminology 27(1):27–55

Skolnick J (1966) Justice without trial: law enforcement in democratic society. John Wiley and Sons, New York

Smith D (1986) The neighborhood context of police behavior. In: Reiss A, Tonry M (eds) Communities and crime. University of Chicago Press, Chicago, pp 313–341

Terrill W (2003) City of San Antonio, Texas Police Department use of force analysis July 1, 2001–December 31, 2002. San Antonio Police Department, San Antonio

Terrill W, Reisig M (2003) Neighborhood context and police use of force. J Res Crime Delinq 40:291–321

Terrill W, Alpert G, Dunham R, Smith M (2003) Evaluating police use of force: an application of the force factor. Police Q 6:150–171

Weisburd D (2008) Place-based policing. Police Foundation, Washington, DC

Weisburd D, Morris N, Groff E (2009) Hotspots of juvenile crime: a longitudinal study of arrest incidents at street segments in Seattle, Washington. J Quant Criminol 25(4):443–467

Williams J, Hester G (2003) Sheriff law enforcement officers and the use of force. J Crim Justice 31:373–381

Wilson J (1968) Varieties of police behavior. Harvard University Press, Cambridge

Wolfgang M, Figlio R, Sellin T (1972) Delinquency in a birth cohort. University of Chicago Press, Chicago

Chapter 10
Mapping Spatiotemporal Patterns of Liquor Law Violation Citations During Oktoberfest in College Town of La Crosse, Wisconsin

Gargi Chaudhuri, Steven Oxley, and Scott Wenzlaff

Abstract Annual events like Oktoberfest, Mardi Gras, and Halloween are very popular among young crowds, especially in college towns, and a significant part of the celebration consists of alcohol consumption. Binge drinking during such celebrations often results in alcohol-related crime, assaults, vandalism, and even fatal accidents. Yet, in crime studies, little research has been conducted to investigate the impact of such events on dynamics of crime and underage drinking in the local communities. The present study aims to analyze the influence of Oktoberfest on spatiotemporal pattern of underage drinking in college town of La Crosse in southwest Wisconsin. The study uses 5 years (2008–2012) of liquor law citation records and GIS techniques to explore the spatiotemporal pattern of underage drinking during the week of Oktoberfest and a week before and after the festival. Analysis conducted using grid thematic maps and local information showed that the local celebration of Oktoberfest results in an increased number of liquor law violations citations which coincided with increased number of fatal accidents, alcohol-related crimes, and public nuisance. Knowledge gained from results of grid thematic mapping was used to create a multi-criteria evaluation (MCE) that helped to identify a probability surface for high concentration of liquor law violation and thus underage drinking in certain parts of the town. Validation of the resultant map shows that 85–97 % of the citation location in last 5 years falls within the high probability zone delineated by our map. This type of mapping approach is useful to the local law enforcement officials and volunteer watch groups to provide focused deployment of intervention measures and increased vigilance to restrict alcohol consumption of underage youths and prevent associated crime and accidents.

G. Chaudhuri (✉)
Department of Geography and Earth Science, University of Wisconsin-La Crosse, 1707 Pine Street, 2031 Cowley Hall, La Crosse, WI 54601, USA
e-mail: gchaudhuri@uwlax.edu

S. Oxley • S. Wenzlaff
Department of Geography and Earth Science, University of Wisconsin-La Crosse, 1707 Pine Street, 2004 Cowley Hall, La Crosse, WI 54601, USA
e-mail: oxley.steve@uwlax.edu; wenzlaff.scot@uwlax.edu

G.A. Elmes et al. (eds.), *Forensic GIS: The Role of Geospatial Technologies for Investigating Crime and Providing Evidence*, Geotechnologies and the Environment 11, DOI 10.1007/978-94-017-8757-4_10,

Keywords Underage drinking • Campus policing • Crime mapping • Multi-criteria evaluation

10.1 Introduction

Since the mid-1990s, there has been increasing use of GIS in crime analysis and mapping for allocation of resources, focused deployment, and strategic planning in the United States (Hoover et al. 2010). Application of GIS is very effective in visualizing patterns of criminal activity and identifying social and physical factors that might encourage such activity. This type of application can be best categorized under strategic crime mapping and analysis. According to US Department of Justice, strategic crime analysis is defined as "the study of crime and law enforcement information integrated with socio-demographic and spatial factors to determine long term 'patterns' of activity, to assist in problem solving, as well as to research and evaluate responses and procedures." (Boba 2001, p. 13). The present chapter focuses on one such application of GIS to study the most common problem of any college town of United States: underage drinking.

At present, crime mapping is a well-established field of study. Yet, effective application of GIS in mapping alcohol-related crimes or underage drinking in college campuses or communities is less popular than other types of crime mapping. Most of the existing applications focus at neighborhood level that used GIS to map areas of alcohol outlets, density of liquor stores, and alcohol-related crimes (Block and Block 1995; Gorman et al. 2001; Scribner et al. 1999; Zhu et al. 2004). Spatial distribution of crime research has been mostly cross-sectional since the locations depend on opportunities to commit crime (Groff et al. 2009; Chainey et al. 2008). Hotspot mapping and its variants have been the predominant method to study spatial distribution of crime. Hotspot mapping has been also used to predict future locations of crime (Chainey et al. 2008). The temporal pattern of crime mapping has been relatively less well explored compared to its spatial counterpart (Townsley 2008; Felson and Poulsen 2003; Ratcliffe 2004a). Temporal mapping within the hot spots of crime helps to determine the variability or stability of crime patterns (Townsley et al. 2000; Bennett 1995; Weisburd et al. 2004; Spelman 1995; Taylor 1999). Depending on the type of crime, the pattern of crime occurrences varies greatly by hour of the day (Felson and Poulsen 2003). Yet if time fragmentation is applied, the number of crime declines greatly (Weisburd et al. 2004; Felson and Poulsen 2003).

Two aspects are generally taken into consideration for temporal mapping: first, at what time interval the crime incidences are mapped (viz., hourly, daily, weekly, monthly, annually, and regular/irregular interval) and secondly, the temporal length of the data record. Both aspects of temporal mapping are important to capture a comprehensive pattern of crime. Multi-temporal mapping shows past trend of crime and therefore can help to determine possible future trends in the area. Thus, predictive mapping serves as an excellent resource for risk analysis and prioritizing police resources (Johnson and Bowers 2004a, b). Application of such informative GIS mapping and predictive analysis is popular in liquor-related crime analysis to understand crime patterns and design intervention mechanisms with the use of spatial decision support systems (Block and Block 1995; Scribner et al. 1999; Goodchild et al. 2000; Zhu et al. 2004). The present study extended this theme and used GIS and predictive mapping to understand the spatiotemporal pattern of underage drinking in the college town of La Crosse, WI, during Oktoberfest. The chapter also highlighted how a social event like Oktoberfest provokes binge drinking among underage youths and results in alcohol-related crimes, assaults, and accidents.

The study used 5 years (2008–2012) of liquor law citation records from the local (La Crosse Police Department) and campus police department (University of Wisconsin-La Crosse Police Department) as a representative of underage drinking pattern. The objective of this study was to analyze the impact of Oktoberfest in increasing liquor law violation in the college town of La Crosse. Specifically, the study used grid thematic mapping to analyze the multi-temporal pattern of the liquor law violation citations during the last week of September and first 2 weeks of October (the week before Oktoberfest, the week of Oktoberfest, and the week after Oktoberfest) from 2008 to 2012. Analysis was conducted for the weeks before and after the Oktoberfest week as a representative trend of liquor law violation during regular non-festive weeks of the year. The comparison of number of citations among these 3 weeks over a span of 5 years helped to measure the magnitude of increase in number of citations during Oktoberfest from regular weeks. Knowledge gained from results of grid thematic mapping was used to create a multi-criteria analysis that helped to identify a probability surface for high concentration of liquor law violation and thus underage drinking in certain parts of the town. The chapter first introduces the study area and then provides a brief discussion about alcohol-related problems associated with local Oktoberfest celebration, followed by methodologies used for the study, and finally reports results and a discussion of the findings.

10.2 Underage Drinking in College Campuses

Alcohol is one of the most harmful drugs, which causes major harm to both individual and the society, and yet the policies and restrictions of its use are not as regulated as some other harmful drugs such as cocaine, heroin, and marijuana (ICL 2010). Thus, alcohol problems in college towns are a matter of serious concern both for campus administrators and law enforcement agencies. There has been growing evidence of ecological association between alcohol and violence especially in college towns (Snowden 2012). According to the Center of Disease Control and Prevention's 2011 Youth Risk Behavior Survey, people aged between 12 and 20 years drink 11 % of all alcohol consumed in the United States (Eaton et al. 2012). The CDC Fact sheets also report that alcohol consumption of youths may result in:

- School problems, such as higher absence and poor or failing grades
- Social problems, such as fighting and lack of participation in youth activities
- Legal problems, such as arrest for driving or physically hurting someone while drunk
- Physical problems, such as hangovers or illnesses
- Unwanted, unplanned, and unprotected sexual activity

- Disruption of normal growth and sexual development
- Physical and sexual assault
- Higher risk for suicide and homicide
- Alcohol-related car crashes and other unintentional injuries, such as burns, falls, and drowning
- Memory problems
- Abuse of other drugs
- Changes in brain development that may have lifelong effects
- Death from alcohol poisoning (CDC 2012)

Thus, prevention of alcohol consumption needs a community-based effort along with the school administration and law enforcement officials to determine success and improve effectiveness.

In recent years, a number of studies have focused on the impact of exceptional events (natural disasters, sporting events, festivals, concerts, political rallies, etc.) on spatiotemporal distribution of crime (Leitner and Guo 2013; Andresen and Tong 2012; Leitner and Helbich 2011; Leitner et al. 2011). The list of events categorized as exceptional events by Leitner and Guo (2013) covers a wide spectrum in terms of both severity and characteristics, but can be easily subdivided into organized events (such as sporting events, festivals, concerts, and political rallies) (Andresen and Tong 2012) and natural events (such as natural disasters) (Leitner and Helbich 2011; Leitner et al. 2011). Empirical observations suggest that locally organized annual social events (like Mardi Gras, Halloween, Oktoberfest, etc.) have a large influence on underage binge drinking and associated crime in college or university towns. The social dynamics of drinking in events like these are similar to any social ecological model of drinking related to an alcohol outlet and problems (Gruenewald 2007) but at a larger scale and for a limited time period. A number of studies have shown positive correlation between physical availability of alcohol and increased consumption (Watts and Rabow 1983; Gruenewald et al 1996; Alaniz et al. 1998; Weitzman et al. 2003; Treno et al 2003) and between outlet densities and alcohol-related crimes violence, drink driving, and traffic crashes (Britt et al 2005; Zhu et al. 2004; Norström 2000; Huckle et al. 2008). Thus, events like these may be economically beneficial to the local community but create an opportunity for young people to indulge into binge drinking in the form of social celebration.

10.3 Study Area

10.3.1 La Crosse, Wisconsin

Situated on the right bank of Mississippi river, the city of La Crosse has a resident population of approximately 51,320 people (US Census 2010) with predominantly white population (89.8 %, US Census 2010). The city is divided into two parts by the La Crosse River Marsh and connected by three major roads (State Highways

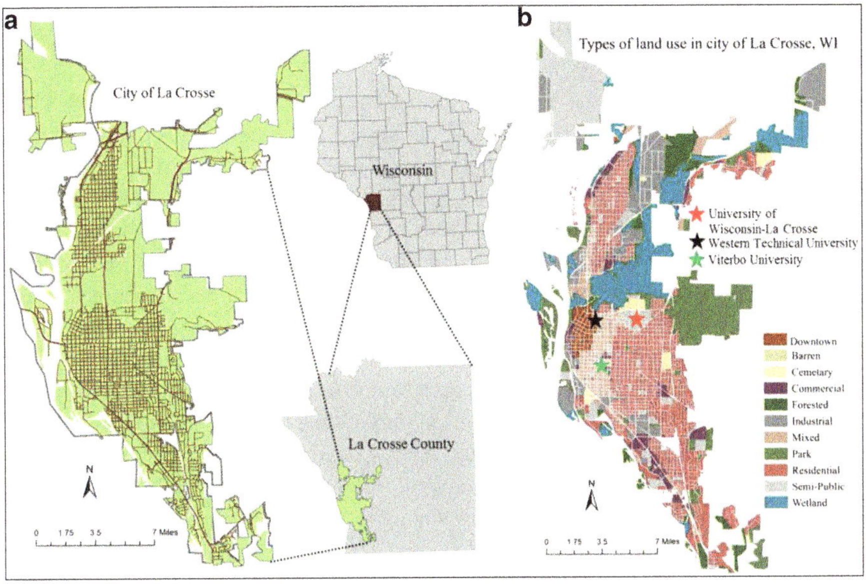

Fig. 10.1 (**a**) Study area: La Crosse, Wisconsin; (**b**) Land use map of La Crosse City and location of the three colleges

Table 10.1 Estimate of student population of three colleges in La Crosse, WI

College	Total students	Undergraduate	Postgraduate
UW-L	10,227	9,441	786
WTC	5,122	5,122	0
VU	3,192	2,279	919
Total	18,541	16,842	1,705

Source: Individual college websites (UW-L: www.uwlax.edu; WTC: www.westerntc.edu; VU: www.viterbo.edu)

53, 35, and 16). The majority of the land use in the city is residential (Fig. 10.1) followed by mixed land uses, including traditional neighborhood development. The small town is home to three undergraduate colleges: University of Wisconsin-La Crosse (UW-L), Western Technical College (WTC), and Viterbo University (VU). The majority of the population in these colleges is traditional undergraduate students. Table 10.1 provides an estimate of student population of the three colleges.

Demographically, 11 % of the city population is within the age group of 15–19 years, and 18.2 % is within 20–24 years of age (U.S. Census Bureau 2010). Table 9.1 shows that among the total number of students, 91 % are undergraduates. Assuming that a majority of these undergraduate students are 21 years and below, a large number of these students are potential candidates for underage drinking especially during Oktoberfest.

10.3.2 Oktoberfest in La Crosse, Wisconsin

La Crosse hosts one of the biggest Oktoberfest celebrations in the United States, attracting around 100,000–150,000 visitors during that week and adding $6 million to the local economy (La Crosse Tribune 2011 and StarTribune 2012). Being home to three universities with a majority of undergraduate students, a good proportion of the visitors consist of young people from neighboring towns. The number of law enforcement officials deployed in town during those days is notably more than usual to keep matters under control. Thus, with an increased number of people, a large number of law enforcement officials in the streets, and the intrinsic nature of the festival celebration, there is a huge number of alcohol-related citations, especially for underage drinking. There have been several cases of fatal accidents, including drowning in the local river, not to mention public assaults and vandalism during the week of Oktoberfest. Since 2004, after several deaths, accidents, and disorder created by drunken people in the town, law enforcement officials and city administrators started taking serious action to change the situation. A local newspaper *StarTribune* (2012) reported that after such incidents, police started responsible server education and took precautions to stop alcohol availability to minors in local bars, and city administrators limited keg purchases and imposed keg registration rules, recriminalized public intoxication, offered alcohol education classes in lieu of fines, and started restricting license issuance to reduce bar density. Eventually, a number of community and student volunteered groups started organizing watch groups along the parks near the river to assure safety of drunken people in the downtown where there is a high density of pubs and to stop public violence especially in student neighborhoods (StarTribune 2012). Every year there has been increased number of police to control the crowd, but the problem still persists.

Nationally, the university administrators and law enforcement officials are working towards a coordinated effort to fight against underage drinking and alcohol abuse in college campuses. Yet popular festivals like Oktoberfest celebrate alcohol consumption that helps to boost the local economy.

10.4 Data

At present there are no specific data for underage drinking in La Crosse. After exploring the database provided by the police departments and discussion with police officials in UWL campus, we decided to focus our analysis on liquor law violation citations to best capture the underage drinking trend. Data was collected from the University of Wisconsin-La Crosse Police Department and La Crosse Police Department for the last week of September and the first 2 weeks of October of every year from 2008 to 2012. To be specific, the time period of our data spans from Friday of the week before Oktoberfest week to Thursday of the week after (21 days). The weeks before and after the Oktoberfest were assumed to represent regular non-festival

TABLE 11.1 Diagnostic Structure for Substance Use Disorder in DSM, ICD-11, and ASEBA

DSM *A problematic pattern of substance use leading to clinically significant impairment or distress, as manifested by at least two of the following, occurring within a 12-month period*		*ICD-11*	*ASEBA*
DSM-IV-TR (2000)	*DSM-V (2013)*	*Example: Alcohol*	*Items*
Substance Abuse • Recurrent substance-related legal problems* • —————— • Recurrent use resulting in a failure to fulfill major role obligations at work, school, or home • Recurrent use in situations in which it is physically hazardous • Continued use despite having persistent or recurrent social or interpersonal problems caused or exacerbated by the effects of a substance	**Substance Use Disorder** • —————— • Craving, or a strong desire or urge to use+ • Recurrent use resulting in a failure to fulfill major role obligations at work, school, or home • Recurrent use in situations in which it is physically hazardous • Continued use despite having persistent or recurrent social or interpersonal problems caused or exacerbated by the effects of a substance	**Harmful Pattern of Substance Use** A pattern of alcohol use that has caused damage to a person's physical or mental health or has resulted in behavior leading to harm to the health of others. The pattern of alcohol use is evident over a period of at least 12 months if substance use is episodic or at least one month if use is continuous. Harm to health of the individual occurs due to one or more of the following: (1) behavior related to intoxication; (2) direct or secondary toxic effects on body organs and systems; or (3) a harmful route of administration. Harm to health of others includes any form of physical harm, including trauma, or mental disorder that is directly attributable to behavior related to alcohol intoxication on the part of the person to whom the diagnosis of Harmful pattern of use of alcohol applies.	**Ages 1.5–5** • Not included **Ages 6–18** • Alcohol • Tobacco • Uses Drugs

TABLE 11.1 (Continued)

DSM *A problematic pattern of substance use leading to clinically significant impairment or distress, as manifested by at least two of the following, occurring within a 12-month period*		*ICD-11*	*ASEBA*
DSM-IV-TR (2000)	*DSM-V (2013)*	*Example: Alcohol*	*Items*
Substance Dependence • Substance is often taken in larger amounts or over a longer period than was intended • Persistent desire or unsuccessful attempts to cut down or control use • A great deal of time is spent in activities necessary to obtain a substance, use a substance, or recover from its effects • Important social, occupational, or recreational activities are given up or reduced because of use • Use is continued despite knowledge of having a persistent or recurrent physical or psychological problem that is likely to have been caused or exacerbated by substance use • Tolerance • Withdrawal **Severity Categories:** ________ ________ ________	• Substance is often taken in larger amounts or over a longer period than was intended • Persistent desire or unsuccessful attempts to cut down or control use • A great deal of time is spent in activities necessary to obtain a substance, use a substance, or recover from its effects • Important social, occupational, or recreational activities are given up or reduced because of use • Use is continued despite knowledge of having a persistent or recurrent physical or psychological problem that is likely to have been caused or exacerbated by substance use • Tolerance • Withdrawal **Severity Categories:** Mild (2–3 symptoms) Moderate (4-5 symptoms) Severe (6 or more symptoms)	**Substance Dependence** Alcohol dependence is a disorder of regulation of alcohol use arising from repeated or continuous use of alcohol. The characteristic feature is a strong internal drive to use alcohol, which is manifested by impaired ability to control use, increasing priority given to use over other activities and persistence of use despite harm or negative consequences. These experiences are often accompanied by a subjective sensation of urge or craving to use alcohol. Physiological features of dependence may also be present, including tolerance to the effects of alcohol, withdrawal symptoms following cessation or reduction in use of alcohol, or repeated use of alcohol or pharmacologically similar substances to prevent or alleviate withdrawal symptoms. The features of dependence are usually evident over a period of at least 12 months but the diagnosis may be made if alcohol use is continuous (daily or almost daily) for at least 1 month.	**Ages 1.5–5** • Not included **Ages 6–18** • Not included

... in DSM-5. *Craving was added as a criterion in DSM-5. Sources: DSM-IV-TR (APA, 2000); DSM-5 (APA, 2013).

Prevalence

PREVALENCE DURING ADOLESCENCE

Substance use is common among adolescents and teens, particularly use of alcohol and marijuana. Use generally increases with age (Johnston et al., 2019) and it is therefore important to examine prevalence rates across age. Research estimates the *lifetime prevalence* rate of alcohol use is 23.5% among U.S. eighth graders and 58.5% among 12th graders (Johnston et al., 2019). Similarly, *30-day prevalence* rates for alcohol have been estimated at 8.2% and 30.2% in these ages. Estimates of marijuana use are similar, with 13.9% of eighth graders and 43.6% of 12th graders reporting lifetime use, as well as 5.6% and 22.2%, respectively, endorsing use within the last 30 days (Johnston et al., 2019). Problematic use is also rather common, with 13.8% of 12th graders reporting binge drinking (i.e., consuming five or more drinks) within the previous two weeks and 11.3% reporting at least one month-long period of daily marijuana use at some point in their lifetime (Johnston et al., 2019).

Regarding other substances, research by the Centers for Disease Control and Prevention (CDC, 2018) suggests about 14.0% of high schoolers endorse lifetime use of illicit drugs besides marijuana and alcohol (e.g., cocaine, methamphetamine.), with 30-day prevalence rates for these substances ranging from 3.0% to 6.0% among adolescents (Johnston et al., 2019). Of note, however, there has been a general, downward trend in lifetime use rates of these drugs among youth, decreasing from 22.6% in 2007 to 14.0% in 2017 (CDC, 2018). This trend is consistent with data suggesting decline in most substance use since peaks in the late twentieth century, with the notable exception of vaping (i.e., electronic smoking), which recently increased rapidly (Johnston al., 2019).

Although some data suggest adolescent substance use has improved over time, some substances remain priority concerns. Specifically, current estimates indicate that more than two million people in the US have an opioid use disorder (OUD) (Collins, Koroshetz, & Volkow, 2018), representing a national health epidemic (Volkow, Frieden, Hyde, & Cha, 2014). While most with OUD are mature adults, opioid misuse and its sequelae are also at crisis levels for youth, including older adolescents (ages 16–18) and emerging adults (ages 18–26). Among adolescents, during the past decade about 150,000 persons per year reported OUD, with about 80% addicted to prescription medications (Minozzi, Amato, Bellisario, & Davoli, 2014; Subramaniam, Fishman, & Woody, 2009; Wenner & Gigli, 2016). The rate of lethal overdoses attributable to opioids in adolescents increased 268% between 1999 and 2016, and a total of 8,986 teens died from overdoses over this 17-year period, with heroin accounting for 23% of deaths (Gaither, Shabanova, & Leventhal, 2018).

PREVALENCE BY RACE AND ETHNICITY

African American youth consistently report lower levels of illicit drug use than White or Hispanic youth; however, marijuana use among these groups appears to have become comparable within recent years (Johnston et al., 2019). Research also suggests African American adolescents exhibit lower rates of SUDs compared to non-Hispanic White adolescents (Merikangas et al., 2010).

Comorbidity

Findings from the National Comorbidity Survey Replication—Adolescent Supplement (NCS-A) (Merikangas et al., 2010), a nationally representative survey of comorbid psychiatric and substance use disorders, suggest that adolescents who engage in problematic substance use are also likely to exhibit a mood, anxiety, or behavioral disorder. Specifically, 11.4% of adolescents (ages 13–18) meet diagnostic criteria for an SUD, with approximately 4.4% of the total sample evidencing at least one other diagnosis as well (Merikangas et al., 2010). Similarly, Conway and colleagues (2016) found that adolescents reporting a prior mental health diagnosis also endorsed high rates of alcohol abuse (10.3%) and illicit drug abuse (14.9%) and were at an increased risk of developing more problematic use patterns (e.g., transitioning to regular use). The highest prevalence rates for alcohol and drug abuse were among youth reporting anxiety disorders (17.3% and 20.0%), as well as disruptive behavior disorders (15.6% and 24.0%) (Conway, Swendsen, Husky, He, & Merikangas, 2016).

Developmental Psychopathology Models of Substance Use Disorder

We present multiple developmental pathways resulting in substance use, reflecting that SUDs reflect *equifinality* in the real world (see Figure 11.1). We will discuss these disorders as related to four separate etiological pathways posited in the literature: (1) an *externalizing pathway*, related to an individual's difficulty in controlling their behavior (e.g., Giancola & Tarter, 1999; Tarter, Kirisci, Habych, Reynolds, & Vanyukov, 2004), (2) an *internalizing pathway*, where substance use serves to diminish internal distress (e.g., Khantzian, 1997; Kirisci et al., 2015), (3) a *repeated use pathway*, which highlights the changes of brain mechanisms that result from substance use itself (e.g., Volkow, Wang, Fowler, & Tomasi, 2012), and, finally, (4) a *contextual pathway*, composed of broader environmental factors that create an environment of risk related to all other pathways. Of note, these trajectories consist of interrelated, cross-cutting weeks during a year. For ease of understanding, from here on, the week before Oktoberfest week will be called week 1, Oktoberfest week will be called week 2, and the week after Oktoberfest week will be called week 3. For each day, citations were divided into three time windows, 4 AM–12 PM, 12 PM–8 PM, and 8 PM–4 AM, to understand the daily temporal distribution of incidents.

The data collected from both the city and the campus police department consisted of date and time of citations, location of the citation in the form of nearest street address, case number, and statute description. The types of offenses taken into consideration are underage alcohol (1st–4th offense), public consumption, underage liquor violation, public intoxication, false ID, selling liquor to minors/intoxicated persons, underage drinking (procures, possess, consumption), fermented malt beverage license, and sale of malt beverage to an underage person. We understand that liquor law violation citations may not be the best data to study the effect of Oktoberfest on drinking behavior, but currently there is no other suitable existing data that can represent this issue. So in this study, we assumed that citation data represented the spatiotemporal distribution of underage drinking in the city where locations with a higher number of citations represented locations with large number of underage drinkers and vice versa.

The police records available for the study did not record information regarding the residency of the violators, so it was not known whether the violators were residents of La Crosse or visitors. The data was geocoded using the US street geocode service by ESRI. The geocoding processes generated a 97–99 % success rate, which was above the acceptable success rate generated by Ratcliffe (2004b). Information regarding the location of student housing and pubs was gathered by local survey using a land use map of La Crosse city. Other types of data used for mapping and analysis include the La Crosse City administrative boundary, street network data, demographic data from the US census (tract boundary), and land parcel data from the La Crosse local authority (parcel boundary).

10.5 Methodology and Results

10.5.1 Multi-temporal Gridded Thematic Mapping

For this study, grid thematic mapping hotspot analysis was used in order to avoid the problem of misaligned administrative boundaries of different data sources and make them comparable to the sociodemographic and land use data of the city. Grid thematic mapping (Chainey et al. 2008; Eck et al. 2005) helped to deal with this area unit problem by interpolating the data into regular grids (or quadrants). In this study, a square grid size of 120 m by 120 m is used, which almost matches with the average city block size of the area. Since a number of citations have the same location, the grid thematic mapping approach helped to avoid the stacked data problem.

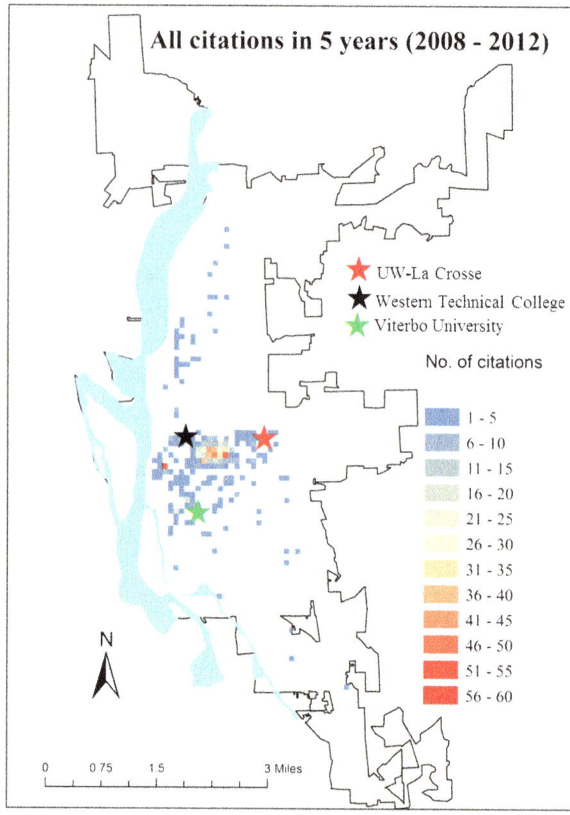

Fig. 10.2 Gridded thematic hotspot map of citations during 2008–2012. Legend represents numbers citations in each grid cell

Using the grid thematic mapping hotspot analysis, multiple spatiotemporal hotspot maps were produced to analyze the spatial distribution of number of citations and its change over time. Figure 10.2 shows a composite thematic map of spatial distribution of citations during the 5-year period (2008–2012). The grid cells with high number of citations were mostly within the downtown area, especially surrounding the three colleges, and sparsely distributed in the northern and the southern parts of the city.

Figure 10.3 shows thematic hotspot maps of citations for every year from 2008 to 2012. This series of maps showed that over the years not only overall number of citations increased but also the incidents of citations became more localized in the neighborhoods surrounded by three colleges, especially in 2012.

Figure 10.4 shows the comparison of spatial distribution of citations among 3 weeks. The results show that the number of citations during week 2 was high compared to the number of citations during week 1 and 2, for every year. The maps also show that, from 2008 to 2012, the number of citations in week 1 decreased, whereas in week 2 they increased, but were more localized, and finally in week 3, except in 2011, the number of citations decreased and were sparsely distributed

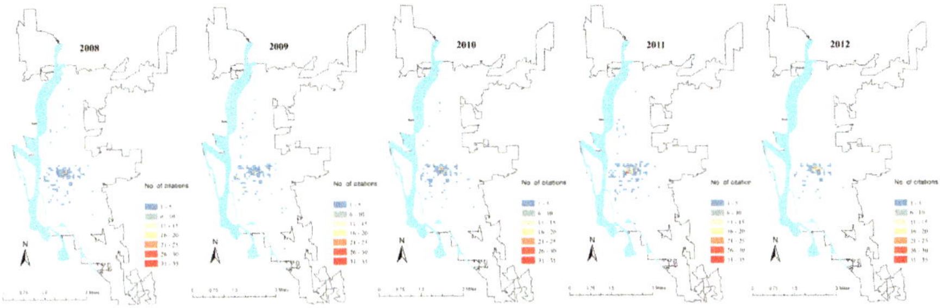

Fig. 10.3 Liquor law violation citations from years 2008 to 2012. Legend represents numbers of citations in each grid cell

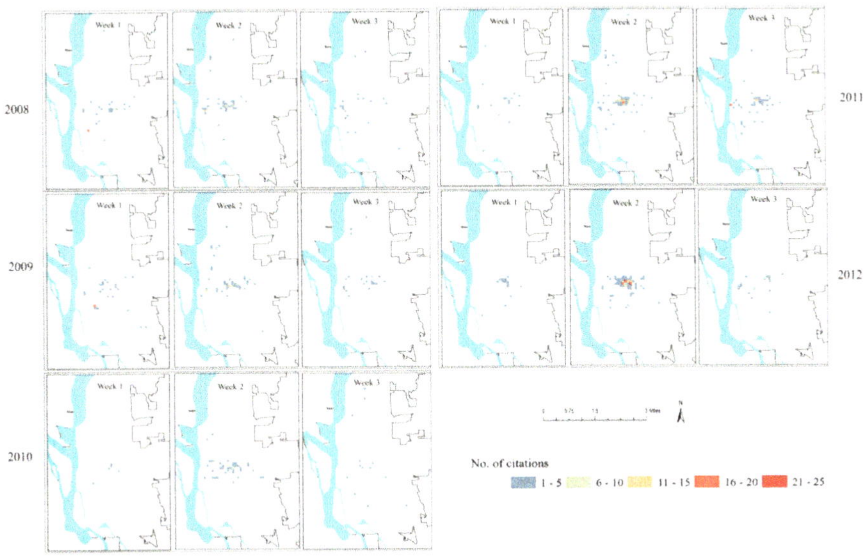

Fig. 10.4 Spatiotemporal distribution of liquor law violation citation by weeks in each year. Legend represents numbers of citations in each grid cell

across the city. It should be noted that, in addition to increased number of binge drinkers and large number of visitors, the increase in the number of citations during the Oktoberfest week may be also due to the large number of police patrols around the city (La Crosse Tribune 2011). Figure 10.5 shows the daily trend for each of the 3 weeks (starting from Friday of week 1 and ending on Thursday of week 3). For all 5 years, the incidence of citations was more or less concentrated during the weekends (Friday–Sunday), with the highest on Saturdays. There were some citations on other days of the week, but the pattern was too inconsistent to draw any conclusion.

To understand the daily temporal distribution of the citation incidents, each day of the week was divided into three time windows: (a) 4:00 AM–12:00 noon, (b) 12:00 noon–8:00 PM, and (c) 8:00 PM–4:00 AM. The results (Fig. 10.6), showed

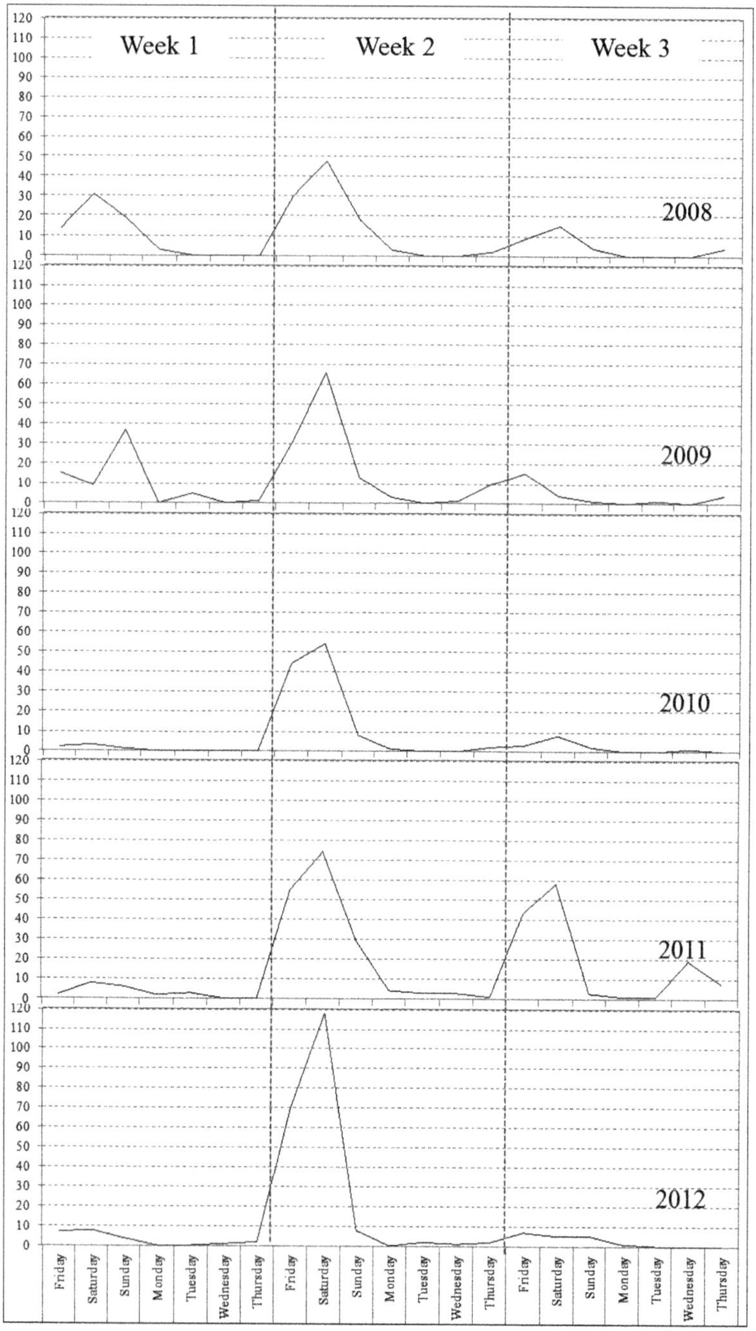

Fig. 10.5 Number of citations for each day of the week. The vertical axis shows the number of citations in each day. The horizontal axis shows the day of the week. The corresponding dates for each of the week differ with every year

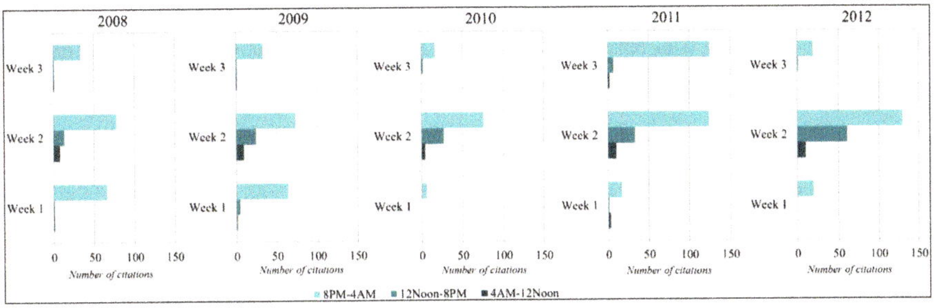

Fig. 10.6 Number of citations for each time period of each week for each year

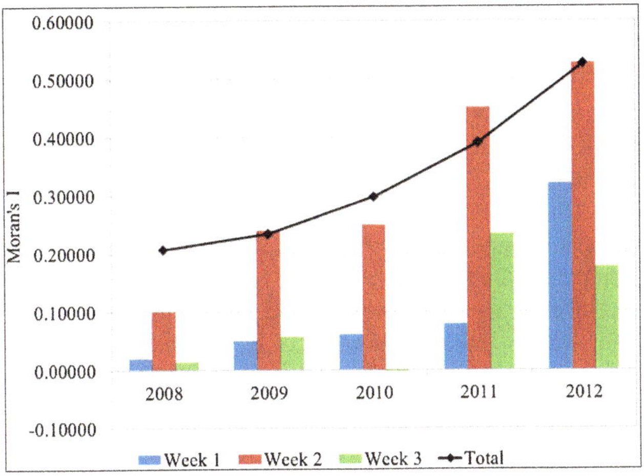

Fig. 10.7 Moran's *I* for years 2008–2012

that the maximum numbers of citations were recorded between 8:00 PM and 4:00 AM during week 2. The trend remained the same for all 5 years, though the number of citations almost doubled from 2008 to 2012. Similar trends were found for weeks 1 and 3, and the fewest citations were recorded during those 2 weeks between 4:00 AM and12:00 noon (Fig. 10.6).

Moran's *I* values were calculated to measure the spatial autocorrelation of incidents for every week of every year. The results (Fig. 10.7) suggested that over the span of 5 years, the incidents became more significantly clustered with strong positive autocorrelation. Comparison among 3 weeks, for every year, suggested that week 2 had higher values of Moran's *I* than week 1 and 3. So it can be concluded that during regular non-festival weeks, the citation incidents were not clustered in the neighborhoods surrounded by the three colleges.

Thus multi-temporal grid thematic mapping of the citation record showed that there was high variability in the spatial distribution of citation incidents among the 3 weeks of study. The number of citations recorded was highest during the Oktoberfest week compared to the weeks before and after. Among the days of the weeks, Saturdays had the highest number of citations, and finally among hours of the day, the highest number of citations was recorded between 8:00 PM and 4:00 AM. Overall the total number of liquor law violation citations almost doubled from 2008 to 2012.

10.5.2 Spatiotemporal Clustering

The gridded thematic hotspot analysis helped us to understand the spatiotemporal variation of the liquor law violation citations during the 3 weeks of the study. But the hotspot analysis failed to provide any information regarding space-time correlation of the citation incidents. A spatiotemporal clustering of the citations will be able to inform about the correlation between distance and time interval for pairs of incidents (Mantel 1967). In recent years, a number of spatiotemporal techniques and space-time statistical tests have been developed within the realm of GIS (Kulldorff et al. 1998, 2005; Kulldorff 2001; Chang et al. 2008; Mountrakis and Gunson 2009; Khan et al. 2009; Pei et al. 2010; Mennis 2010; Nakaya and Yano 2010; Eckley and Curtin 2013).

In the domain of crime mapping and analysis, the Mantel index has been applied to understand spatiotemporal clustering of different types of crimes (Johnson and Bowers 2004a, b; Bowers and Johnson 2005; Malizia 2012). The Mantel index was originally developed to measure levels of disease transmission, based on the correlation between two dissimilarity matrices (e.g., distance and time), and is still a popular measure used in epidemiology (Mantel 1967; Mantel and Bailar 1970; Johnson and Bowers 2004a, b; Eckley and Curtin 2013). The Mantel index (*r*) is based on the cross product of two interval variables (such as distance and time) and summarizes comparisons between pairs of points (Levine 2010). The Mantel index can be written as

$$r = \sum_{i=1}^{N} \sum_{j=1}^{N} Z_x * Z_y / (-1) \tag{10.1}$$

where $Z_x = \dfrac{X_{ij} - \bar{X}}{S_x}$; $Z_y = \dfrac{Y_{ij} - \bar{Y}}{S_y}$;

X_{ij} and Y_{ij} are the two variables distance (in meters) and time (in hours) for comparing two observations *i* and *j*, and Z_x and Z_y are the normalized variables.

For this study the citation data was used to test the existence of spatiotemporal clustering of underage drinkers. Spatiotemporal clustering will help us to understand whether the underage drinking pattern was clustered in particular places in the city and time of the day or dispersed. Simulated confidence intervals were used for significance testing to avoid error due to high interdependency among the variables (Levine 2010). Since it's a two-tailed test, a lower threshold of 2.5 % and upper threshold of 97.5 % were adopted, so if the observed Mantel index is smaller than the lower threshold or larger than the upper threshold, then it can be concluded that the underage drinking patterns were clustered in space and time (Levine 2010).

Table 10.2 Mantel index to test spatiotemporal clustering of citation locations from 2008 to 2012

Year	*N*	No. of comparisons	*r*	Simulation 2.5 %	Simulation 97.5 %	Approx. *p*-level
2008	200	19,900	0.02698	−0.03142	0.03239	n.s
2009	213	22,578	0.00049	−0.02914	0.03166	n.s
2010	132	8,646	0.10912	−0.03885	0.04492	n.s
2011	323	52,003	0.00676	−0.02262	0.02539	n.s
2012	234	27,261	0.15091	−0.02812	0.03070	0.021

N number of points, *n.s* not significant, *r* observed Mantel index

For this study, the Mantel index was computed for all the years, and the results (Table 10.2) show that for years 2008, 2009, and 2011, the observed Mantel indices lie between their respective upper and lower threshold values, which leads to the conclusion that the citations were not spatiotemporally clustered for those years. But in 2010 and 2012, the observed Mantel indices were above their respective upper threshold values, from which we conclude that only for those years the spatiotemporal clustering of citations was most likely greater than what would be expected on the basis of a chance distribution.

Though the Mantel index is a useful tool for space-time analysis, it has certain limitations such as that extreme values of either space or time can distort the relationship and that smaller sample sizes or narrow time windows can make the index unstable (Levine 2010). For this study the data did not have extreme values for space or time and the data size was fairly large, so it can be claimed that the Mantel index results were unbiased and stable.

10.5.3 Multi-criteria Evaluation for Mapping Probability Surface of Citation Location

The analysis of the grid thematic mapping suggested that the majority of the incidents were clustered in and around the neighborhoods of the three colleges. So in order to restrict underage drinking, various alcohol-related accidents, and nuisance by drunken crowds, law enforcement officials need to focus deployment of intervention measures in those neighborhoods. But at present there is no authoritative model or map that exactly delineates the areas where there is a high probability of such occurrences. No doubt that such map will be highly valuable for implementation of effective preventive policies.

The knowledge gained from the thematic maps helped us to narrow down the sociodemographic aspects of the neighborhoods where high clusters of the liquor law violation were cited within the last 5 years. Among others, four major factors were identified which, if mapped, will help to define areas where underage drinking are more likely to occur. These four factors were distance from the three colleges, distance from the pubs and bars, distance from student housings (nonuniversity owned), and the population aged between 15 and 24 years in each block group.

Table 10.3 Classification of parameters

Factors	Distance in miles	Probability zone
Distance from colleges	0.00–0.50	High
	0.50–1.00	Medium
	1.00–1.50	Low
	Beyond 1.50	Very low
Distance from pubs and bars	0.00–0.25	High
	0.25–0.50	Medium
	1–1.50	Low
	Beyond 1.50	Very low
Distance from student housing (nonuniversity owned)	0.00–0.25	High
	0.25–0.50	Medium
	1–1.50	Low
	Beyond 1.50	Very low
Population aged between 15 and 24 years (in each block group)	1,000–7,000	High
	400–1,000	Medium
	100–400	Low
	0–100	Very low

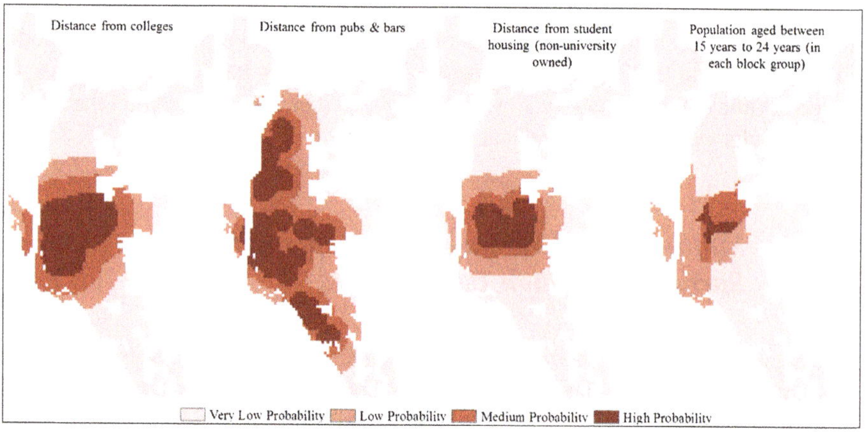

Fig. 10.8 Spatial zonation of four parameters for multi-criteria evaluation

Assuming that the past is the best predictor of the future, these four factors can help to predict locations with high likelihood of underage drinking occurrences during future Oktoberfest celebrations.

All four factors were spatially divided into four zones, and each zone was named as follows: high probability zone, medium probability zone, low probability zone, and very low probability zone, according to the given thresholds. Table 10.3 provides detailed zonation information of each parameter.

Figure 10.8 classifies the four factors into four probability zones. Multi-criteria evaluation (MCE) was used to create a map of potential areas of liquor law violation using the classified parameter maps. The resultant composite map was named a

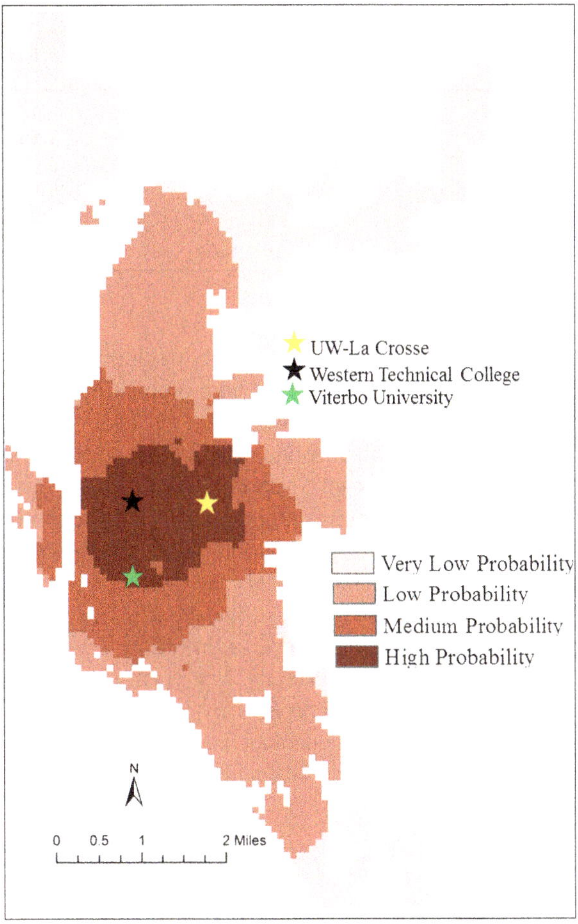

Fig. 10.9 Probability surface map

"probability surface map." Multi-criteria evaluation is a type of analysis that is easily implemented in a GIS environment using map algebra operations (Malczewski 2004). Different variants of the MCE method have been developed over the years and have been applied in GIS environment, especially for land or suitability analysis. Details about various forms of MCE have been well documented by Malczewski in his review papers (2004, 2006). In this study, weighted linear combination was used (Malczewski 2004; Jiang and Eastman 2000) where weights were assigned to the parameters according to their relative importance. For this study, equal weights were assigned to all the parameters as they were assumed to have equal importance in delineating a successful probability surface for location of liquor law violation. The resultant composite map was also recategorized into zones of high, medium, low, and very low probability of incident occurrence. The final map (Fig. 10.9) shows that the high probability zone coincides with the downtown area, mixed land use areas, and residential areas around the three college campuses. The medium

Table 10.4 Percentage of citation data points within each of the probability zones

Year	High probability (%)	Medium probability (%)	Low probability (%)	Very low probability (%)
All 5 years	89.02	7.17	3.81	0.00
2008	85.50	7.00	7.50	0.00
2009	84.98	8.45	6.57	0.00
2010	93.18	3.03	3.79	0.00
2011	85.76	11.76	2.48	0.00
2012	97.86	2.14	0.00	0.00

probability zone coincided mostly with mixed land use areas with bars and residential areas encircling the high probability zone. The low and very low probability zones cover the rest of the city.

In order to establish confidence in the final MCE map, the locations of the actual citation incidents for all 5 years were compared with the predicted probability surface map. Since there was no existing scale to determine the strength of the analysis, we assumed that if more than 80 % of the citation incidents coincided with high probability zone, then the final map would be acceptable for practical application. The result (Table 10.4) shows that 89.02 % of citations from all 5 years coincide with the high probability zone, followed by 7.17 % and 3.81 % in medium and low probability zones, respectively. For each year, the percentage of actual citations located within the high probability zone varied from 84.98 % (2009) to 97.86 % (2012). Thus, it can be concluded that the factors used to create the MCE map were successful to determine the potential areas of liquor law violation citation locations in La Crosse.

10.6 Discussion

Oktoberfest is not only a popular festival, but for a small town like La Crosse, it is also an important contributor to the local economy. The present study used a multi-temporal grid thematic mapping approach to understand the effect of Oktoberfest on underage drinking in the city. Analysis of the last 5 years of liquor law violation citations helped to map the magnitude of change in number of citations during Oktoberfest compared to regular non-festival times of the year. The results of mapping also showed the number of violations has doubled in last 5 years, but it was not known whether the increase was due to an increased number of underage drinkers, or large number of visitors or increased patrolling, or a combination of all these three factors. Detailed temporal analysis showed that maximum numbers of citations were recorded on Saturday of Oktoberfest week and between the 8:00 PM and 4:00 AM time window.

The lack of residential information of the violators restricted the study from determining the proportion of citations that can be attributed to the visitors. The Mantel index was used to determine the correlation between distance and time relationship between pair of citation incidents. Results showed that for years 2008, 2009, and 2011, there was no spatiotemporal clustering of the citation incidents, but for years 2010 and 2012, the citation incidents were spatiotemporally clustered. But the index was not able to identify the locations and time window of the clustering.

The MCE map delineated potential areas of liquor law violation incidents, and the validation of the map showed that in the last 5 years on average 89 % of the citation incidents overlap with the high probability zone delineated by the probability surface map. Though the MCE results were successful for this study, it had certain limitations. Firstly, only four factors were used with equal weights to determine the probability surface, so the analysis neither identified which factors had more influence than the others nor captured the effect of any other factor which may have influenced the spatiotemporal pattern of citation incidents. Secondly, the results may be biased due to the distance threshold assumption for each of the factors and the size of the grids used.

The data used in this study was not the best representation of underage drinking in La Crosse; nevertheless, the present approach not only provided useful information to determine potential locations of future incidents but also showed the need for further research on the topic that will help to better develop intervention strategies, with more data related to alcohol-related crimes and accidents and the social dynamics of the event. The mapping effort in this study will be helpful to the local police and community volunteers to deploy watch groups to restrict underage drinking and liquor-related crimes during Oktoberfest. The MCE application used in this study is also applicable in any other college towns having social events like Oktoberfest, where alcohol consumption is an important part of celebration. The multi-temporal analysis of citation incidents showed that existing preventive actions and the yearly increase in the number of police patrolling in La Crosse are not enough to curb the negative social impact of this popular festival. Thus, it is the responsibility of the city administrators to determine a constructive course of actions to better control the situation.

Acknowledgments We would like to thank the University of Wisconsin-La Crosse Police Department and La Crosse Police Department for providing us data and insights about policing during Oktoberfest. We would also like to extend our gratitude to Professors Cynthia Berlin (Dept. of Geography and Earth Science) and Kim Vogt (Dept. of Sociology) of University of Wisconsin-La Crosse for their valuable suggestions and guidance.

References

Alaniz ML, Cartmill RS, Parker RN (1998) Immigrants and violence: the importance of neighborhood context. Hisp J Behav Sci 20:155–176

Andresen MA, Tong W (2012) The impact of the 2010 Winter Olympic games on crime in Vancouver. Can J Criminol Crim Justice 54(3):333–361

Bennett T (1995) Identifying, explaining and targeting burglary “hot spots”. Eur J Crim Policy Res 3(3):113–123

Block RL, Block CR (1995) Space, place and crime: hot spot areas and hot places of liquor-related crime. Crime Prev Stud 4

Boba R (2001) Introductory guide to crime analysis and mapping. Office of Community Oriented Policing Services, Washington, DC

Bowers KJ, Johnson SD (2005) Domestic burglary repeats and space-time clusters the dimensions of risk. Eur J Criminol 2(1):67–92

Britt H, Carlin B, Toomey T et al (2005) Neighborhood level spatial analysis of the relationship between alcohol outlet density and criminal violence. Environ Ecol Stat 12:411–426

Centers for Disease Control and Prevention (CDC) (2012) Atlanta, GA: fact sheets: http://www.cdc.gov/alcohol/fact-sheets/underage-drinking.htm. Accessed Sept 22 2013

Chainey S, Tompson L, Uhlig S (2008) The utility of hotspot mapping for predicting spatial patterns of crime. Secur J 21:4–28

Chang W, Zeng D, Chen H (2008) A stack-based prospective spatio-temporal data analysis approach. Decis Support Syst 45:697–713

Eaton DK, Kann L, Kinchen SA et al (2012) Youth risk behavior surveillance—United States, CDC. CDC Morb Mortal Surveill Summ 61(SS-04):1–162

Eck JE, Chainey S, Cameron JG, et al (2005) Mapping crime: understanding hot spots. National Institute of Justice Special Report, Office of Justice Programs, U.S. Department of Justice

Eckley DC, Curtin KM (2013) Evaluating the spatiotemporal clustering of traffic incidents. Comput Environ Urban Syst 37:70–81

Felson M, Poulsen E (2003) Simple indicators of crime by time of day. Int J Forecast 19:595–601

Goodchild MF, Anselin L, Appelbaum RP et al (2000) Towards spatially integrated social science. Int Reg Sci Rev 23(2):139–159

Gorman DM, Speer PW, Gruenewald P et al (2001) Spatial dynamics of alcohol availability, neighbourhood structure and violent crime. J Stud Alcohol 62(5):628–36

Groff E, Weisburd D, Morris NA (2009) Where the action is at places: examining spatio-temporal patterns of juvenile crime at places using trajectory analysis and GIS. In: Weisbard D, Bernasco W, Bruinsma GJN (eds) Putting crime in its place. Springer, New York, pp 61–86

Gruenewald P (2007) The spatial ecology of alcohol problems: niche theory and assortative drinking. Addiction 102:870–878

Gruenewald P, Millar PJ, Treno AJ et al (1996) The geography of availability and driving after drinking. Addict Sci 91:967–983

Hoover L, Zhang Y, Zhao J (2010) Geographic Information Systems and their effects on police efficacy: Campbell Systematic Review. Accessed from http://campbellcollaboration.org/lib/project/146/

Huckle T, Huakau J, Sweetsur P et al (2008) Density of alcohol outlets and teenage drinking: living in an alcogenic environment is associated with higher consumption in a metropolitan setting. Addiction 103:1614–1621

Imperial College London (ICL) (2010) Alcohol most harmful drug based on multicriteria analysis, November 1. http://www3.imperial.ac.uk/newsandeventspggrp/imperialcollege/newssummary/news_1-11-2010-15-43-18. Accessed 25 Dec 2012

Jiang H, Eastman JR (2000) Application of fuzzy measures in multi-criteria evaluation in GIS. Int J Geogr Inf Sci 14(2):173–184

Johnson SD, Bowers KJ (2004a) The burglary as clue to the future the beginnings of prospective hot-spotting. Eur J Criminol 1(2):237–255

Johnson SD, Bowers KJ (2004b) The stability of space-time clusters of Burglary. Br J Criminol 44(1):55–65

Khan G, Santiago-Chaparro K, Qin X et al (2009) Application and integration of lattice data analysis, network K-functions, and geographic information system software to study ice-related crashes. Transp Res Rec 2136:67–76

Kulldorff M (2001) Prospective time periodic geographical disease surveillance using a scan statistic RID H-4282-2011. J R Stat Soc A Stat Soc 164:61–72

Kulldorff M, Athas W, Feuer E et al (1998) Evaluating cluster alarms: a space-time scan statistic and brain cancer in Los Alamos, New Mexico RID H-4282-2011. Am J Public Health 88(9):1377–1380

Kulldorff M, Heffernan R, Hartman J et al (2005) A space-time permutation scan statistic for disease outbreak detection RID H-4282–2011. PLoS Med 2(3):216–224

La Crosse Tribune (2011) Oktoberfest roundup: Mob will mean more police for 2012, reported by Anne Jungen, September 27. http://lacrossetribune.com/news/local/oktoberfest-roundup-mob-will-mean-more-police-for/article_c5b3c2a4-e89f-11e0-ac71-001cc4c03286.html. Accessed 31 Dec 2012

Leitner M, Guo D (2013) The impact of hurricanes on crime using a data mining and visual analytics approach. In Proceedings of 26th International Cartographic Conference, Dresden, Germany, August 25–30

Leitner M, Helbich M (2011) The impact of Hurricanes on crime: a spatio-temporal analysis in the city of Houston, TX. Cartogr Geogr Inf Sci 38(2):214–222

Leitner M, Barnett M, Kent J et al (2011) The impact of Hurricane Katrina on reported crimes in Louisiana: a spatial and temporal analysis. Prof Geogr 63(2):244–261

Levine N (2010) CrimeStat: a spatial statistics program for the analysis of crime incident locations (v 3.3). Ned Levine & Associates, Houston; National Institute of Justice, Washington, DC

Malczewski J (2004) GIS-based land-use suitability analysis: a critical overview. Prog Plann 62:3–65

Malczewski J (2006) GIS-based multicriteria decision analysis: a survey of the literature. Int J Geogr Inf Sci 20(7):703–726

Malizia N (2012) The effect of data inaccuracy on tests of space-time interaction (Working paper # 10). GeoDa Center for Geospatial Analysis and Computation, Arizona State University, Phoenix

Mantel N (1967) The detection of disease clustering and a generalized regression approach. Cancer Res 27(2 Part 1):209–220

Mantel N, Bailar JC (1970) A class of permutational and multinomial tests arising in epidemiological research. Biometrics 26(4):687–700

Mennis J (2010) Multidimensional map algebra: design and implementation of a spatio-temporal GIS processing language. Trans GIS 14(1):1–21

Mountrakis G, Gunson K (2009) Multi-scale spatiotemporal analyses of moose–vehicle collisions: a case study in northern Vermont. Int J Geogr Inf Sci 23:1389–1412

Nakaya T, Yano K (2010) Visualising crime clusters in a space-time cube: an exploratory data-analysis approach using space-time kernel density estimation and scan statistics. Trans GIS 14(3):223–239

Norström T (2000) Outlet density and criminal violence in Norway 1960–95. J Stud Alcohol 61

Pei T, Zhou C, Zhu A-X et al (2010) Windowed nearest neighbor method for mining spatio-temporal clusters in the presence of noise. Int J Geogr Inf Sci 24:925–948

Ratcliffe JH (2004a) The hotspot matrix: a framework for the spatio-temporal targeting of crime reduction. Police Pract Res 5(1):05–23

Ratcliffe JH (2004b) Geocoding crime and a first estimate of a minimum acceptable hit rate. Int J Geogr Inf Sci 18(1):61–72

Scribner R, Cohen D, Kaplan S et al (1999) Alcohol availability and homicide in New Orleans: conceptual considerations for small area analysis of the effect of alcohol outlet density. Alcohol Drugs 60(3)

Snowden A (2012) Alcohol outlets and assaults in a College Town: accounting for social organization and land use. Paper presented at the ASC Annual Meeting, Washington Hilton, Washington, DC (http://www.allacademic.com/meta/p517139_index.html)

Spelman W (1995) Criminal careers of public places. In: Eck JE, Weisburd D (eds) Crime and place: crime prevention studies. Willow Tree, Monsey, p 4

StarTribune (2012) Oktoberfest in La Crosse: a party town tries to sober up, reported by Larry Oaks, October 1. http://www.startribune.com/local/171948771.html. Accessed 31 Dec 2012

Taylor R (1999) Crime, grime, fear, and decline: a longitudinal look, Research in brief. National Institute of Justice, Washington, DC

Townsley M (2008) Visualizing space time patterns in crime: the hotspot plot. Crime Patterns Anal 1(1):61–74

Townsley M, Homel R, Chaseling J (2000) Repeat burglary victimisation: spatial and temporal patterns. Austr N Z J Criminol 33(1):37–63

Treno A, Grube J, Martin S (2003) Alcohol availability as a predictor of youth drinking and driving: a hierarchical analysis of survey and archival data. Alcohol Clin Exp Res 27:835–840

U.S. Census Bureau (2010) Census. Profile of general population and housing characteristics. http://factfinder2.census.gov/faces/tableservices/jsf/pages/productview.xhtml?pid=DEC_10_DP_DPDP1. Accessed 31 Dec 2012

Watts RK, Rabow J (1983) Alcohol availability and alcohol related consumption in 213 Californian cities. Alcohol Clin Exp Res 7:47–58

Weisburd D, Bushway S, Lum C et al (2004) Trajectories of crime at place: a longitudinal study of street segments in the city of Seattle. Criminology 42(2):283–321

Weitzman E, Folkman A, Folkman K et al (2003) The relationship of outlet density to heavy and frequent drinking and drinking-related problems among college students at eight universities. Health Place 9:1–6

Zhu L, Gorman DM, Horel S (2004) Alcohol outlet density and violence: a geospatial analysis. Alcohol Alcohol 39(4):369–375

Chapter 11
A Web-Based GIS for Crime Mapping and Decision Support

Guiyun Zhou, Jiayuan Lin, and Xiujun Ma

Abstract Timely mapping of crime locations and accurate detection of spatial concentrations of crime help to identify where crimes concentrate in space and time and thus provide important information for crime reduction efforts of law enforcement. The main objective of this chapter is to design and implement a Web-based GIS for crime mapping and decision support. The prototype system allows users to detect and view crime hotspots in a Web environment in the form of thematic layers overlaid on background data. Four hotspot mapping techniques, i.e., choropleth mapping, grid mapping, spatial ellipse mapping, and kernel density mapping, are implemented. The system can be used for mapping crime hotspots, predicting the locations of future crime, and optimizing crime reduction efforts. It is a rich Internet application and is much more efficient than script-based clients in hotspot detection, map manipulation, and rendering. It is entirely based on open-source software, making it affordable for many small- and medium-sized police departments in developing countries. Results from the prototype development demonstrate that for a Web-based crime mapping and decision support system, rich Internet application technology in combination with open-source software is an effective solution in terms of both system performance and financial cost.

G. Zhou (✉)
School of Resources and Environment, University of Electronic Science and Technology of China, Chengdu 611731, People's Republic of China
e-mail: zhouguiyun@uestc.edu.cn

J. Lin
Institute of Mountain Hazards and Environment, Chinese Academy of Sciences, Chengdu 610041, People's Republic of China
e-mail: linjiayuan@gmail.com

X. Ma
Key Laboratory of Machine Perception of Ministry of Education, School of Electronics Engineering and Computer Science, Beijing University, Beijing 100871, People's Republic of China
e-mail: maxj@cis.pku.edu.cn

G.A. Elmes et al. (eds.), *Forensic GIS: The Role of Geospatial Technologies for Investigating Crime and Providing Evidence*, Geotechnologies and the Environment 11,
DOI 10.1007/978-94-017-8757-4_11,

Keywords Crime mapping • Web-based GIS • Flash • HDML5 • Rich Internet applications (RIAs)

11.1 Introduction

Crime has an inherent geographic component, and the locations of crime are not randomly distributed (Phillips and Lee 2011). Many police departments map crime data as a part of their routine strategic and tactical activities (Levine 2006). Optimizing the spatial and temporal allocation of law enforcement resources can contribute significantly to crime reduction and prevention. Timely mapping of crime locations and accurate detection of spatial concentrations or clusters of crime help to identify where crime tends to concentrate in space and time and thus provide important information for law enforcement resource allocation. Crime mapping can be used for responding to incidents, generating intelligence products, reassurance policing, performance monitoring, and information sharing and partnership working (Chainey and Ratcliffe 2005; Home Office 2005). Much of a crime mapping effort is devoted to the identification of crime hotspots. The areas of concentrations or clusters of crime are referred to as hotspots. Hotspots are scale-dependent (Wu and Grubesic 2010). Various hotspot mapping techniques such as choropleth mapping and kernel density mapping are available to provide different insights into the spatial patterns of crime occurrences. Hotspots are of particular interests to law enforcement agencies since they can be used to help predict the locations of future crime and to optimize crime reduction efforts (Chainey et al. 2008).

Various software products are available for crime mapping and analysis, and a selected list of them is given in Table 11.1. Among them, ArcGIS for Desktop Basic, previously called ArcView, is a generic GIS software whereas other products are mainly developed for dedicated purposes such as spatial modeling and crime mapping. Many commonly used crime mapping tools such as CrimeStat, ArcView, and GeoDa (Eck et al. 2005) are desktop applications. A Web-based system offers many advantages over desktop applications. The availability of a Web-based crime mapping and decision support system helps law enforcement agencies to conduct timely crime mapping and hotspot detection and to facilitate decision support in a crime crackdown. It delivers mapping tools to a wide range of police staff more easily and faster. It ensures that the most up-to-date crime data are used for mapping and analysis by accessing a centralized and regularly updated crime database. A Web-based system can easily be integrated with other police systems such as command and control centers for responding to incidents much more efficiently. Some Web-based crime mapping systems are available to the public. The website operated by the British government utilizes Google Maps and offers public access to monthly updated street level crime data in England and Wales (Police UK 2013). All crime incidents reported to the police are expected to be shown on the website. The Omega Group has a Web-based product which can be used through a police department's intranet. A version of the product is up and running on the Internet, helping law

Table 11.1 Selected applications used for crime mapping and analysis

Software	URL	Application type	Based on commercial products?	Functionalities
CrimeStat	http://www.icpsr.umich.edu/CrimeStat/	Desktop	No	Spatial distribution, hotspot analysis, interpolation statistics
GeoDa	http://geodacenter.asu.edu/	Desktop	No	Statistical graphics, spatial autocorrelation, spatial regression
ArcGIS for Desktop Basic[a]	http://www.esri.com/software/arcgis/arcgis-for-desktop	Desktop	Yes	Choropleth mapping, density mapping
Hotspot Detective	http://jratcliffe.net/hsd/	Desktop plug-in	Yes, MapInfo	Hotspot surface mapping, timeline graphs, repeat location finder
Police, UK	http://www.police.uk	Web	Yes, Google Maps	Crime location mapping
CrimeMapping	http://www.crimemapping.com/	Web	Yes, ArcGIS Server	Crime location mapping, trend report
CrimeReports	https://www.crimereports.com/	Web	Yes, Google Maps	Crime location mapping
Oakland Crimespotting	http://oakland.crimespotting.org/	Web	Unknown	Crime location mapping
WebCAT	http://ieeexplore.ieee.org/xpls/abs_all.jsp?arnumber=1242424	Web	No	Kernel density mapping, choropleth mapping

[a]ArcGIS for Desktop Basic is the new name given by ESRI. Previously it was called ArcView

enforcement agencies throughout North America to provide the public with valuable information about recent crime activities in their neighborhoods (Crime Mapping 2013). The application is based on ESRI mapping engine and maps crime locations based on their types. CrimeReports runs a similar crime mapping service as the Omega Group on the Internet, with crime data coverage extended outside North America (CrimeReports 2013). Its mapping service is based on Google Maps. Oakland Crimespotting is an interactive map of crimes in Oakland and a tool for mapping crime locations (Oakland Crimespotting 2013). Its client side is based on Adobe Flash, a rich Internet application technology that will be discussed in more detail in the latter part of the chapter. WebCAT is a Web-based crime analysis toolkit available to all Virginia crime agencies (Hawkins et al. 2003; Calhoun et al. 2008). It uses open-source MapServer as its GIS Web server (MapServer 2013). In terms of crime mapping, it implements choropleth mapping and kernel density mapping functionalities on the server side.

Law enforcement agencies generally require a Web-based mapping application on their intranets, which makes public mapping services such as Google Maps unavailable. Compared with the development and maintaining costs of crime mapping applications, commercial Web-based mapping products such as ArcGIS Server are generally expensive and add to total financial costs significantly, which makes their wide use impractical for many small- and medium-sized police departments, particularly in developing countries. In addition, most current Web-based GIS applications for crime mapping primarily use JavaScript-based technology for rendering point, line, or polygon dynamic data on the client side, which is efficient only when the amount of dynamic data is limited, as is often the case with public mapping services on the Internet. However, for crime hotspot mapping the amount of dynamic data is relatively large, which may comprise hundreds or thousands of polygons. In order to effectively process these dynamic data on the client side, the client side will need to use rich Internet applications (RIAs). An RIA is a Web-based application that has the typical characteristics of desktop applications. Platforms that support RIAs include Adobe Flash, Microsoft SilverLight, and JavaFX. RIAs can shift a large part of processing burdens from the server to the client, allowing the server to provide services to more clients. RIAs provide efficient client-side rendering mechanisms that can present responsive and graphically rich user interfaces, lowering the amount of traffic between the client and the server, as compared to the HTML-based client–server architecture (O'Rourke 2004; Li et al. 2011). For a Web-based GIS for crime mapping and decision support, RIAs can implement computation-intensive algorithms on the client side and present the mapping output in an efficient and responsive way, allowing the Web system to respond to more user requests very quickly.

This chapter aims to present the design and implementation of a Web-based GIS for efficient crime mapping and decision support, which is based on open-source software and rich Internet client. The remainder of the chapter is organized as follows. The next section reviews some commonly used hotspot mapping techniques. Section 11.3 gives an overview of Web-based GIS techniques and compares the performances of various techniques for client side rendering of geospatial data. The architecture and development of a prototype system of the Web-based crime mapping and decision support system is introduced in Sect. 11.4. Section 11.5 presents some example application results of the system. Section 11.6 concludes the chapter with a discussion on future improvements.

11.2 Hotspot Mapping Methods

Many researchers have reviewed crime hotspot mapping techniques (Eck et al. 2005; Chainey et al. 2008). In this section, we give a review of four hotspot mapping techniques commonly used by law enforcement agencies and researchers. Figure 11.1 shows some commonly used methods for crime hotspot mapping.

11.2.1 Choropleth Mapping

Choropleth mapping techniques are widely used to represent spatial distributions of geographic phenomena, and it is no exception with crime data (Harries 1999; Home Office 2005). To create a choropleth map, the study area is divided into various geographic areal units, and each areal unit is shaded based on the number of crimes committed within it. The commonly used geographic areal units for crime mapping are administrative regions and police districts such as census tracts and police precincts. Areal units are classified based on its crime count using a classification scheme such as equal interval and natural breaks (Dent et al. 2008) and shaded using a coloring scheme. Choropleth maps help to identify areal units of high crime concentration and can be used to detect areal units of abnormal changes in crime occurrences by mapping the difference in crime counts of comparable time periods. Figure 11.2 shows the flowchart to generate a choropleth crime map, and an example of choropleth maps for crime mapping is shown in Fig. 11.1b.

11.2.2 Grid Mapping

One salient problem with choropleth maps for crime mapping is the varying sizes and shapes of the geographic areal units. In order to overcome this problem, equally sized grids can be generated across the study area and be shaded depending on the number of crimes falling within each of them (Bowers et al. 2001; LeBeau 2001). The uniform size of grids facilitates the identification and comparison of hotspots. The main issue with grid mapping is the determination of grid sizes. In addition, grid mapping output is often not aesthetically appealing. An example of grid thematic maps is shown in Fig. 11.1c.

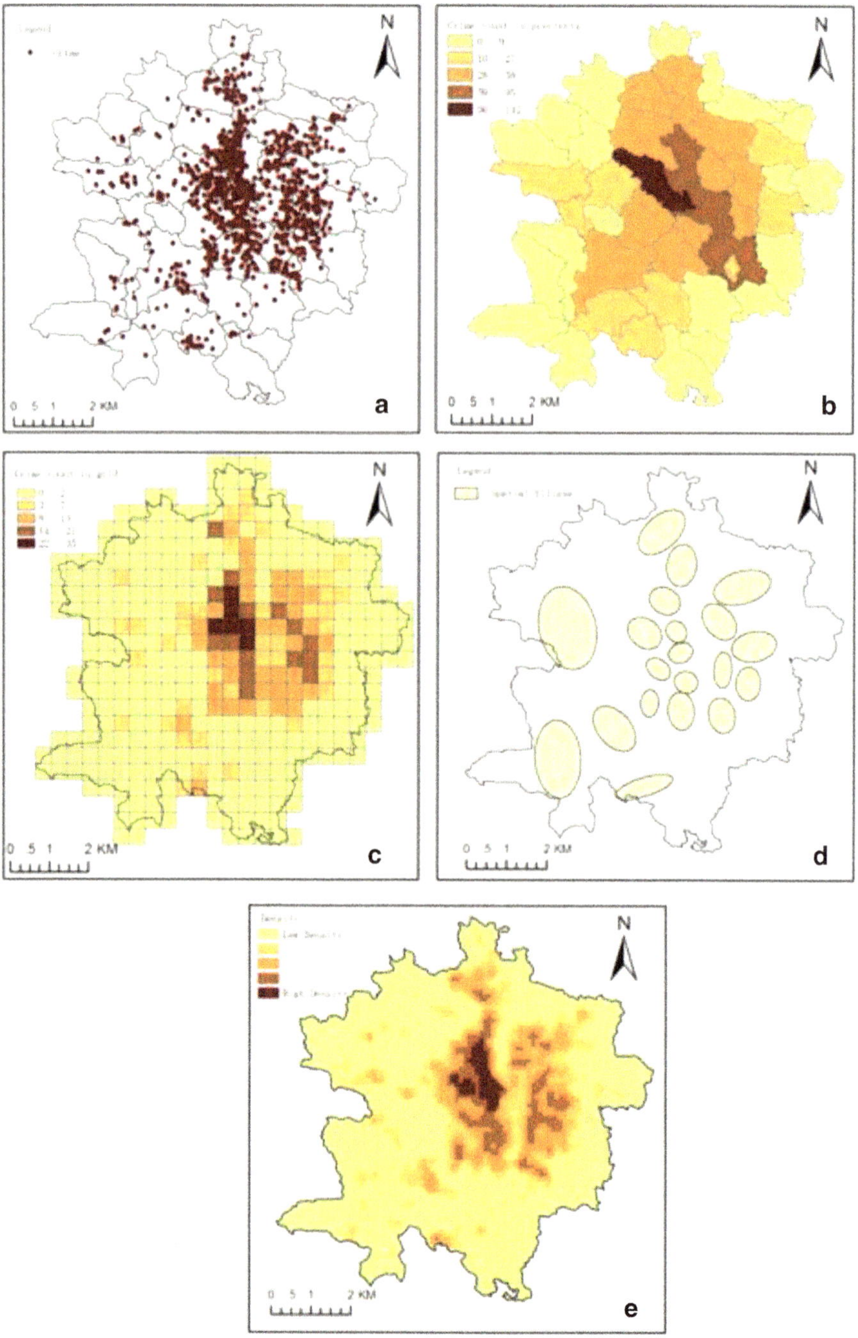

Fig. 11.1 Common crime mapping techniques: (**a**) discrete point mapping, (**b**) choropleth mapping, (**c**) grid thematic mapping, (**d**) spatial ellipse mapping, and (**e**) kernel density mapping

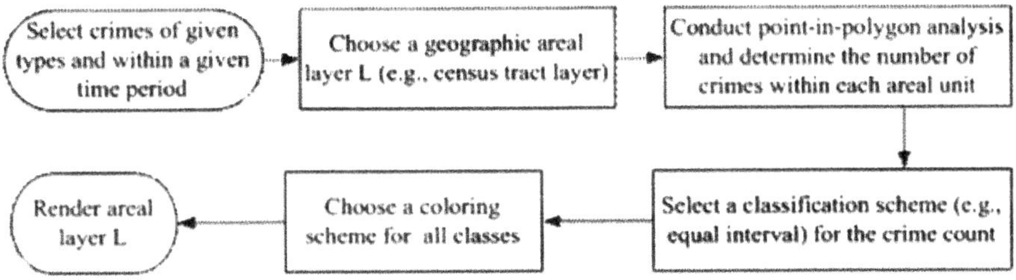

Fig. 11.2 Flowchart of choropleth crime mapping

11.2.3 Spatial Ellipses

In this hotspot mapping method, points are grouped into clusters based on their proximity and a standard deviational ellipse is fitted to each cluster to indicate the dispersion of crime points in the cluster through the ellipse's size and alignment (Cromley 1992). Figure 11.1d shows an example of spatial ellipse maps.

The K-means clustering algorithm is one of the most commonly used clustering methods for generating spatial ellipse maps (Levine 2010). Users specify the number of clusters to be grouped and a separation value between initial seed locations. The algorithm makes initial estimates of seed locations for a specified number of clusters, assigns each crime to its nearest seed, recalculates a new seed for each cluster, and repeats the procedure. The procedure stops when there are no or few changes to each cluster.

The derivation of spatial ellipses depends only on the spatial positions of crimes in a cluster and does not rely on geographic boundaries, as is the case with choropleth mapping and grid mapping (Martin et al. 1998).

11.2.4 Kernel Density Mapping

Kernel density is probably the most commonly used hotspot mapping technique for visualizing hotspots and has higher prediction abilities than other techniques (Chainey et al. 2008; Levine 2010). Its popularity is partly due to its ability to produce aesthetically appealing maps. The flowchart to create a kernel density surface is shown in Fig. 11.3. To create a kernel density map, a grid is generated to cover the study area, then a kernel function is centered on each crime point, and density values are calculated for each grid cell by aggregating contributions from all crimes. Two commonly used kernel functions are normal distribution function and triangular function.

For kernel functions, the bandwidth and grid dimensions (or grid cell size) are specified by the users. For novice users, guidelines exist for choosing the bandwidth and grid dimensions (Williamson et al. 1999; Eck et al. 2005). For experienced users, changing these parameters allows them to investigate the spatial effects of these parameters on the density patterns.

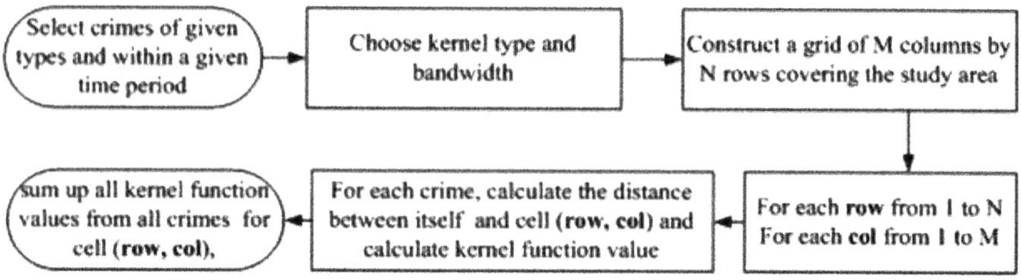

Fig. 11.3 Flowchart to construct a kernel density surface

The resulting density surface represents the point density of crime in the study area and can be rendered using some classification and coloring schemes, as with choropleth mapping and grid mapping. An example of kernel density map is presented in Fig. 11.1e. Examples demonstrating the use of kernel density mapping techniques can be found in the work of many researchers and governmental reports (Chainey and Ratcliffe 2005; Eck et al. 2005).

11.3 Web-Based GIS Techniques

11.3.1 An Overview of Web-Based GIS Techniques

Depending on the requirements of individual systems, different architectures of Web-based GIS have been presented in the literature. Many early Web GIS systems are based on the widely used commercial product ESRI ArcIMS (Su et al. 2000; Chang and Park 2004; Sugumaran et al. 2004; Tsou 2004; Kontoes et al. 2005; Rao et al. 2007), and some are based on open-source platforms such as MapServer and GeoServer (Tuama and Hamre 2007; Kruger et al. 2007). These systems, which are usually implemented in Dynamic HyperText Markup Language (DHTML) and have limited interactive capabilities, deliver dynamically generated maps to the clients. Generating maps dynamically on the server posts a huge load demand, resulting in long response time and limited number of concurrent requests. Using cached tiles or pre-generated tiles has become a common practice to improve system response in various public mapping services such as Google Maps and Microsoft Bing Maps. Intranet applications have also started adopting such techniques (Kulawiak et al. 2010), and open-source script-based libraries such as OpenLayers provide efficient tile rendering mechanisms. Caching or pre-generating tiles is a technique appropriate for background maps and images which do not need frequent updates, and other more dynamic data such as points and lines are rendered on the client using DHTML techniques. DHTML is suitable for rendering limited amount of data, which is usually the case in public mapping services. We have been developing Web GIS applications for crime mapping for nearly a decade for many municipal police departments across China, and the applications have followed similar technological routes as described above.

The above described architectures use thin-client script-based technology, which rely heavily on the server for processing and do not take advantage fully of client-side computing capabilities. To overcome this problem, in early systems, rich clients use technologies such as Java Applet or custom-made browser plug-ins (Al-Sabhan et al. 2003). Custom-made browser plug-ins are rarely used in today's applications because of such issues as compatibility, security, and initial downloading and installation. Java Applet, for various reasons, is seldom used as well. On the other hand, Adobe Flash, previously called Adobe Flex, is compatible with Windows, Macintosh, and Unix browsers; additionally, Flash starts much faster than Java Applet, and Flash can easily add animation, video, and interactivity to Web pages. It has extensive features and consistent appearance in major browsers. Flash comes with a variety of built-in user interface and other components. For Web GIS systems, Flash provides effective vector and raster manipulation and rendering operations. More importantly, since popular Internet video-sharing websites such as YouTube prefer Flash to display video contents, the Flash plug-in is the world's most pervasive software reaching 99 % of Internet-enabled desktops in major western countries (Adobe Flash 2011), which Microsoft SilverLight cannot compete with. Commercial products like ArcGIS Server Flash APIs are commonly used to build Flash-based Web mapping systems (Li et al. 2011; Defne et al. 2012).

11.3.2 Comparison of Client-Side Rendering Mechanisms

A Web-based GIS for crime mapping and decision support requires an efficient client side rendering mechanism to support the presentation of large number of crime points, polygonal features, and other spatial data types. In this section, DHTML-rendering methods and rich-client methods are compared in terms of their performances.

There are three methods in DHTML to render geographic features in a browser. The three methods are scalable vector graphics (SVG), vector markup language (VML), and HTML5. SVG is a standard proposed by the World Wide Web Consortium supported by major browsers, such as Mozilla Firefox, Internet Explorer 9, Google Chrome, Opera, and Safari, to render two-dimensional vector graphics. VML is an XML-based language to render vector graphics and has been supported by Internet Explorer between versions 5 and 9. Before Internet Explorer 9, VML was the only available method to render vector graphics in Internet Explorer. Microsoft has encouraged the use of SVG over VML in Internet Explorer 9. Despite that, VML is still widely used in many websites because of the large number of legacy users of Internet Explorer 6 and Internet Explorer 8. HTML5 is a new version of HTML that supports rendering graphics using the Canvas tag. HTML5 graphics rendering capability is supported in all mobile browsers and latest versions of desktop browsers.

Flash is the most popular rich-client technology compared with SilverLight and JavaFX and is chosen to compare with DHTML. To compare the performances of DHTML and RIA for mapping geospatial features, two commonly used mapping

Table 11.2 Test vector datasets

	Dataset name	No. of polygons	No. of vertices	Sources
DS1	Countries – 100 m	177	10,589	Natural Earth http://www.naturalearthdata.com/downloads/110m-cultural-vectors/110 m-admin-0-countries/
DS2	Countries – 50 m	242	99,606	Natural Earth http://www.naturalearthdata.com/downloads/50m-cultural-vectors/50m-admin-0-countries-2/
DS3	Countries – 10 m	253	536,399	Natural Earth http://www.naturalearthdata.com/downloads/10m-cultural-vectors/10m-admin-0-countries/

Table 11.3 Time (in milliseconds) used from zooming 20 times

		Browsers		
Dataset	Rendering methods	IE9	Chrome	Firefox
DS1	SVG	1,345	858	1,622
	HTML5	1,909	1,277	1,598
	VML	9,156	N/A	N/A
	Flash	71	72	74
DS2	SVG	10,331	5,342	12,441
	HTML5	13,606	8,813	10,318
	VML	120,588	N/A	N/A
	Flash	371	379	372
DS3	SVG	55,730	21,734	60,480
	HTML5	115,048	58,959	67,665
	VML	No response[a]	N/A	N/A
	Flash	1,831	1,890	1,887

[a]More than 10 min for the operation to be completed

operations, i.e., zooming and panning, are compared. Two open-source mapping frameworks, OpenLayers and OpenScales, are used to conduct the comparison. OpenLayers is an open-source JavaScript-based client-side mapping library that supports SVG, VML, and HTML5. OpenScales is an open-source Flash-based mapping library. Three browsers, Internet Explorer 9, Google Chrome, and Mozilla Firefox, are used for the experiments. The experiments are conducted on a Windows 7 system, with 4GB of RAM and an I3 3.30 GHz processor. In all experiments, the entire dataset is loaded before the operations are conducted. Table 11.2 lists three datasets of various sizes used for the comparison. Tables 11.3 and 11.4 list the time used for zooming and panning 20 times, respectively.

For panning operations, all rendering methods take a negligible amount of time. This is due to the fact that the entire dataset is preloaded and the panning operations involve the relocation of the previously generated images, which can be done in an efficient way by translating the container's location only. For the zooming operations, the performance of each rendering method varies considerably. Flash is basically independent of the browsers and it takes no more than one tenth of the time used by the three DHTML methods. As an RIA technology, the performance

Table 11.4 Time (in milliseconds) used from panning 20 times

		Browsers		
Dataset	Rendering methods	IE9	Chrome	Firefox
DS1	SVG	1	2	4
	HTML5	1	5	8
	VML	6	N/A	N/A
	Flash	3	3	3
DS2	SVG	2	3	4
	HTML5	3	5	8
	VML	6	N/A	N/A
	Flash	3	3	2
DS3	SVG	3	6	8
	HTML5	3	5	4
	VML	10	N/A	N/A
	Flash	3	3	1

of Flash depends on its internal rendering mechanism, and the browsers it runs within have little effect. Among the three DHTML methods, SVG takes the least amount of time and VML takes largest amount of time for all three browsers when applicable. The poor performance of VML is possibly partly due to the fact that Microsoft is on the way to phase it out. Among the three browsers, Chrome performs the best for all rendering methods, and the efficient V8 JavaScript Engine used by Chrome contributes to its excellent performances. As the volume of the dataset increases, the time used increases. For the largest dataset (DS3) which contains 253 polygons and 536,999 vertices, SVG takes at least 21 s, HTML5 58 s, and VML 10 min, while Flash takes around 2 s.

A Web-based crime mapping and decision support system needs to render large amount of dynamic data, particularly line and polygon features. The excellent performance of Flash, particularly for zooming operations, makes it a good choice for client-side geographic feature rendering.

11.4 Prototype Design and Development

11.4.1 Users and System Requirements

The Web-based crime mapping and decision support system primarily targets law enforcement agencies. The system can be deployed on their proprietary intranets. Law enforcement staff can perform sophisticated hotspot analysis and mapping using various methods. We are particularly concerned with small and medium-sized police departments in developing countries, which cannot afford expensive commercial GIS software in addition to the development and maintenance costs of the system. The system can also be useful to the public when deployed on the Internet, enabling them to view, browse, and search crime points of different types and time periods.

Drawing upon our previous requirement analyses with many law enforcement agencies, we conclude that the Web-based GIS system for crime mapping and decision support should implement such fundamental requirements as basic mapping operations and hotspot mapping. Basic mapping operations should enable users to zoom and pan to areas of interest; manage layer properties such as visibility, opacity, and display order; retrieve spatial data through Open Geospatial Consortium (OGC) protocols such as Web Feature Service (WFS) and Web Mapping Service (WMS); and display tiled maps for background display through Web Map Tile Service (TMS). Hotspot mapping functions should enable users to specify mapping methods and related parameters, define classification and coloring schemes, etc. In addition, crime database should be kept up to date with original master dataset within a reasonable time lag. Existing and incoming crime records should be geocoded accurately and stored in the spatial database before being mapped for hotspot analysis.

11.4.2 Choice of Open-Source Products

Since one objective of this study is to make the Web-based GIS system affordable for small- and medium-sized police departments in developing countries, we choose to use open-source products instead of commercial GIS systems. Open-source software products such as OpenLayers and MapServer have been widely used in Web mapping and Web-based GIS systems (Tuama and Hamre 2007; Kulawiak et al. 2010; Rinner et al. 2011). On the server side, we choose PostgreSQL as the database server, PostGIS as the spatial database engine, and GeoServer as the GIS server. PostgreSQL is a powerful, open-source object-relational database system and runs on all major operating systems. It has a proven architecture that has earned a strong reputation for reliability, data integrity, and consistency. PostGIS is an open-source software product that adds support for geographic objects to the PostgreSQL object-relational database. GeoServer is an open-source Java-based software server that allows users to share and edit geospatial data. Designed for interoperability, it publishes data from major spatial data sources such as Shapefile, PostGIS, GeoTIFF, ArcSDE, GML, MySQL, Oracle, etc. It supports standard protocols including WFS, WMS, and WCS (Web Coverage Service) and can be configured through its user-friendly Web administration interface, which enables the rapid and easy publishing of spatial data.

On the client side, OpenScales is chosen as the mapping engine. OpenScales is an open-source mapping framework based on Adobe Flash and used for building rich Internet mapping applications. It is a pure client-side product and can work with a suite of open-source and commercial products through standard protocols and supports different types of layers such as tiled map layers, common OGC protocols like WFS and WMS, efficient and fast vector rendering capabilities, powerful style APIs, popup support, and XML runtime configuration. It comes with a range of controls and components including essential map control, layer manager control, zoom/pan control, overview map, and scale bar. Together with good

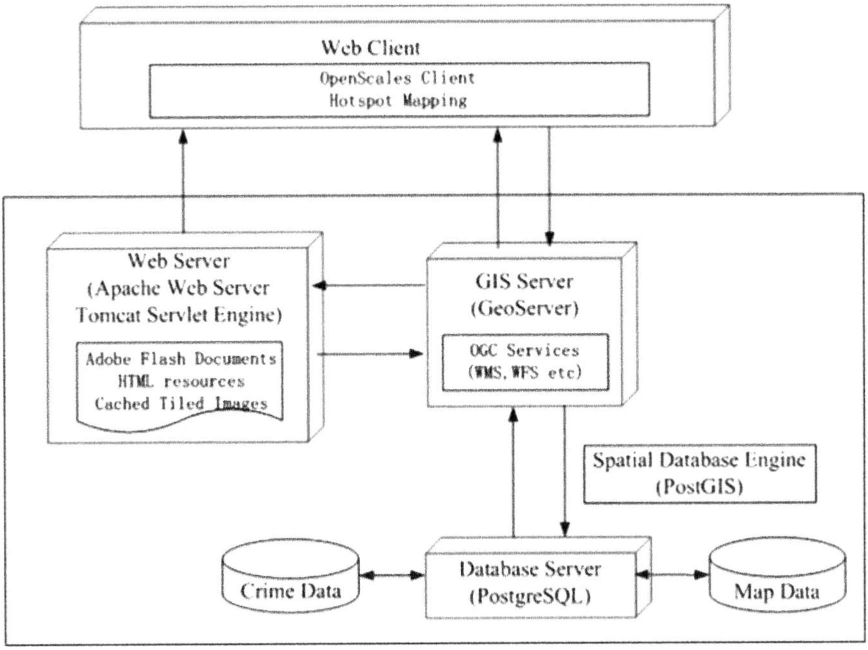

Fig. 11.4 Architecture of Web-based crime mapping and decision support system

documentation and a vast library of examples and tutorials, these controls and components enable the rapid and easy development of mapping clients with advanced features. As an open-source framework, OpenScales can be easily extended to accommodate custom requirements by allowing developers to delve into the source codes and to better understand the inner mechanisms of the framework.

11.4.3 System Architecture

Figure 11.4 illustrates the architecture of the Web-based crime mapping and decision support system. This multi-tier, client-server architecture is composed of a database server, a GIS server, a spatial database engine, a Web server, and a Web client. All server-side software is open-source products. The database server hosts crime data and other map data such as administrative boundaries and police precincts. The database server is spatially enabled through the spatial database engine. The GIS server publishes such OGC services as WFS and WMS for mapping, querying, and updating the spatial database through the spatial database engine. The Web server hosts OpenScales documents, custom-made Flash controls, HTML resources, and the cached tiled images.

The Web client is installed with Adobe Flash Player and accesses the Web resources on the Web server and GIS functionalities through the GIS server. The

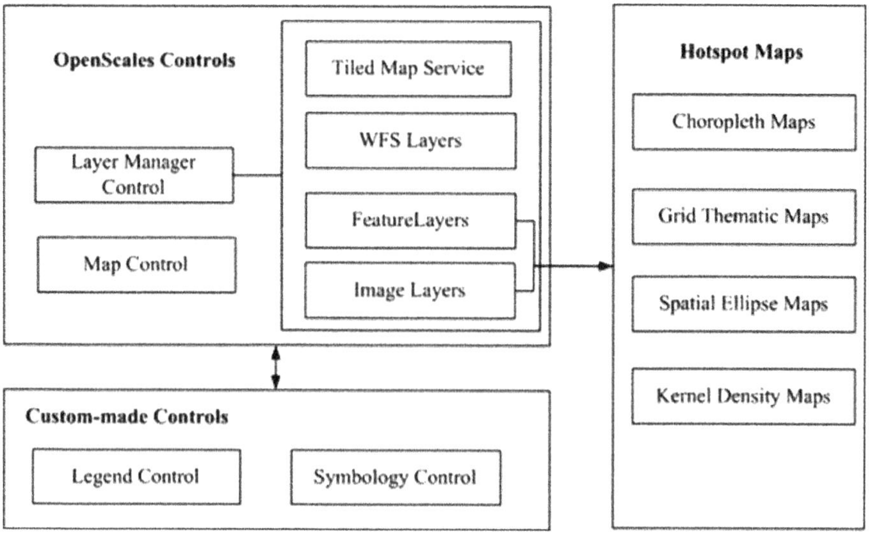

Fig. 11.5 Client-side architecture of the system

Web client provides basic GIS functionalities through OpenScales framework and various crime hotspot mapping options. The architecture of the client side is shown in Fig. 11.5. The application uses OpenScales controls such as map control and layer manager control for basic presentation and manipulation operations. The layer manager control manages such layers as tiled map service, WFS layers, feature layers, and image layers. The tiled map service is for the presentation of cached background data. WFS layers are used to retrieve spatial data such as crime points and administrative boundaries from the GIS server. Feature and image layers are used to render hotspot maps. In addition, custom controls including legend and symbology controls are implemented. The legend control shows the legend of the current hotspot map. The symbology control allows users to define classification schemes, color ramps, and symbology properties for each class.

11.4.4 Prototype Development

A prototype of the Web-based GIS for crime mapping and decision support has been developed according to the requirements and design described in the previous sections. The prototype serves to demonstrate that open-source software can meet the requirements of Web-based GIS systems for crime mapping and rich Internet applications give users more freedom than thin-client applications to manipulate hotspot mapping options. The prototype also serves to refine the requirements of law enforcement agencies.

The main user interface of the prototype system is shown in Fig. 11.6. Spatial data are retrieved and rendered by using OpenScales built-in feature layers and

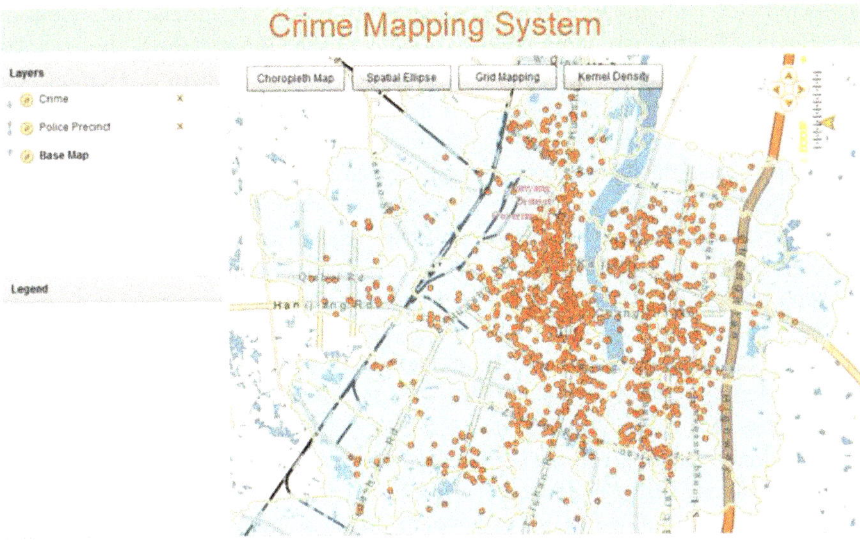

Fig. 11.6 The user interface of the system prototype

OGC layers. Hotspot map generation methods are implemented in ActionScript on the client side. Hotspots are arranged in thematic layers, whose properties such as opacity, visibility, and display order can be changed in the layer manager control. Spatial data are published from GeoServer using standard OGC protocols.

Background maps provide important reference information such as roads, buildings, rivers, parks, and recreational locations for the spatial distribution of crime, and they are an inherent part of a crime mapping and decision support system. Background maps generally do not change frequently and can be cached at different scales in tiled images to provide efficient Web mapping, which is a common practice in Internet mapping. As compared with on-the-fly map generation, cached images provide an additional benefit, i.e., visually more aesthetic maps. This is because cached images are usually prepared with a lot of effort at each display scale, sometimes with proprietary tools, whereas on-the-fly map generation generally does not satisfactorily deal with such factors as feature selection and text labeling.

11.5 Example Application Results

In this section, we demonstrate the application results of the prototype system. In the examples, the spatial data are originally obtained from a police department we worked with before and used here for demonstration purpose only. The crime dataset include such information as address, crime type, crime time, etc. Police precinct boundaries are used for choropleth mapping. Other spatial data are used primarily for background reference data. At each display scale, our background maps are prepared with care and tiled images are generated with a proprietary tool built on ArcEngine. The background tiled images are published through the Web server. All data are processed to be in a local projected coordinate system to facilitate the calculation of geometric measures such as distance.

For many police departments, geocoding crime records is the first step toward building a spatial database for crime mapping. Various geocoding methods and tools are available to provide satisfactory positional accuracy for the majority of records in western countries such as the United States (Zandbergen 2008). This is not the case for developing countries such as China (Li et al. 2010). In China, geocoding existing crime records is a daunting issue. This is due to the huge number of historical crime records to be processed and the lack of satisfactory automatic geocoding methods. Because of historical, cultural, and other reasons, Chinese addresses lack uniform formats, and the street numbers do not provide information on the relative positions of addresses in many cases (Li et al. 2010). Automatic tools generally yield address matching with an accuracy of less than 70 % according to our experiences with many Chinese police departments. The remaining unmatched addresses are geocoded manually. For our example crime dataset, the majority of the crime points are already geocoded previously, and the remaining addresses are geocoded manually. For daily operational application of the system, incoming crime records should be geocoded and stored in the spatial database using a dedicated tool. The dedicated tool should provide an automatic tool to geocode incoming crime locations and provide functionalities to manually adjust wrongfully geocoded addresses.

11.5.1 Hotspot Mapping Implementation

Figure 11.7 exemplifies choropleth mapping in the prototype system using all crimes for the year of 2010. As a rich-client application, the prototype needs to conduct point-in-polygon algorithm, which is implemented by OpenScales, in order to determine the number of crimes falling within each areal unit. Different classification and coloring schemes are provided to facilitate rendering.

Figure 11.8 exemplifies grid mapping in the prototype system. The grid mapping function allows users to specify an area of interest and grid size and number of grids along the horizontal and vertical directions. Once the grid dimensions are defined, the number of crimes falling inside each grid cell can be easily calculated, and then it becomes choropleth mapping with all necessary functions available to users.

Figure 11.9 shows an example of spatial ellipse mapping in the prototype system. The spatial ellipse function requires spatial clustering of crime points. The system implements K-means clustering method. Once the clusters are defined, a standard deviational ellipse is calculated for each cluster and displayed to users.

Figure 11.10 shows an example of kernel density mapping in the prototype system. With the kernel density mapping function, users can specify the dimensions of the grid to be generated for holding the kernel density value at each point in the

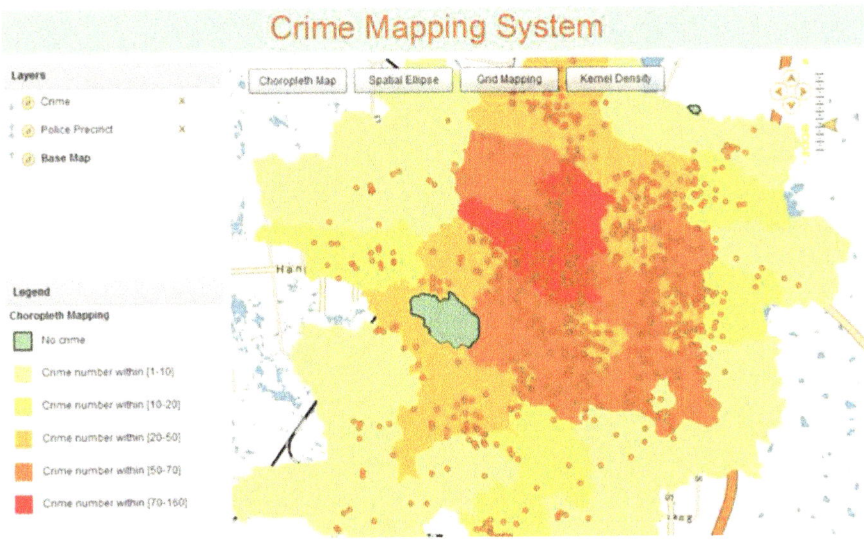

Fig. 11.7 Choropleth mapping overlaid with crime points

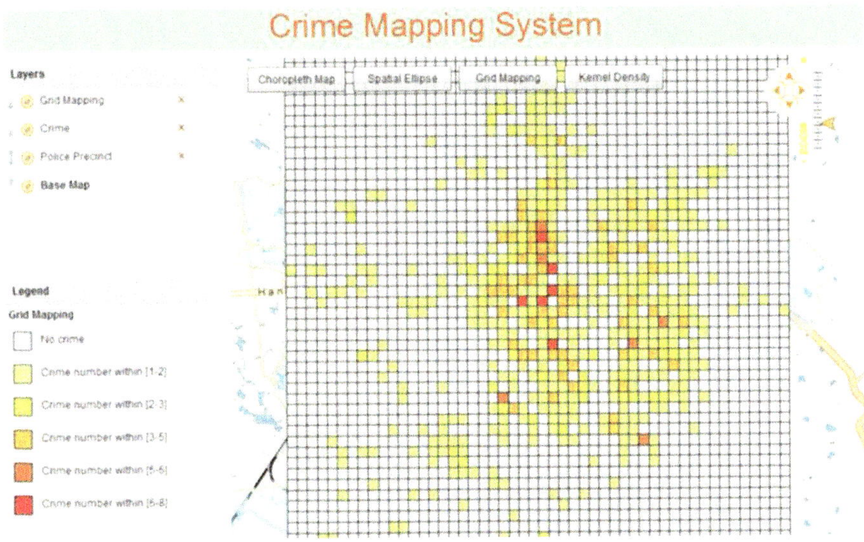

Fig. 11.8 Grid thematic mapping for crime hotspots

study area. Kernel functions such as the normal distribution function and triangular function can be specified and kernel density values contributed from all crime points are summed up for each grid cell. The bandwidth of the kernel function is defined by users. Once each grid cell receives its kernel density, the grid can be rendered in a way similar to choropleth mapping.

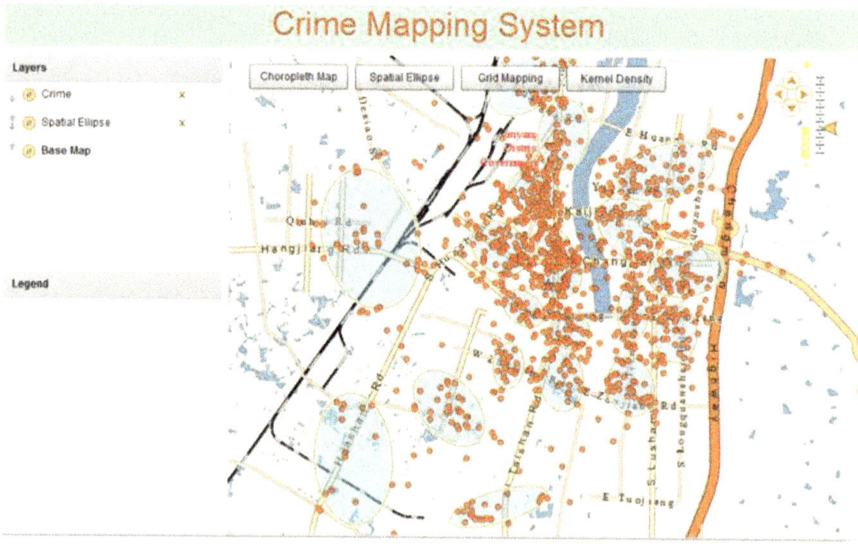

Fig. 11.9 Spatial hotspot ellipses overlaid with crime points

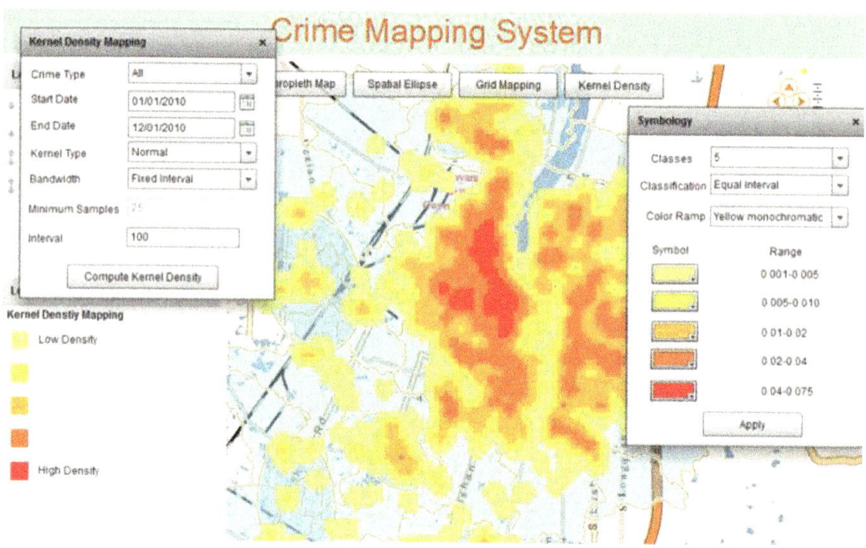

Fig. 11.10 Kernel density hotspot mapping

11.5.2 Spatial and Temporal Scale Effects

Spatial and temporal scales are key factors in many geographical phenomena, and crime hotspot detection is no exception (Wu and Grubesic 2010). To explore the spatial scale effects on hotspot detection and mapping, users can vary geographical

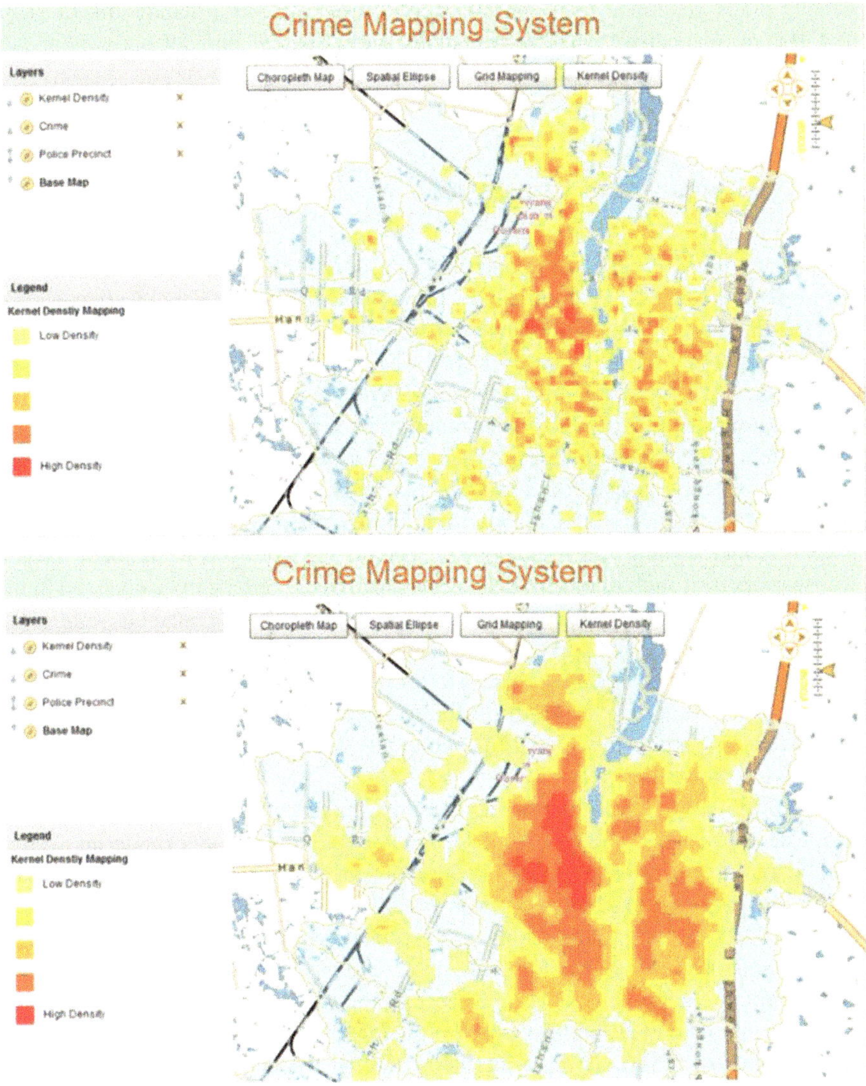

Fig. 11.11 Spatial scale effects on kernel density mapping. *Top*: normal kernel function with a bandwidth of 50 m. *Bottom*: normal kernel function with a bandwidth of 100 m. These two maps use the same classification and coloring schemes

areal units in choropleth mapping, grid size in grid mapping, the number of clusters to be grouped in spatial ellipse mapping, and the bandwidth of kernel function in kernel density mapping. In all four mapping methods, users can also vary the time span of crime to study the effect of various temporal scales. Users can study a subset of crime points by filtering the crime type as well. Figure 11.11 shows two kernel density maps generated using normal kernel function with a bandwidth of 50 m and 100 m, respectively. These two maps show distinct hotspot patterns at two spatial scales.

11.6 Summary and Conclusion

Existing applications for crime mapping and decision support are either desktop applications or based on commercial GIS products. The main objective of this research is to design and implement a crime mapping and decision support system based on open-source software packages. The prototype allows users to detect and view crime hotspots in the form of thematic layers overlaid on background data in a Web environment. The system is designed to be a part of a comprehensive police GIS system which basically encapsulates major aspects of routine police operations and can be effectively used for decision making. The prototype utilizes four hotspot mapping techniques, i.e., choropleth mapping, grid mapping, spatial ellipse mapping, and kernel density mapping. Hotspots are arranged in thematic layers, whose properties such as opacity, visibility, and display order can be changed in the layer manager control. The prototype can be used to map crime hotspots, predict the locations of future crime, and optimize crime reduction efforts for law enforcement agencies. The prototype allows users to specify various options for each hotspot mapping technique and to apply different classification and coloring schemes. Users can also study the spatial and temporal effects on crime hotspot patterns through the prototype.

The prototype is a rich Internet application, making it much more efficient than script-based clients in hotspot detection, map manipulation, and rendering. Large number of polygons such as police precincts and grid cells can be easily manipulated within a Flash client, whereas a script-based client is extremely cumbersome to handle a decent number of polygons. The prototype is entirely based on open-source software, making it affordable for many small- and medium-sized police departments in developing countries. GeoServer is used as the GIS server to publish crime data and other spatial data through standard protocols like WFS and WMS. Spatial data are stored in PostgreSQL and spatially enabled through spatial database engine PostGIS. The rich-client mapping framework OpenScales, as an open-source product, can be easily extended to meet custom requirements by allowing developers to delve into the source codes and to better understand the inner mechanisms of the software.

Although the prototype system has been tested and used for mapping property theft on public transit systems in Beijing metropolitan area, it is still in preliminary development stage. For future improvements, classification and coloring schemes may be saved on the server to allow users to retrieve and reapply them easily. Statistical charts such as bar charts and pie charts may be made available for better visualization and comparison. Other clustering methods such as hierarchical nearest neighbor clustering will be implemented to offer different options for clustering and to generate spatial ellipses for hotspots. More kernel types will be made available for kernel density mapping. Default bandwidth will be calculated for novice users.

Issues such as geocoding accuracy of addresses in China have broader implications, and advances in solving them will greatly benefit GIS-based crime mapping systems.

The prototype focuses on hotspot mapping. Other common GIS functionalities can be added to it to provide forensic cues for law enforcement agencies. One direction is to map hotel check-in records of suspects of motorcycle theft. Law enforcement agencies in many nations like China have all the check-in records of all hotel guests. Geospatial analytic tools can be applied to find those motorcycle theft suspects who have hotel check-in records outside their regular living areas during the period of the theft. A rule of thumb is that those suspects are more likely to have committed the crimes. Web-based GIS rich Internet applications can provide such functionalities on the client side more efficiently than script-based GIS applications.

Despite the fact that the prototype still needs further refinements, results obtained from the system development and testing stages demonstrate that rich Internet application technology, in combination with open-source software, is an effective solution for a Web-based GIS crime mapping and decision support system in terms of both system performance and financial cost.

Acknowledgments The work was supported by the Fundamental Research Funds for the Central Universities (No. ZYGX2010J089), National Natural Science Foundation of China (No. 41371399, 41101380, 41371341, 41301459), and National Science & Technology Pillar Program of China (No. 2012BAJ05B04).

References

Adobe Flash (2011) Adobe Flash Platform runtimes/Statistics: PC penetration. http://www.adobe.com/products/flashplatformruntimes/statistics.html. Accessed 12 Nov 2011

Al-Sabhan W, Mulligan M, Blackburn GA (2003) A real-time hydrological model for flood prediction using GIS and the WWW. Comput Environ Urban Syst 27(1):9–32

Bowers K, Newton M, Nutter R (2001) A GIS-linked database for monitoring repeat domestic burglary. In: Hirschfield A, Bowers K (eds) Mapping and analysing crime data – lessons from research and practice. Taylor & Francis, London, pp 120–136

Calhoun C, Stobbart C, Thomas D, Villarrubia J, Brown D, Conklin J (2008) Improving crime data sharing and analysis tools for a Web-based crime analysis toolkit: WebCAT 2.2. In: Proceedings of the 2008 IEEE systems and information engineering design symposium, University of Virginia, Charlottesville, 25 Apr 2008

Chainey S, Ratcliffe JH (2005) GIS and crime mapping. Wiley, London

Chainey S, Tompson L, Uhlig S (2008) The utility of hotspot mapping for predicting spatial patterns of crime. Secur J 21(1–2):4–28

Chang Y-S, Park H-D (2004) Development of a web-based Geographic Information System for the management of borehole and geological data. Comput Geosci 30(8):887–897

Crime Mapping (2013) Crime mapping – building safer communities. http://www.crimemapping.com/. Accessed 1 May 2013

CrimeReports (2013) Crime reports for US and Canada. https://www.crimereports.com/. Accessed 1 May 2013

Cromley RG (1992) Digital cartography. Prentice Hall, Englewood Cliffs

Defne Z, Haas KA, Fritz KM, Jiang L, French SP, Shi X, Smith BT, Neary VS, Stewart KM (2012) National geodatabase of tidal stream power resource in USA. Renew Sustain Energy Rev 16(5):3326–3338

Dent B, Torguson J, Hodler T (2008) Cartography: thematic map design, 6th edn. McGraw-Hill, Boston

Eck JE, Chainey S, Cameron JG, Leitner M, Wilson RE (2005) Mapping crime: understanding hot spots. National Institute of Justice, Washington, DC. http://www.nij.gov/pubs-sum/209393.htm. Accessed 22 Aug 2013

Harries K (1999) Mapping crime: principle and practice. National Institute of Justice, Washington, DC. https://www.ncjrs.gov/html/nij/mapping/. Accessed 19 Nov 2011

Hawkins C, Pittman J, Prats F, Wheeler W, Brown D, Dalton J (2003) WebCAT: the design and implementation of the web-based crime analysis toolkit. In: Jones MH, Tawney BE, White KP (eds) Proceedings of the 2003 IEEE systems and information engineering design symposium, University of Virginia, 24–25 April 2003

Home Office (2005) Crime mapping: improving performance – a good practice guide for front line officers. Home Office, London

Kontoes CC, Sykioti O, Paronis D, Harisi A (2005) Evaluating the performance of the space-borne SAR sensor systems for oil spill detection and sea monitoring over the south-eastern Mediterranean Sea. Int J Remote Sens 26(18):4029–4044

Kruger A, Khandelwal SG, Bradley A (2007) AHPSVER: a Web-based system for hydrologic forecast verification. Comput Geosci 33(6):739–748

Kulawiak M, Prospathopoulos A, Perivoliotis L, łuba M, Kioroglou S, Stepnowski A (2010) Interactive visualization of marine pollution monitoring and forecasting data via a Web-based GIS. Comput Geosci 36(8):1069–1080

LeBeau J (2001) Mapping out hazardous space for police work. In: Hirschfield A, Bowers K (eds) Mapping and analysing crime data – lessons from research and practice. Taylor & Francis, London, pp 139–155

Levine N (2006) Crime mapping and the Crimestat program. Geogr Anal 38(1):41–56

Levine N (2010) A spatial statistics program for the analysis of crime incident locations. http://www.icpsr.umich.edu/CrimeStat/. Accessed 22 Aug 2013

Li B, Zhang X, Chen Y (2010) Automatic construction and visualization of address models. 2010 Sixth International Conference on Natural Computation (ICNC), Yantai, Shandong, 10–12 August 2010

Li S, Xiong C, Ou Z (2011) A Web GIS for sea ice information and an ice service archive. Trans GIS 15(2):189–211

MapServer (2013) MapServer-open source web mapping. http://mapserver.org/. Accessed 1 May 2013

Martin D, Barnes E, Britt D (1998) The multiple impacts of mapping it out: police, geographic information systems (GIS) and community mobilization during Devil's Night in Detroit, Michigan. In: La Vigne N, Wartell J (eds) Crime mapping case studies: successes in the field. Police Executive Research Forum, Washington, DC

O'Rourke C (2004) A look at rich Internet applications. Oracle Magazine, July/August 2004 issue

Oakland Crimespotting (2013) http://www.crimemapping.com/. Accessed 1 May 2013

Phillips P, Lee I (2011) Crime analysis through spatial areal aggregated density patterns. GeoInformatica 15(1):49–74

Police UK (2013) Local crime and policing information for England and Wales. http://www.police.uk/. Accessed 1 May 2013

Rao M, Fan G, Thomas J, Cherian G, Chudiwale V, Awawdeh M (2007) A Web-based GIS decision support system for managing and planning USDA's conservation reserve program (CRP). Environ Model Softw 22(9):1270–1280

Rinner C, Moldofsky B, Cusimano MD, Marshall S, Hernandez T (2011) Exploring the boundaries of Web map services: the example of the online injury atlas for Ontario. Trans GIS 15(2):129–145

Su Y, Slottow J, Mozes A (2000) Distributing proprietary geographic data on the World Wide Web – UCLA GIS database and map server. Comput Geosci 26(7):741–749

Sugumaran R, Meyer JC, Davis J (2004) A Web-based environmental decision support system (WEDSS) for environmental planning and watershed management. J Geogr Syst 6(3):307–322

Tsou M-H (2004) Integrating Web-based GIS and image processing tools for environmental monitoring and natural resource management. J Geogr Syst 6(2):155–174
Tuama ÉÓ, Hamre T (2007) Design and implementation of a distributed GIS portal for oil spill and harmful algal bloom monitoring in the marine environment. Mar Geod 30(1–2):145–168
Williamson D, McLafferty S, Goldsmith V, Mallenkopf J, McGuire P (1999) A better method to Smooth Crime Incident Data. ESRI ArcUser Magazine, January–March, 1–5. http://www.esri.com/news/arcuser/0199/crimedata.html. Accessed 1 Sept 2012
Wu X, Grubesic TH (2010) Identifying irregularly shaped crime hot-spots using a multiobjective evolutionary algorithm. J Geogr Syst 12(4):409–433
Zandbergen P (2008) A comparison of address point, parcel and street geocoding techniques. Comput Environ Urban Syst 32(3):214–232

Chapter 12
Use of Geographically Weighted Regression on Ecology of Crime, Response to Hurricane in Miami, Florida

William C. Walker, Sunhui Sim, and Lisa Keys-Mathews

Abstract Research has illuminated the complex natural and social processes that occur after a natural disaster. Despite emergent efforts given to understanding the relationship between natural disasters and crime, few geographers have studied the effect that natural disasters have on the space-time behavior of crime patterns using local-level data. This research highlights aspects of change in patterns of crime as a result of a hurricane disaster; the underlying social, economic, and demographic characteristics may contribute explanations of the changes. Specifically, this study analyzes multiple types of crime as a response to Hurricane Wilma in Miami, Florida, 2005. The results reveal that more accurate predictions of crime for specific crime types in specific cities with use of geographically weighted regression are possible.

Keywords Natural disasters • Crime • Space-time analysis • Geographically weighted regression (GWR)

12.1 Introduction

Natural disasters and severe weather phenomena are remarkable, yet terrifying displays of power, capable of producing widespread destruction and loss of life due to their large scale and unpredictability. According to the National Hurricane Center (2011), Hurricane Wilma in 2005 was the fourth costliest and twelfth most intense tropical cyclone in the United States since 1975. Despite the human technocratic domain, events like Hurricanes Ivan and Wilma humble society's capacity to calculate chaos, specifically that of crime and deviance.

W.C. Walker • S. Sim (✉) • L. Keys-Mathews
Department of Geography, University of North Alabama, UNA Box 5064, Florence, AL 35632, USA
e-mail: walkerdc5@gmail.com; ssim@una.edu; lkeysmathews@una.edu

G.A. Elmes et al. (eds.), *Forensic GIS: The Role of Geospatial Technologies for Investigating Crime and Providing Evidence*, Geotechnologies and the Environment 11, DOI 10.1007/978-94-017-8757-4_12,

Despite emergent efforts given to understanding the relationship between natural disasters and crime, few geographers have studied the effect that hurricanes have on the space-time behavior of crime patterns. Research methods predominantly involve time series analyses of purely temporal crime data or the use of temporally or spatially aggregated data, giving limited consideration to the spatial distribution of crime events or underlying characteristics of crime locations. To better understand the relationship, more recent studies have incorporated the dimensions of space, time, and underlying characteristics (e.g., LeBeau 2000; Hagenauer et al. 2011; Leitner and Helbich 2011).

The primary objective of this research is to examine underlying characteristics of crime locations. The relationship between natural disasters, the space-time distribution of criminal activity, and potential causal characteristics has only recently been investigated (Hagenauer et al. 2011; Leitner and Helbich 2011). This research highlights aspects of change in patterns of crime as a result of a hurricane disaster, arguing that the underlying social, economic, and demographic characteristics may contribute explanations of the changes. Specifically, this study analyzes the variation in multiple types of crime as a response to Hurricane Wilma in Miami, Florida.

12.2 Disasters and Societal Response

According to the Centre for Research on the Epidemiology of Disasters (CRED 2011), the United States has experienced, on average, 7.23 natural disasters per year since 1900. These disasters are responsible for roughly 368 deaths, affect over 236,000 people, and cause $5.1 billion in damages ever year. Hurricanes represent the largest disaster force, accounting for only 12 % of these disasters, but producing 39 % of all deaths, 49 % of all people affected, and 60 % of all damage costs related to natural disasters.

There has been much debate about the definition of disaster characteristics. These include speed of onset, length of forewarning, magnitude of physical processes, geographical scope, temporal duration of their effects, frequency and temporal regularity, extent of environmental cues, and familiarity to disasters by the community and disaster response personnel (Tierney et al. 2001). Alexander (1993) suggests that temporal phases of disasters should be considered a disaster characteristic. Timelines of human crisis are needed to gauge progress, assess recovery, and evaluate the effectiveness of policy. Kates and Pijawka (1977) list activities that cities undergo during the disaster recovery process in a four-stage model, which includes:

1. The emergency period
2. The restoration period
3. A replacement reconstruction period
4. A period of commemorative, betterment, and development reconstruction (Fig. 12.1)

The emergency period, which may last from a few days to much longer periods of time, focuses on coping activities related to destruction of capital and loss of life,

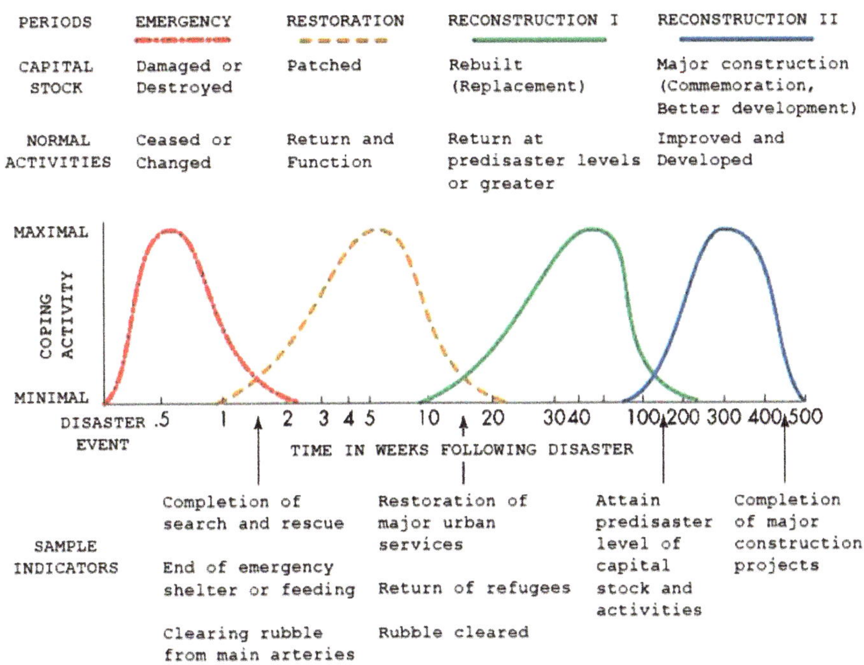

Fig. 12.1 Model of disaster recovery, illustrating the temporal phases of a disaster (Adapted from Kates and Pijawka 1977)

when normal activities cease. The restoration period involves the restoration of building stock and infrastructure and covers the time when activities return to relatively normal operation. This period may last from several months to more than a year. The replacement reconstruction period is highlighted by capital stock and activities returning to pre-disaster levels. The fourth stage involves large projects, usually financed by the government, to memorialize, indicate betterment, or aid future growth (Kates and Pijawka 1977).

12.3 The Relationship Between Disaster and Crime

Until recently, there has been a limited amount of empirical research which investigates the relationship between disasters and crime. On one hand, there is a perspective that has confidence in the occurrence of cohesion and altruism after disasters, while on the other, there is a view of torn social fabric and loss of social control (Drabek 1975; Quarantelli and Dynes 1970; Perry and Lindell 1978). Despite general acceptance of the former in the academic realm, disaster myth proliferation through mass media coverage (Wenger and Friedman 1986) and pervasive misconceptions about looting (Quarantelli and Dynes 1970) influence what is generally perceived about the prevalence of crime after disasters.

Conflicting findings give no conclusive insight into the reactions of criminal activity following a natural disaster. Researchers have found increased crime rates (VanLandingham 2008, 2007; Thornton and Voigt 2007), decreased crime rates (Suar and Kar 2005), varying rates among different types of crime (Bailey 2009; Zahran et al. 2009; Varano et al. 2010), and unchanged crime rates (Bass 2008; Cromwell et al. 1995) in the wake of a disaster. However, it is generally accepted that crime decreases immediately following a disaster, apart from domestic violence (Enarson 1999) and fraud (Davila et al. 2005).

Cohen and Felson (1979) propose a routine activity approach for analyzing crime rates, which transcends fundamental human ecological theories. The authors argue that structural changes in "recurrent and prevalent" (Cohen and Felson 1979, 593) activity patterns can alter crime rates by influencing space-time convergence of three minimal elements required for direct-contact predatory violations: (1) motivated offenders, (2) suitable targets, and (3) the absence of capable guardians against a violation. Crime is created by the intersection of suitable targets and motivated offenders in time and space and is arbitrated by capable guardians. Any considerable disruption to any of these elements at the macro level will cause changes in crime rates.

Exceptionally large-scale events, such as hurricanes, are capable of affecting all elements of the routine activity theory. Shifting populations and changes to the physical environment as a result of these events are capable of altering the dynamics of prevalent activities (Decker et al. 2007). The findings of Cromwell et al. (1995) concur that under extreme conditions, official guardianship systems may break down, an influx of outsiders increases the number of potential offenders, and potential targets, both people and places, are more vulnerable. LeBeau (2002) suggests that target vulnerability increased after a hurricane, as did attempts to increase guardianship.

12.4 Ecology of Crime and Geographically Weighted Regression

Stark (1987) proposes an ecology of crime theory, which is comprised of 30 propositions as explanation for deviance of place, as a means of interpreting characteristics of neighborhood level crime. These pertain to characteristics of place factors (density, poverty, mixed use, transience, and dilapidation), impacts these have on moral order as people respond to them (moral cynicism, opportunities for crime, motivation to deviate, and diminished social control), and how these responses increase deviance through consequences (attracting deviant people and activities, driving out the least deviant, and further reducing social control). These elements offer insight into the complex underpinnings of criminal activity following a disaster. Empirical research focusing on determinants or underlying factors of crime has generally included demographic, socioeconomic, housing, disaster, and proximity variables in their analyses. Osgood (2000) includes at-risk information, housing, demographic, income, unemployment, and proximity variables in his regression analysis of crime.

Zahran et al. (2009) investigate socioeconomic, social order, and disaster variables in a longitudinal study of county-level crime. Leitner and Helbich (2011) analyze socioeconomic, demographic, housing, and proximity variables at the census tract level in a space-time study of crime.

Regression analysis involves creating a model that "fits" the data or an empirical approximation of true relationships between variables. Part of this model-building process involves the estimation of unknown variables, with techniques such as ordinary least squares (OLS) (Montgomery et al. 2001). While conventional regression techniques, such as ordinary least squares, may provide insight into global trends in data, their inherent statistical assumptions may be inappropriate for spatial data. This is due to the assumption of first-order variation, without respect to second- or higher-order variation (Bailey and Gatrell 1995). Resulting issues of spatial autocorrelation, spatial dependency, and spatial heterogeneity violate assumptions necessary for regression modeling (Anselin et al. 2000).

In response to the shortcomings of global regression models, methods have been proposed to estimate local parameters of spatial data while considering spatial nonstationarity, or process not constant across space. In geographically weighted regression (GWR), the model creates a continuous surface of parameter estimates that vary over space (Fotheringham et al. 2002, 2000). In crime analysis, regression analysis has played a role in attempts to identify explanatory aspects of crime (Anselin et al. 2000). Tita and Radil (2010) provide a review of exploratory regression techniques related to criminology. GWR techniques have been incorporated in crime research focusing on local relationships. Cahill and Mulligan (2007) concluded that GWR techniques explained more variance about the relationship between elements of opportunity and crime in Portland, Oregon. Graif and Sampson (2009) found GWR superior to global techniques in determining the effects of neighborhood characteristics, namely, immigration and cultural differences, on homicide rates in Chicago, Illinois. GWR contributes to "spatially informed criminology by exploring the geographic dimensions of immigration and by modeling the heterogeneous spatial patterns underlying the risk" of crime (Graif and Sampson 2009).

Theoretical perspectives, including environmental criminology, have often been considered in selecting variables for regression analysis. In analyzing aggregate rural youth violence among 264 non-metropolitan counties in four states, Osgood (2000) considers social disorganization when selecting variables for OLS and Poisson-based regression techniques. Selected variables include measures of residential instability, ethnic heterogeneity, family disruption, poverty, unemployment, and a dummy variable representing proximity to metropolitan counties. Social disorganization and segmented assimilation theories are considered in selecting variables for OLS and GWR techniques for analyzing the effects of neighborhood characteristics on homicide at the census tract level in Chicago (Graif and Sampson 2009). Independent variables considered for regression include measures of population density, poverty, female-headed households, households with public assistance, residential stability, and language diversity. Leitner and Helbich (2011) use several easily obtainable variables in their spatial regression analysis in their spatiotemporal analysis of crime in Houston, Texas, following Hurricanes Rita and Katrina. Using

2000 census data, the authors derived 11 demographic and socioeconomic variables. These include (a) 2000 % Whites, (b) 2000 % African Americans, (c) 2000 % Asians, (d) 2000 % civilian labor force that is unemployed, (e) 1999 % persons below the poverty level, (f) 1999 median household income in $1,000, (g) 2000 % housing units vacant, (h) 2000 rental vacancy rate, (i) 2000 % owner-occupied housing units, and (j) 2000 % renter-occupied housing units. Additionally, a proximity variable – distance to the nearest police (sub)station – was also derived from spatial data.

These studies have generally found a consistent set of explanatory variables associated with crime clustered in space. Consistency among relevant literature reveals a theoretically based selection of potential predictor values for regression modeling. However, it is stressed that ecological studies of crime can be explored for local processes that are not supported by crime theory (Cahill and Mulligan 2007); that is, casual activity may vary over space and exploring local spatial pattern can be useful for providing local context of crime which are not accounted for Global OLS model. Furthermore, different types of crime may require different theories and subsequently different variables (Tita and Radil 2010). The present study will contribute to an understanding of ecological aspects of crime events in Miami, Florida.

12.5 Hurricane Wilma and Miami, Florida

Hurricane Wilma was originally formed as a tropical depression in the Caribbean Sea near Jamaica on October 15, became a hurricane on October 18, became a Category 5 hurricane with winds of 15 mph, and struck south Florida on October 24, 2005, as a Category 3 hurricane, before producing ten tornadoes over the peninsula (Fig. 12.2). Wilma is attributed with causing five directly related deaths and a substantial number of downed trees and damaged houses. Wilma also caused the largest disruption of electrical service damage ever to occur in Florida. The extent of this disruption included outages in 42 Florida counties, equal to 98 % of South Florida. Insurance claims for Wilma totaled $10.3 billion, of an estimated $20.6 billion in total damages (National Hurricane Center 2006).

The jurisdiction of the Miami Police Department comprises all unincorporated areas and some incorporated areas of Miami-Dade County, one of Florida's numerous coastal counties. The Miami-Dade Police Department provides basic law enforcement services to the more than 1.2 million residents living in the unincorporated areas, in addition to those living within contracted incorporated areas (Miami-Dade County 2004). Because the Miami-Dade Police Department provides contractual services to incorporated areas outside of its official jurisdiction while letting other municipalities police their own jurisdictions, a complete contiguous study area could not be obtained, but contiguity was required in order to eliminate island effects in the data. We excluded city of Miami, South Miami, Pinecrest, West Miami, Miami Springs, Hialeah, Medley, North Bay Village, Miami Beach and Key Biscayne. Consequently, the study area comprised 345 contiguous US Census Bureau block groups in Miami-Dade County (2010), an area of approximately 741 km.

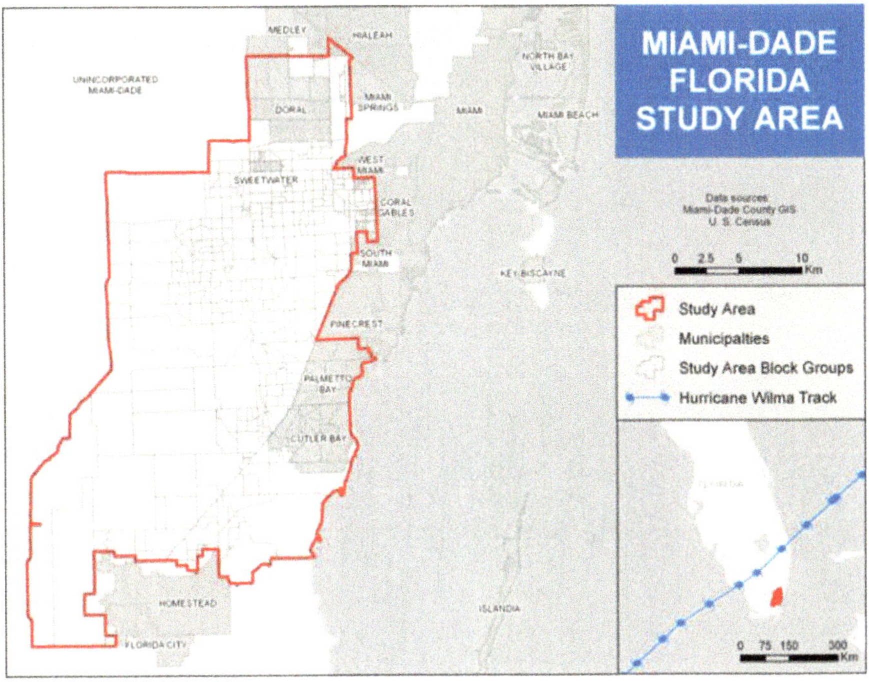

Fig. 12.2 Study site of Miami-Dade, Florida. A hurricane track for Wilma has been included for reference

12.6 Data and Methods

12.6.1 Crime Data

Crime data were collected for a study period 3 months prior to and after Hurricane Wilma. This includes the dates July 23, 2005, to January 24, 2006. Types of crime in the datasets were converted to Uniform Crime Reporting (UCR) classifications (Uniform Crime Reporting Program 2004) (Table 12.1).

Time series were created for daily crime totals for selected types of offenses for 6 months for the study site (Fig. 12.3). They reveal that reports of larceny-theft generally decline during hurricane landfall, while reports of burglary generally increase. Miami-Dade experienced highest totals of 29 burglaries on October 25 and 87 larceny-thefts on October 18. These are considerably higher than the respective 6-month average totals of 15 and 53.

The spatial distribution of burglary and larceny-theft during the study period appears to be heterogeneously distributed and occurs at higher densities in certain areas (Fig. 12.4). Kernel density estimates (KDE) are widely accepted techniques for estimating a surface density from crime point data, as they can easily illustrate density or clustering among points (Fotheringham et al. 2000; Eck et al. 2005;

Table 12.1 Study area, counts, and percentages

UCR crime type	Count	%
(1) Criminal Homicide	140	0.08
(2) Robbery	512	3.10
(3) Aggravated Assault	1,013	6.14
(4) Burglary	2,849	17.27
(5) Larceny-theft	9,907	60.05
(6) Motor vehicle theft	2,133	12.93
(7) Arson	69	0.42
Total:	16,497	

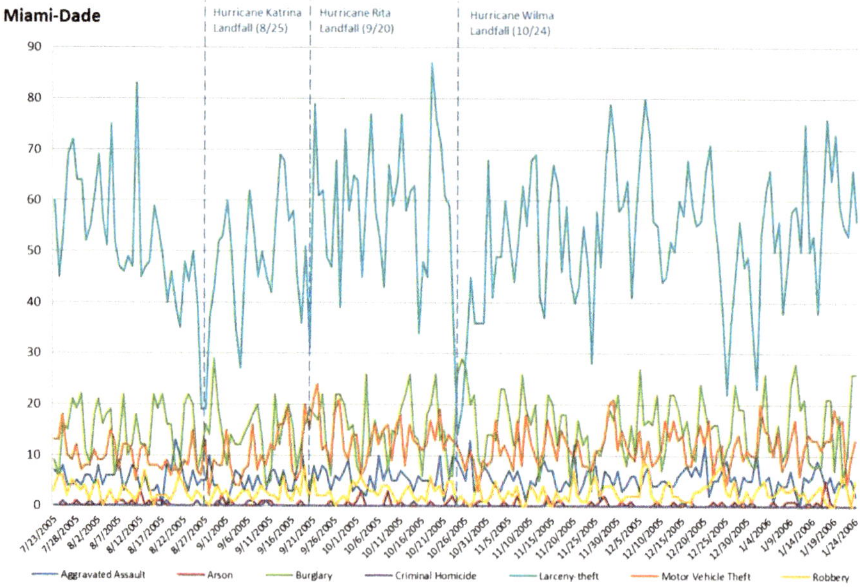

Fig. 12.3 Daily crime totals by crime. Time series for 6 months of crime counts in Miami

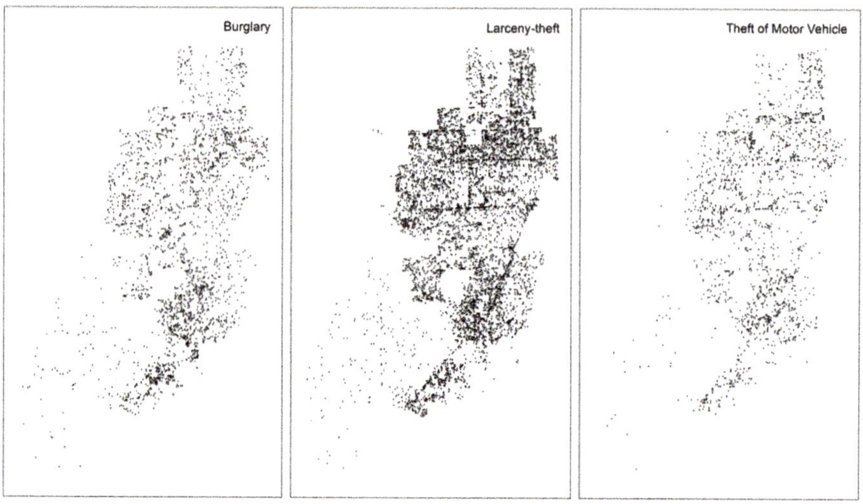

Fig. 12.4 Point map of the selected crimes across Miami

Fig. 12.5 Kernel densities of the selected crimes across Miami

Chainey and Ratcliffe 2005). In Miami, burglary hotspot concentrated over a relatively larger area to the south; the three major hotspots of larceny-theft are scattered out in small pockets across Miami, and the relatively weak hotspots of theft of motor vehicle are located in small pockets to the north and south (Fig. 12.5).

A general consensus exists within the literature about concepts that influence crime including guardianship, motivated offenders, suitable targets, socioeconomic status (SES) (also defined as disadvantage, affluence, and poverty), transience, accessibility, and levels of formal social control (i.e., police concentration) (Cahill and Mulligan 2007; Stark 1987; Cohen and Felson 1979). These concepts are broad and general, and researchers often disagree about the selection of proxy variables to represent these factors. A set of ecological concepts were identified within the literature (Table 12.2), along with theoretical descriptions and the sources of these descriptions, and proxy variables that characterize the broad concepts.

A racial heterogeneity index, residential stability, and family disruption are proxy variables for guardianship. Population density represents motivated offenders. Median household income, % unemployed, % persons below the poverty level, and % college graduates are proxy variables for socioeconomic status, disadvantage, affluence, and poverty. To represent transience, % housing units vacant and rental vacancy rates constitute proxy variables. Suitable targets are represented by % owner-occupied housing units and % renter-occupied housing units. Distance to the nearest municipal police (sub)station and distance to the nearest county police (sub) station are proxy variables for formal social control or police concentration. Distance to the nearest interstate/expressway exit and distance to the nearest major road constitute proxies for accessibility. Stability is measured as the proportion of residents who have lived in the same residence since 1995. The heterogeneity index

Table 12.2 Variables for regression

Variable	Expected correlation	Interpretation	Source
Racial heterogeneity index	+	Guardianship	2000 census
Residential stability	–		
Family disruption	+		
Population density	–	Potential offenders	
Median household income	–	Disadvantage, affluence, poverty	
% unemployed	+		
% persons below poverty level	+		
% college graduates	–		
% housing units vacant	+	Transience	
Rental vacancy rate	+		
% owner-occupied housing units	+	Potential targets	
% renter-occupied housing units	+		
Dist. to the nearest municipal police (sub)station	+	Guardianship	GIS derived
Dist. to the nearest county police (sub)station	+		
Dist. to the nearest interstate/ expressway exit	–	Accessibility	
Dist. to the nearest major road	–		

was calculated from seven race classifications (White, African-American, American Indian, Asian, Pacific Islander, Hispanic, and others) and considers the relative size and number of racial groups in an area. This measure has a range of 0–1 with 1 representing maximum heterogeneity, or least variation in the racial groups of an area. Family disruption is measured by the proportion of divorced or separated persons of all married persons in the population. The college graduate variable is the proportion of all individuals 25 years or older who have obtained a college education.

12.6.2 Methods

Following the technique presented by Leitner and Helbich (2011), a scan statistic was applied to each crime type using a circular search window, with 1 day as the temporal unit. The upper limit of the geographical size of crime clusters was set to the area covered by 50 % of the crime cases, and the maximum temporal length was set to 50 % of the study period. As a result, a "typical" daily crime and "highest" daily crime emerged and are used for a dual kernel density estimate by subtracting a typical crime total from a high period crime total. Miami had its highest daily totals of burglary (29) on October 25, larceny-theft (87) on October 18, and theft of motor vehicle (24) on September 22. Typical daily totals for burglary (15), larceny-theft (53), and theft of motor vehicle (11) occurred on September 29, October 6, and August 12, respectively. The dates of the highest totals for burglary and larceny are during Hurricane Wilma, and the date of highest total for theft of motor vehicle is

1 month before Hurricane Wilma and right during Hurricane Rita (September 20, 2005). Dual KDEs were created using CrimeStat 3.3 (Levine 2010). Parameters included the use of a quartic kernel function with an adaptive bandwidth of 15 crime points. The cell sizes were determined by dividing the shorter of the two sides of the minimum bounding rectangle for the study area by 150 (Eck et al. 2005). This step resulted in cell sizes of approximately 607 ft. for Miami. A dual KDE was calculated for three different types of crime – burglary, larceny-theft, and theft from a motor vehicle. Density values from this process were then aggregated to the block group level for regression modeling.

The final techniques may help illuminate the characteristics that influence criminal activity during hurricane disasters. This research analyzed 11 socioeconomic and demographic variables and four proximity variables at the census block group level for each of the cities. Aspatial and spatial regression techniques were used to gain a better insight into the possible causes for crime clusters and general increases in crimes associated with hurricane landfall. The 15 socioeconomic, demographic, and proximity variables discussed previously were used as independent variables. For each selected crime, a dependent variable was calculated from the dual kernel density estimations from the previous section, representing an overall increase in the selected crime between the base and high periods, which were subsequently aggregated to the block group level. Initially, an ordinary least squares (OLS) regression was used to regress all 15 independent variables onto each of the dependent variables. A step-wise algorithm was implemented to reduce the number of independent variables in the model. The selection was stopped until all regression coefficients were significant at $p=0.1$. The statistical software package SPSS was used during the step-wise process. The final OLS identified variables that influenced increases in crime at the global level. Ultimately, a geographically weighted regression (GWR), which accounts for local variation, was used to regress onto each dependent variable, the corresponding independent variables identified in the OLS process. GWR was performed using ArcMap 10.

12.7 Results

In order to analyze a general increase in crime during the study periods, a dual kernel density estimate (KDE) was calculated to identify areas of increased crime density. The crime density surface represents a change between a period of many crimes and a period of typical crimes. Subsequent surfaces reveal homogeneity among the different types of crime. The range of values varies among each location and type of crime, but darker areas represent increases in crime per unit area. Miami-Dade experienced increased burglary in a small pocket to the south, larceny-theft in small pockets to the north and south, and theft of motor vehicle over a relatively larger area to the north (Fig. 12.6). This pocket coincides with the established burglary hotspot, but larceny-theft and theft of motor vehicle clusters do not overlap areas of the established hotspots, but rather areas of lower densities or areas experiencing less than typical crime.

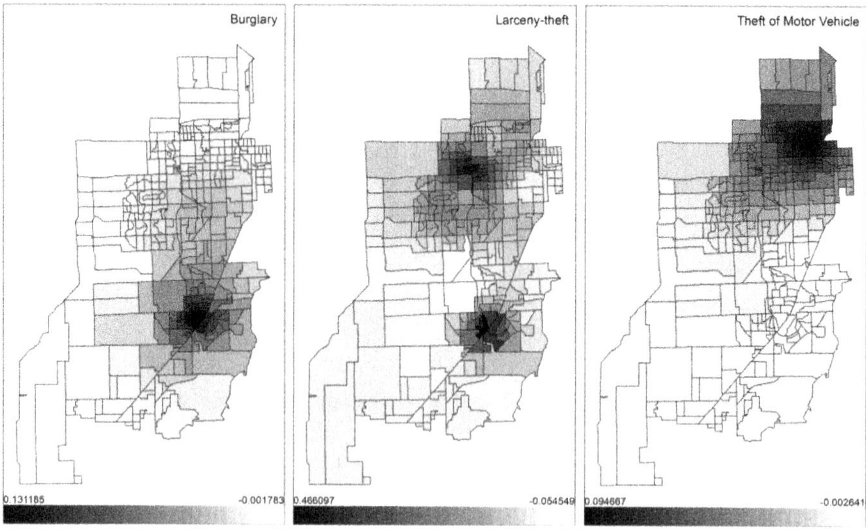

Fig. 12.6 Dual kernel densities of the selected crimes across Miami

As a global model, a parameter is estimated for each variable and is assumed to be equal across the study area. A GWR analysis was subsequently used to explore these findings and to identify local variations in the estimated parameters. An initial GWR model was developed to identify parameter estimates for variables that are close to zero. These variables are eliminated as they tend to not influence the dependent variable. A final GWR model is identified that has significant regression coefficients. Seven variables were identified by the OLS model of burglary at the study area level. Three of these were found to have contributed to the increased burglary at the local level (Table 12.3). These included the heterogeneity index, percent college graduates, and percent housing units vacant. The relationship between the independent variables and the general increase in burglaries is revealed by a spatial distribution of regression coefficients (Figs. 12.7, 12.8 and 12.9).

12.7.1 Miami-Dade Burglary

The global model identified seven significant parameter estimates (Table 12.3), of which three estimates were in the expected direction. The heterogeneity index and housing unit vacancy were positively correlated, and college graduates negatively correlated to the general increase in burglary. These relationships were expected, but those of the remaining four estimates were inverse to what was anticipated. This finding indicates that areas farther from municipal and county police stations experienced a higher level of burglary. Additionally, areas farther from interstate exits and areas with higher population densities experienced more burglaries.

Table 12.3 Final models of ordinary least squares regression and GWR

Variable	OLS results		
	Crime type		
	Burglary	Larceny-theft	Theft of vehicle
Intercept	0.055086	0.224976	0.051222
Racial heterogeneity index	0.065096		−0.080394
Residential stability			0.000254
Population density	−0.000002		0.000002
% persons below poverty level			−0.000529
% college graduates	−0.000511	−0.001666	
% owner-occupied housing units		0.000557	
% renter-occupied housing units			0.000209
Dist. to the nearest municipal police (sub) station	−0.000001	−0.000002	
Dist. to the nearest county police (sub)station	−0.000001	−0.000005	
Dist. to the nearest interstate/expressway exit	0.0000004		−0.000001
Dist. to the nearest major road		−0.000006	
Adjusted R^2	0.429	0.292	0.513
AICc	−1,738.5	−725.5	−1,773
	GWR results		
Adjusted R^2	0.869	0.872	0.972
AICc	−2,200.2	−1,223.5	−2,684.9

AICc Corrected Akaike Information Criterion

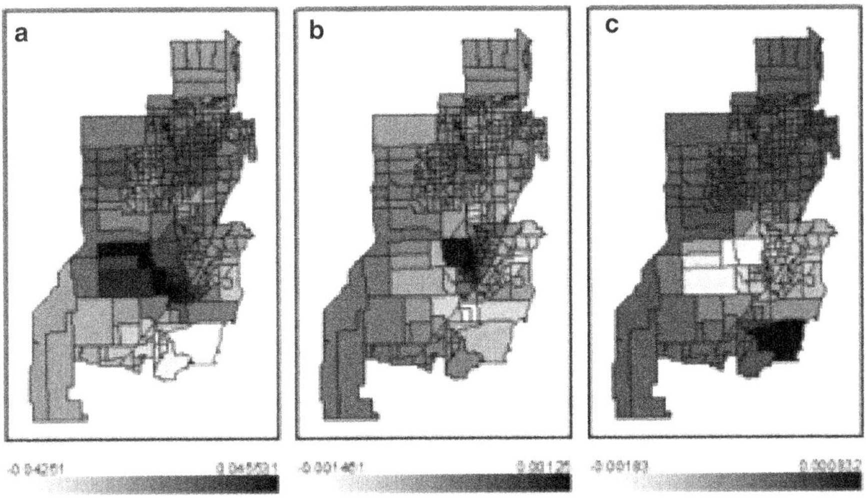

Fig. 12.7 Spatial distribution of regression coefficients of independent variables in GWR model for burglary in Miami-Dade: (**a**) heterogeneity index, (**b**) % housing units vacant, and (**c**) % college graduate

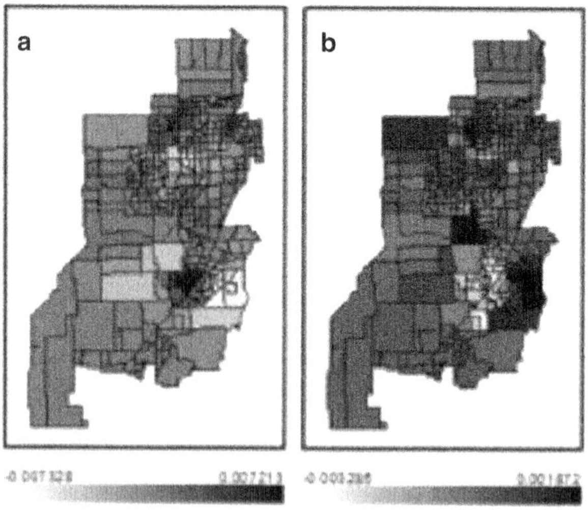

Fig. 12.8 Spatial distribution of regression coefficients of independent variables in GWR model for larceny-theft in Miami-Dade: (**a**) % college graduates and (**b**) % owner-occupied housing units

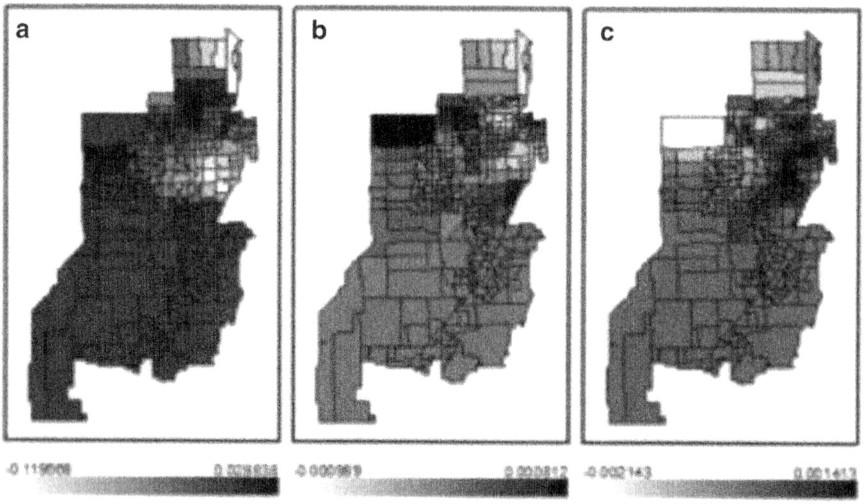

Fig. 12.9 Spatial distribution of regression coefficients of independent variables in GWR model for theft of motor vehicle in Miami-Dade: (**a**) heterogeneity index, (**b**) % persons below poverty line, and (**c**) % housing units vacant

The final OLS model explained 43 % of the variance in the general increase of burglary and produced an AICc of −1739. The final GWR model estimated coefficients for the significant variables – heterogeneity index, % college graduates, and % housing units vacant – which exerted the strongest influence on the density of burglaries. This model explained 87 % of the variance and produced an AICc of −2200, both improvements over the global model. The spatial distribution of regression coefficients (Fig. 12.6) reveals the relationship between the independent variables and the general increase in burglaries is heterogeneous across the study area. A positive relationship between the independent variables and the general increase in burglaries is revealed by the area near the location of two significant burglary clusters – toward the northeast and along the coast (Fig. 12.7). This means that an increase in the heterogeneity index and % housing units vacant and a decrease in the percentage of college graduates result in more burglaries.

12.7.2 Miami-Dade Larceny-Theft

The global model identified five significant variables, of which two were in the expected direction (Table 12.3). Estimates for % college graduates and distance to the nearest major road were both negative correlated to the increase in larceny-theft, while the other estimates did not have the expected direction. This finding means that areas closer to police stations and areas with higher % owner-occupied housing units experienced more larceny-thefts. The final OLS model explained 29 % of the variance, a relatively low explanatory power. The OLS model produced an AICc of −726. The final GWR model estimated coefficients for % college graduates and % owner-occupied housing. The model explained 87 % of the variance and produced an AICc of −1223, a significant improvement over OLS. The spatial distribution of coefficients (Fig. 12.8) shows that the relationship between the independent variables and the general increase in larceny-theft is heterogeneous across the study area. A generally positive relationship between the independent variables and the general increase in larceny-theft is revealed by the area near the location of several significant clusters in the north (Fig. 12.8). This means that a decrease in % college graduates and increase in % owner-occupied housing units results in more larceny-thefts.

12.7.3 Miami-Dade Theft of Motor Vehicles

The OLS model identified seven significant estimates, of which three were in the expected direction (Table 12.3). Estimates for distance to the nearest interstate exit were negatively correlated, while % renter-occupied housing units and % housing units vacant were positively correlated to the general increase in motor vehicle thefts. The remaining estimates, which were not in the expected direction, revealed that a decreased heterogeneity index and % persons below the poverty line and increased % population density and residential stability resulted in more motor vehicle thefts. The final OLS model accounted for 51 % of the variance and produced an AICc of −1773.

The final GWR model estimated coefficients for the heterogeneity index, % persons below the poverty line, and % housing units vacant. The model explained 97 % of the variance and produced an AICc of −2684.87, tremendous improvements over the global model. The heterogeneous spatial distribution of regression coefficients (Fig. 12.9) reveals a generally positive relationship between the independent variables and the increase in burglaries near the location of the significant theft of motor vehicle cluster (Fig. 12.9). This means that a decrease in racial heterogeneity and % persons below the poverty line and an increased % housing units vacant result in more motor vehicle thefts.

12.8 Summary and Discussion

This study investigates the impacts of hurricanes on the spatial and temporal distribution of crime in the cities of Miami, Florida. Societal structures influence criminal opportunity and activity, such as orders for mandatory (or voluntary) evacuations or curfews. The highest daily totals for crimes, which were used to estimate density surfaces, also occurred 1 week prior to landfall or on the day of landfall. Highest daily crime total for theft of motor vehicles occurred during the landfall of Hurricane Rita other than the study event, Hurricane Wilma. This research finds that crime clusters were influenced by hurricanes.

The application of regression techniques to models of change in crime density as a result of tropical storms also produced interesting findings. Initial OLS models used to identify suitable regression variables were not robust, but gave backing to theoretically supported variables. Measures of guardianship (the heterogeneity index), of transience (% housing units vacant), of affluence/poverty (% college), and of potential targets (% owner-occupied housing units) were significant in many of the models and exerted strong influence on crime. Proximity measures (distances to the nearest police stations, interstate exits, and major roads) were significant in nearly every model, but exerted only a weak influence on the number of crimes.

Additionally, in many cases, regression variables did not correlate to increased crime rates as expected. GWR results provided significantly more robust models, but did not give insight into the counterintuitive relationship between explanatory variables and crime. This suggests that crime patterns are influenced by underlying variables differently during hurricane events than what is supported by theory.

Ultimately, findings show that several factors influence criminal activity at different levels following a hurricane. Regression techniques reveal that, during hurricane events, city-wide burglary, larceny-theft, and theft of motor vehicle crimes generally increase because of demographic, economic, and proximity factors. These factors variably influence criminal activity across each city as characteristics of neighborhoods contribute to different aspects of crime. The areas with the increased crimes 1 week prior to the landfall or on the day of the landfall are located in economically disadvantaged neighborhoods. Law enforcement should place some personnel into specific neighborhoods of the city to watch out for potential criminal activities during disaster events.

The findings of this research are important for the underlying socioeconomic, demographic, and proximity variables of crime. A universal model for predicting crime occurrences may be unattainable because of the dynamic nature of urban systems, policy, and criminals, but more accurate predictions of crime for specific crime types in specific cities are possible. The method of this research is also important to assist law enforcement planners and disaster management personnel in preparing for hurricanes and other disasters. Areas of cities may be characterized in advance as at risk for increased crime based on historical data, and law enforcement personnel can be allocated based on future needs.

The current analysis could be expanded with additional variables, such as damage information, land use, distance measure to other facilities, evacuation information, and police plans and activities, to determine crime-specific variables in support of a universal model. Analysis could then be repeated on the same cities, as well as others, to identify temporal and spatial trends of crime. The authors of this chapter are expanding upon the aforementioned research through the comparative study of crime patterns and underlying characteristics of three dissimilarly populated cities affected by significant Atlantic hurricanes. Those cities include Miami, Pensacola in Florida, and Mobile in Alabama.

Acknowledgments This research was funded in part by Department of Justice COPS grant #2008CKWX0245 – AL04108.

References

Alexander D (1993) Natural disasters. Chapman & Hall, New York

Anselin L, Cohen J, Cook D, Gorr W, Tita G (2000) Spatial analysis of crime. Criminal Justice 2000. Office of Justice Programs, U.S. Department of Justice, National Institute of Justice, Washington, DC, pp 212–262

Bailey TC, Gatrell AC (1995) Interactive spatial data analysis. Wiley, New York

Bailey KL (2009) An evaluation of the impact of Hurricane Katrina on crime in New Orleans, Louisiana. Paper 304, applied research projects, Texas State University-San Marcos, San Marcos

Bass L (2008) A pooled time series analysis of Hurricane Katrina evacuees and their effect on crime rates. Paper presented at the 66th Midwest Political Science Association annual national conference, Chicago, 3–6 April

Cahill M, Mulligan G (2007) Using geographically weighted regression to explore local crime patterns. Soc Sci Comput Rev 25:174

Centre for Research on the Epidemiology of Disasters (2011) EM-DAT: the OFDA/CRED international disaster database. http://www.emdat.be/

Chainey S, Ratcliffe JH (2005) GIS and crime mapping. Wiley, London

Cohen LE, Felson M (1979) Social change and crime rate trends: a routine activity approach. Am Sociol Rev 44:588–608

Cromwell P, Dunham R, Akers R, Lanza-Kaduce L (1995) Routine activities and social control in the aftermath of a natural catastrophe. Eur J Crim Policy Res 3(3):57–69

Davila M, Marquart JW, Mullings JL (2005) Beyond mother nature: contractor fraud in the wake of natural disasters. Deviant Behav 26:271–293

Decker SH, Varano SP, Greene JR (2007) Routine crime in exceptional times: the impact of the 2002 Winter Olympics on citizen demand for police services. J Crim Justice 35:89–101

Drabek TE (1975) System shock: immediate responses following impact. In: Mileti DS, Drabek TE, Haas JE (eds) Human systems in extreme environments: a sociological perspective. University of Colorado, Institute of Behavioral Science, Boulder, pp 57–99

Eck JE, Chainey S, Cameron JG, Leitner M, Wilson RE (2005) Mapping crime: understanding hot spots (Special report). National Institute of Justice, Washington, DC

Enarson E (1999) Violence against women in disasters: a study of domestic violence programs in the United States and Canada. Violence Against Women 5:742–768

Fotheringham AS, Brunsdon C, Charlton M (2000) Quantitative geography: perspectives on spatial data analysis. Sage, London

Fotheringham AS, Brunsdon C, Charlton M (2002) Geographically weighted regression: the analysis of spatially varying relationships. Wiley, Chichester

Graif C, Sampson RJ (2009) Spatial heterogeneity in the effects of immigration and diversity on neighborhood homicide rates. Homicide Stud 13(3):242–260

Hagenauer J, Helbich M, Leitner M (2011) Visualization of crime trajectories with self-organizing maps: a case study on evaluating the impact of hurricanes on spatio-temporal crime hotspots. Paper presented at the 25th conference of the International Cartographic Association, Paris, 3–8 July

Kates RW, Pijawka D (1977) From rubble to monument: the pace of reconstruction. In: Haas JE, Kates RW, Bowden MJ (eds) Reconstruction following disaster. MIT Press, Cambridge, MA, pp 1–23

LeBeau JL (2000) Demonstrating the analytical utility of GIS for police operations: a final report. National Criminal Justice Reference Service. https://www.ncjrs.gov/pdffiles1/nij/187104.pdf

LeBeau JL (2002) The impact of a hurricane on routine activities and on calls for police service: Charlotte, North Carolina, and Hurricane Hugo. Crime Prevent Commun Saf 4:53–64

Leitner M, Helbich M (2011) The impact of hurricanes on crime: a spatio-temporal analysis in the city of Houston, Texas. Cartogr Geogr Inf Sci 38(2):214–222

Levine N (2010) CrimeStat v3.3: a spatial statistics program for the analysis of crime incident locations. http://www.icpsr.umich.edu/CrimeStat/index.html

Montgomery DC, Peck EA, Vining GG (2001) Introduction to linear regression analysis, 3rd edn. Wiley, New York

National Hurricane Center (2006) Tropical cyclone report: Hurricane Wilma. http://www.nhc.noaa.gov/pdf/TCR-AL252005_Wilma.pdf

National Hurricane Center (2011) The deadliest, costliest, and most intense United States tropical cyclones from 1851 to 2010 (and other frequently requested hurricane facts). http://www.nhc.noaa.gov/pdf/nws-nhc-6.pdf

Osgood DW (2000) Poisson-based regression analysis of aggregate crime rates. J Quant Criminol 16(1):21–43

Perry RW, Lindell MK (1978) The psychological consequences of natural disaster: a review of research on American communities. Mass Emerg 3:105–115

Quarantelli EL, Dynes RR (1970) Property norms and looting: their patterns in community crisis. Phylon 31(2):168–182

Stark R (1987) Deviant places: a theory of the ecology of crime. Criminology 25(4):893–909

Suar D, Kar S (2005) Social and behavioural consequences of the Orissa supercyclone. J Health Manag 7(2):263–275

Thornton WE, Voigt L (2007) Disaster rape: vulnerability of women to sexual assaults during Hurricane Katrina. J Public Manag Soc Policy 13(2):23–49

Tierney KJ, Lindell MK, Perry RW (2001) Facing the unexpected: disaster preparedness and response in the United States. Joseph Henry Press, Washington, DC

Tita GE, Radil SM (2010) Making space for theory: the challenges of theorizing space and place for spatial analysis in criminology. J Quant Criminol 26:467–479

Uniform Crime Reporting Program (2004) Uniform crime reporting handbook. Federal Bureau of Investigation, Washington, DC

Van Landingham MJ (2007) Murder rates in New Orleans, LA, 2004–2006. Am J Public Health 97(9):1614–1616

Van Landingham MJ (2008) 2007 murder rates in New Orleans, Louisiana. Am J Public Health 98(5):776–777

Varano SP, Schafer JA, Cancino JM, Decker SH, Greene JR (2010) A tale of three cities: crime and displacement after Hurricane Katrina. J Crim Justice 38:42–50

Wenger D, Friedman B (1986) Local and national media coverage of disaster: a content analysis of the print media's treatment of disaster myths. Preliminary paper 99, Disaster Research Center, University of Delaware, Newark

Zahran S, Shelly TO, Peek L, Brody SD (2009) Natural disasters and social order: modeling crime outcomes in Florida. Int J Mass Emerg Disasters 27(1):26–52

Chapter 13
Delineating Legal Forest Boundaries to Combat Illegal Forest Encroachments: A Case Study in Murree Forest Division, Pakistan

Irfan Ashraf, Urooj Saeed, Naeem Shahzad, Javed Gill, Shahid Parvez, and Akram Raja

Abstract The annual deforestation rate of Pakistan is 2.37 %, the highest degradation rate in Asian countries. Illegal encroachment on forest land is one of the major causes of deforestation in Pakistan. This study addresses the issue of forest encroachments in Murree Forest Division (MFD). MFD is a biodiversity-rich popular hill station of Punjab province and hosts thousands of visitors throughout the year. MFD comprises of 19,135 ha (47,285 acres) of state forest land, out of which 1,158.8 ha (2,862 acres) was identified as land encroached upon by the land grabbers, builders, and timber mafia. During the past decade this problem has drastically increased and led to a suo moto initiative by the Lahore High Court (vide Writ Petition No. 1813–2010). For the forensic support to the court, a detailed geographic information system and remote sensing baseline mapping have been carried out in close coordination with the Punjab Forest Department, Survey of Pakistan, Punjab Revenue Department, and WWF-Pakistan. Legal forest boundaries were delineated using high-precision surveying based on a Differential Global Positioning System and total station systems,

I. Ashraf (✉)
GIS Laboratory, WWF-Pakistan (HO), Lahore, Pakistan
e-mail: iashraf@wwf.org.pk

U. Saeed • N. Shahzad
GIS Laboratory, WWF – Pakistan, Lahore, Pakistan
e-mail: usaeed@wwf.org.pk; nsrana@wwf.org.pk

J. Gill
Murree Forest Division, Punjab Forest Department, Lahore, Pakistan
e-mail: dfo_murree@hotmail.com

S. Parvez
Department of Space Science, Punjab University, Lahore, Pakistan
e-mail: shahid.spsc@pu.edu.pk

A. Raja
Survey of Pakistan, Faizabad, Murree Road, Rawalpindi, Pakistan
e-mail: dmp@surveyofpakistan.gov.pk

G.A. Elmes et al. (eds.), *Forensic GIS: The Role of Geospatial Technologies for Investigating Crime and Providing Evidence*, Geotechnologies and the Environment 11, DOI 10.1007/978-94-017-8757-4_13, © Springer Science+Business Media Dordrecht 2014 whereas the encroachments were identified using high-resolution satellite images. The accuracy of these boundaries and encroachments was assessed and validated on the ground as well as on historic maps. The study outcomes were presented to the court as a supporting document of evidence to help them make a better decision against the encroachers and land retrieval. The main goal of the study was to prioritize conversion strategies for sustainable natural resources and segregate them into immediate, mid-term, and long-term process of land retrieval to restoring forests back to their original state. The study prompts the relevant government officials to adopt the latest mapping technologies of GIS, GPS, and RS for efficient forest management practices and sustainable enhancements.

Keywords Forensic GIS • Forest conservation • Differential Global Positioning System (DGPS) • Delineation • Documenting evidence

List of Acronyms

cm	Centimeter
DFO	Divisional forest officer
DGPS	Differential global positioning system
EIA	Environmental impact assessment
EPA	Environmental Protection Agency
FAO	Food and Agriculture Organization of the United Nations
FBM	Forest bench mark
FIRs	First information reports
ft	feet
GIS	Geographic information system
GOP	Government of Pakistan
GPS	Global positioning system
Gt	Gigaton
ha	Hectare
km	Kilometer
m	Meter
MFD	Murree Forest Division
NOC	No objection certificate
PDOP	Positional dilution of precision
PF	Protected forest
PFD	Punjab Forest Department
PRD	Punjab Revenue Department
RF	Reserved forest
RS	Remote sensing
SMC	Study Monitoring Committee
SoP	Survey of Pakistan
WWF	World Wide Fund for Nature
yr	Year

13.1 Introduction

This study is focused on forest land encroachment in Pakistan and proposes an effective solution to help the Judiciary to eradicate the associated problems. The kernel of the study is the baseline data development (a self-contained forest inventory) to make forest monitoring and evaluation easy and efficient on a practical basis. The adoption of a geographic information system (GIS) and remote sensing (RS) along with the global positioning system (GPS) increases the capacity of relevant government officials in updating forest records, continuous monitoring, and evaluation. The major goal of the study was to retrieve the encroached land for forest restoration and to augment forest conservation initiatives. In addition, the outputs will be presented to the Lahore High Court as a supporting document of evidence. Overall the study encompasses:

- Developing reference bench marks for forest boundaries
- Delineating the legal boundaries of state forests in Murree Forest Division by involving line departments
- Identifying the forest sections facing high encroachment threats
- Investigating the delineated forest boundaries with reference to *Patwari* Maps[1] and identifying encroachments

Forests are the lungs of the earth. They play an important role in providing natural habitat to plants, animals, and microorganisms (biotic components) in that area functioning together with all of the nonliving physical (abiotic) factors of the environment, collectively called "a forest ecosystem" and probably one of the oldest surviving ecosystems (Christopherson 2009). FAO studies reveal that the trees and other vegetation globally contribute in the storage of carbon by a number of 289 gigatons (Gt). This carbon storage has decreased by an estimated rate of 0.5 Gt a year between 2000 and 2010 due to a reduction in the total forest cover area (FAO 2010).

Forests in Pakistan are in a critical situation. According to Pakistan Bureau of Statistics for 2010, the total forest land area in Pakistan is 4.23 million ha (almost 5.3 % of its total land) (PADs 2011). Pakistan falls in the lowest category of percentage groups of the total global forest cover (0–9 %), having 2.5 % forest cover reported by Hasan (2001) out of 5.3 % forest designated land. Calculations associate a value of 0.026 ha of forest per capita which is far less than the threshold value of 25 % (a minimum level to meet an ecological balance) of the total land area that should be forested. The previous statistics on forest cover in Pakistan are really alarming (Table 13.1). Reports published in 2010 and 2011 conclude a value of only 2.1 % forest cover land in Pakistan (FAO 2010).

[1] Patwari maps are set of revenue maps usually referred to as village maps or *Patwari* field maps. These include *latha* and *mussavi* maps. A *Patwari* (a lowest rank official in revenue system for two to four villages) is responsible to maintain these maps along other records, i.e., *jamabandi* (record of rights), register *intkaal* (mutations register), *khasra girdawari* (harvest inspection register), *shajra nasb* (genealogy table), field measurement book, and *roznaamcha waqiati* (daily journal).

Table 13.1 Forests in Pakistan (FAO 2010)

Forest cover area				Annual change rate in forest cover					
FY1990	FY2000	FY2005	FY2010	1990–2000		2000–2005		2005–2010	
(000 ha)	(000 ha)	(000 ha)	(000 ha)	1,000 ha/yr	%/yr	1,000 ha/yr	%/yr	1,000 ha/yr	%/yr
2,527	2,116	1,902	1,687	−41	−1.76	−43	−2.11	−43	−2.37

Pakistan is continuously losing its forest and unveils the world's second highest deforestation rate. Major causes of forest degradation in Pakistan are anthropogenic rather than being due to climatic variations (Bhatti 2011). Anthropogenic threats include illegal cutting, commercial overexploitation, illegal land encroachment, overgrazing, poor management, and other man-made ecological changes (Irshad et al. 2011). Hasan, L. (2008) associates these problems to a lack of understanding of Land Tenure Arrangements since British rule in the subcontinent. Implementation of old and obsolete management practices by forest officials is a negative feedback to forest growth in Pakistan. No standardized forest database exists in the country. Register-based records waste much of useful time and are inadequate for forest management activities. Forests land is still measured with the help of *Patwari* maps.

Weak management, a lack of knowledge of forest boundaries, and scarce monitoring activities encourage the encroachers to intrude on forest land. Hasan, L. (2001) reported that due to half-hearted demarcation by the government officials, state forests are facing threats of illegal encroachment. Hasan, L. (2008) associates the deforestation issues to government failure. The ultimate increase in deforestation is also destroying the complex watershed system, ecological biodiversity, and environment (Bukhari 1997). Hence, the "Theory of Himalayan Environmental Degradation" (THED)[2] is proven correct since the last few decades (Ali and Benjaminsen 2004; Hasan 2008). Hussain et al. (2012) found that the participation of the state forests in providing timber and fuel wood is just 14 % and 10 %, respectively.

A GIS, being a decision-making tool, provides an easy-to-understand ground picture of the forest condition and the intensity/level of the encroachment to the Lahore High Court. The study supported the Punjab Forest and Wildlife Departments to manage and protect natural resources of the area through enhancing the capacity of forest officials in forest management resourcefully. The study provides new research avenues for the upcoming researchers, scholars, and academia to take benefit of the baseline data in their studies related to geology, hazard analysis, environment and carbon-stock assessment, etc.

After reading this chapter, the reader will be able to understand the forest situations in Pakistan and the significant use of geospatial technologies to address the forest management issues. The forensic application of GIS, GPS, and RS in forestry is to provide evidence on ground realities about the encroachments and degradation/deforestation in protected and reserved forest land. A step-by-step approach is used to explain the scenarios and methods adopted for this study.

[2]The main tenet of this theory is that increased human population has resulted in increased demands for natural resources, leading to severe resource depletion, especially deforestation.

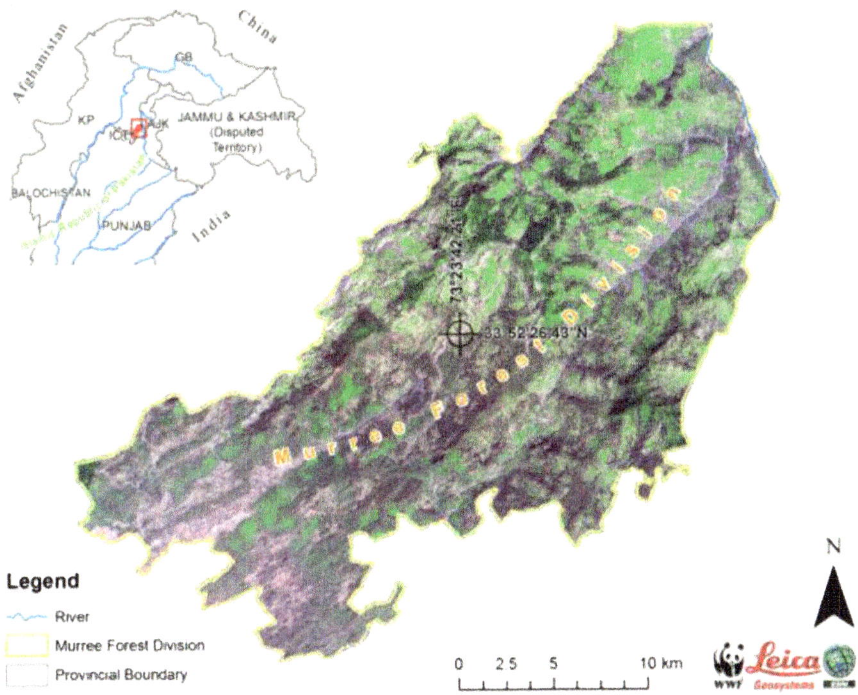

Fig. 13.1 Study area Murree Forest Division – a bird's eye view from SPOT 5 Satellite Sensor

13.1.1 Study Area

The study area is Murree Forest Division, a popular hill station, located along the Islamabad-Kohala highway, 30 km northeast of Islamabad. Geographically centered at 73°23′42.21″ east and 33°52′26.34″ north, MFD is situated in two ecological zones, i.e., "Moist Temperate Coniferous Forests" (Blue pine or Kail zone) and the "Chir Pine Subtropical Forests" (Chir zone), covering a small area of the outer Himalayas (Fig. 13.1) (Roberts 1992). Its altitude ranges from 1,700 to 7,800 ft (520–2,380 m). MFD lies in the outer Himalayas of the sub-tropical continental highlands. Forests in MFD play an important role in sub-watershed of the Indus-Jhelum River systems. It is believed that any damage to the Western Himalayan Eco-region will have negative impacts on the Indus River systems (WWF-Pakistan 2011).

MFD is subdivided into six administrative units, i.e., four subdivisions (Sehr Bagla, Lower Topa, Ghora Gali, and Sambli) and two ranges (Municipal and Ban). MFD comprises of approximately 19,136 ha (47,285 acres) of State Forest land which is almost 44 % of the total area of Murree Tehsil. Management in MFD deals with two major land tenures under the state-owned forest category, i.e., reserved forest and protected forest (Ahmed and Mahmood 1998). Another class of forest that is managed by Guzara Division under privately owned tenure is Guzara Forests (Shamilat Deh). MFD contains 22 protected forests and 23 reserved forests.

13.1.2 Background

Abbasi (2010) highlighted the issue of forest land encroachment in Murree. He reported that land worth more than ten billion rupees has been encroached upon. He added that a total land area of 15,000 kanals[3] has been detected after demarcation of half of the forest land by officials of Punjab Forest Department. The issue was highlighted in media and political pressure increased on demarcation activities. He further stated that the encroachment stress is high in Municipal Forest.

The Lahore High Court took suo moto action against heavy encroachment in MFD on January 27, 2010, which was converted to a Writ Petition no. 1813–2010 by Chief Justice Khwaja Muhammad Sharif (Lahore High Court) on the very next day. Subsequently, WWF – Pakistan became involved in this suo moto case on the principle of forest conservation. WWF – Pakistan offered its services for a detailed study, covering analysis of the forest conversion, retrieval, and a GIS-based mapping to support the prioritization and forest rehabilitation process. This study was jointly funded by the Punjab Government and WWF – Pakistan.

The Murree Forest Division was facing high encroachment pressures due to ambiguous and un-demarcated forest boundary; furthermore, there were no updated forest records. This lack of management opened the gates for local communities to reach protected and reserved forests. This resulted in the conversion of forest land into agriculture land and housing schemes (Hasan, Research Report No. 182, 2001). Interdepartmental conflicts are also a backend driving force for the encroachers and land mafia.

Gill (2013) reported that 2,862 acres (1,158 ha), which is about 6 % of the total reserved and protected forest land, has been detected under illegal encroachments. A total of 140 First Information Reports (FIRs) have been registered in the police station, Murree, in the last 3 years of demarcation process.

One of the largest renowned daily newspapers of Pakistan, *The Dawn* reported (2010) that the Town Municipal Administration had declared 17 housing societies illegal with a warning for investors. An Environmental Impact Assessment (EIA) is a legal requirement as per EIA Rules 2000 under the Pakistan Environmental Protection Act 1997. An Initial Environmental Evaluation is required for any housing colony; however, if it is close to a reserve forest, then an EIA is a legal requirement. The Environment Protection Agency (EPA) reviews the EIA and, after a public hearing, provides a No Objection Certificate (NOC). These illegal housing societies have not gone through such an EIA process.

---

[3] 1 *kanal* = 20 *marlas* = 5,445 sq. ft. = 505.86 Sq. m; 1 acre = 8 *kanals*.

13.2 Concepts

This section will help the reader to understand basic concepts of forest land management in Pakistan; moreover, it will help to develop the technical grounds for the reader to understand the theme of delineation process involved.

13.2.1 Forest Land Tenure in Pakistan

Ahmed and Mahmood (1998) stated that forests in Pakistan can be classified into two tenure types (on the basis of their legal rights), i.e., state owned and privately owned. This legal status also defines the implementation of forest policy on that land. They further stated that the ratios of state-owned and privately owned forests are 66 % and 34 %, respectively. This means that in spite of regionalized forest administration, major areas of forest in Pakistan are still under state control (Hasan 2008; Ahmed and Mahmood 1998). State-owned forest further can be classified into four subclasses (Ahmed and Mahmood 1998): (1) reserved forests, (2) protected forests, (3) un-classed forests, and (4) resumed lands.

Apart from these, municipal and cantonment forests also have state-owned forest land status. This study will mainly focus on the state-owned reserved and protected forest lands in the study area.

The Forest Act 1927 defines reserved forest as "Land designated after settling their ownership and usage rights under Sections 4–26 of The Forest Act 1927. These forests are usually free from rights and concessions and all acts are prohibited unless permitted specifically by the Government through notifications in legal terms" and protected forest as "Land designated after settling their ownership and usage rights under Section 29–34 of The Forest Act 1927. These differ from Reserved Forests in two ways. Firstly, they have not passed through the lengthy process of admittance or extinction of local peoples' rights or concessions. Secondly, in contrast to Reserved Forests, all acts are permitted in Protected Forests unless prohibited by a notification of the Government; the title 'Protected Forests' may therefore be somewhat of a misnomer in practice" (GOP 2002; Ahmed and Mahmood 1998). Murree forests have a longer history under government control and were left temporarily untouched until the annexation of the Punjab (Hasan 2008).

There are four subclasses of the privately owned forests in Pakistan (Ahmed and Mahmood 1998): (1) Guzara Forests (Communal Forests), (2) Chos Act Areas, (3) Section 38 Areas, and (4) Farm Forest Areas. Ahmed and Mahmood (1998) defined Guzara Forest "subsistence" as "Forests since the time of first settlement of land ownership in 1872, and consists sizable patches of wooded lands close to habitations were set aside to meet the bona fide domestic needs of the local communities." Their ownership is vested in local people, either as individual property, or as joint property known as "Village Shamilat/Malkiat Lands."

13.2.2 *Forest Land Map Records in Pakistan*

Forest departments in Pakistan have inadequate technology and expertise to counter the forest management challenges (Hasan 2008). It is difficult to find updated forest maps in Pakistan. Few of the forest units have well managed forest records, but most of them contain outdated forest records and maps due to decentralization and poor management. The ultimate approach to assess forest land is to use the maps of the Revenue Department. Sometimes the forest boundary has to be referred from small-scale Survey of Pakistan (SoP)[4] maps, which have low spatial accuracy. This increases the discrepancies in boundary-related issues on ground.

13.2.3 *Revenue Map Structure*

Pakistan is still using hand-drafted paper-based cadastre maps for revenue assessment. Recently the government of Pakistan took initiative to make land record available in digitized form (Khan 2012). There are two forms of revenue maps in use (*Patwari* maps) in Pakistan which are described below:

13.2.3.1 Mussavi

Mussavi is a "quantifiable paper map." The measuring unit of *mussavi* is *karam*[5] These basically are surveyed paper maps at different scales depending upon village area, normally at a scale of 1 in.:40 *karams*. This scale may vary from area to area depending upon village size. Figure 13.2 (left) gives a glimpse of a *mussavi* Map from Murree *Tehsil*. A general index used to mosaic all the *mussavi* maps is given in Fig. 13.2.

13.2.3.2 Shajra Parcha/Aks Shajra (Latha)

This is a mosaic reflection of all *mussavi* maps of a village on a cotton cloth called *shajra parcha/aks shajra (latha)*. It is a map of smaller scale than the *mussavi* and covers a larger area.

It can be concluded that land records are authentic and highly informative. The tenure system for forest lands is also well established. However, due attention is required to change the practice of an obsolete mechanism of storage and updating of land records in relevant departments.

---

[4] SoP is a national surveying and mapping organization of the country. It is primarily responsible for all sorts of topographical land surveys of the entire country.

[5] 1 *karam* = 5.5 ft; 3 × 3 *karams* = 1 marla = 272.25 sq. ft.

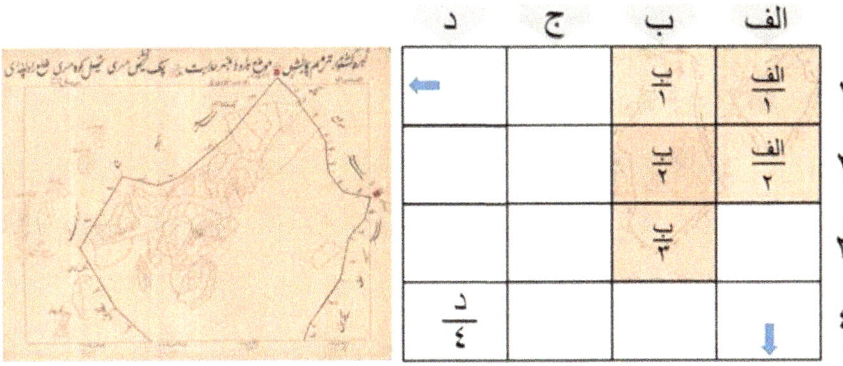

Fig. 13.2 (*Left*) One of *mussavi* maps (scale 30 *karam* equals to 1 in.). (*Right*) General index to *mussavi* map – How to mosaic to *latha/lathas*

13.3 Methods

The study had five general steps: a planning phase, data acquisition and management, data processing, results, and implementation. The planning phase involved the generation of all guidelines to carry out the activities in a smooth and steady manner. Available resources were enlisted and field activities were planned accordingly. The requirements of satellite images and vector data were identified. Timelines were also defined to get the results in time.

13.3.1 Tools Used

For precision surveying, Differential GPS (Leica SR20 Surveying Solution), handheld GPS receivers (Garmin GPSMap 76 CSx), and total station (Sokia Total Solution) were used. A variety of commercial geospatial software and tools including ArcGIS Desktop 9.x, ERDAS Imagine 9.x, Garmin Map Source 6.x (GPS data processing), Leica Geo Office 7, Geo Calculator, and AutoCAD Map 2000I were used to process the data depending upon the nature of the task.

13.3.2 Pre-survey Data Preparation

The baseline maps for detailed field survey were developed by using several reliable data sources. These included latest topographic maps of the Survey of Pakistan (scale 1:50,000) and historic maps of the Survey of India (SoI) of high scale (1:15,840). Both topographic datasets were geo-referenced by using rubber-sheeting

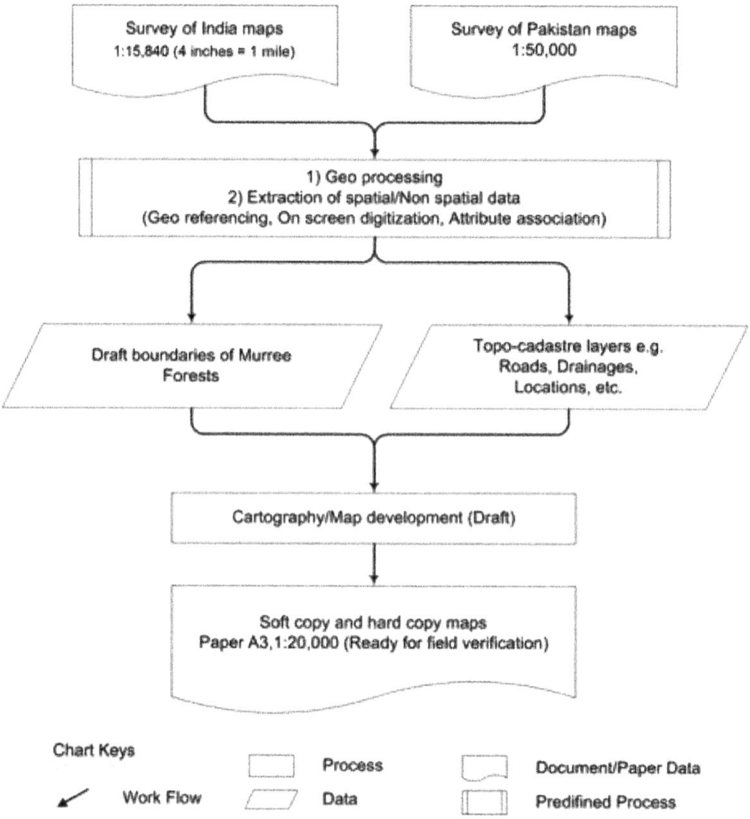

Fig. 13.3 Pre-workflow before boundary delineation survey

technique. Various toponyms including settlements, roads, forest boundaries, and water channels from these topographic maps were extracted in vector format. A total of 2,149 pillar marks (monuments) covering the entire State Forest in MFD were extracted from SoI maps. These extracted boundary marks helped in the estimation of forest boundary lines. On the basis of these draft boundaries, large-scale (i.e., 1:10,000) maps were produced for reference in boundary delineation survey (Figs. 13.3 and 13.4).

Furthermore, for boundary delineation and mapping, high spatial resolution (46 cm) ortho-rectified images of GeoEye-1 (acquisition date October 15, 2009, and June 4, 2010) were procured. "High-resolution merge" with multiplicative and bilinear interpolation techniques was used to merge multispectral and panchromatic bands. This improved the visual interpretability of the dataset. Moreover, images from Google Earth were also acquired and geo-rectified for temporal change detection.

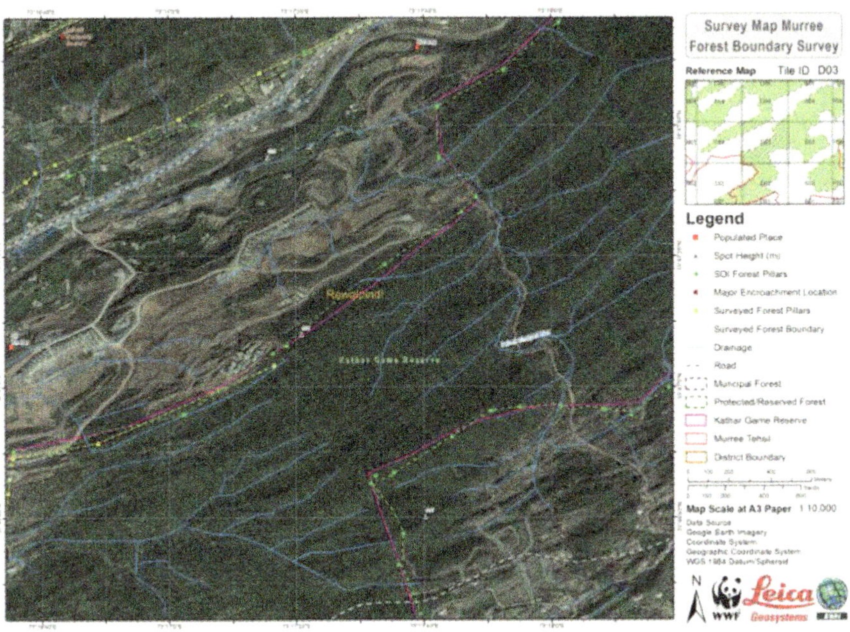

Fig. 13.4 A tile from the map series used for field survey

Fig. 13.5 Temporal images (2002, 2006, and 2012) of Bahria Town – Kathar RF (Google Earth)

13.3.3 Delineating State Forest Boundary Lines

13.3.3.1 Preliminary Assessment of the Problems in the Area

A preliminary assessment of the study area identified major encroachers through a couple of field surveys, pictorial records, multi-temporal images from Google Earth, and consultative meetings with MFD field staff. Maps of temporal changes highlighted the forest destruction in time domain (2002–2012) due to the development and expansion of housing societies inside and along the State Forest land. Figure 13.5 shows the forest degradation in and along Kathar reserved forest.

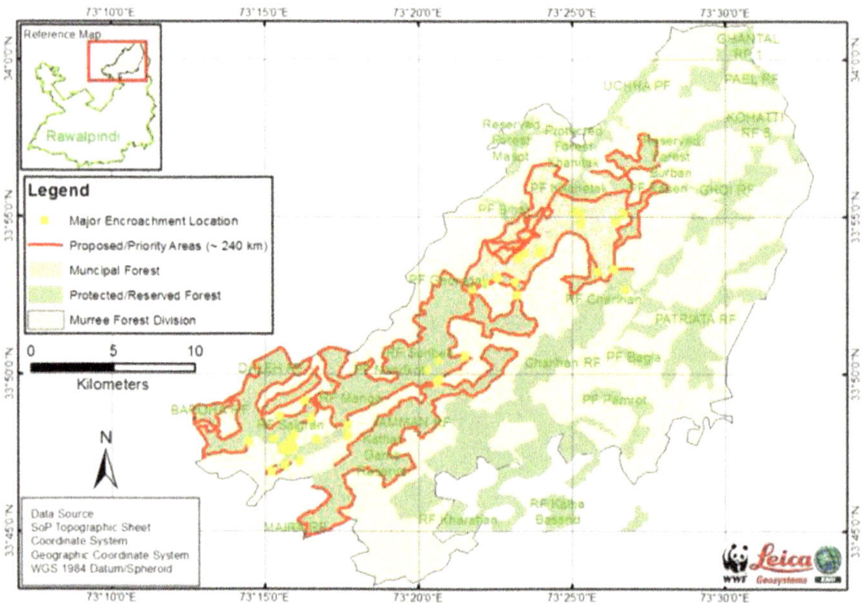

Fig. 13.6 Priority areas for the boundary delineation in MFD

Major societies and their respective encroachment areas were tabulated and small-scale maps were compiled to identify encroachment pressure areas. On the basis of the preliminary assessment, a 240 km-long stretch of RF/PF boundary line was selected as priority for delineation as shown in Fig. 13.6. Updates regarding the encroachments were submitted to the Lahore High Court on every court hearing of the case.

13.3.4 Formulation of Study Monitoring Committee (SMC) and Survey Team

To monitor the survey and mapping activities on a regular basis (quarterly), a committee comprising of representatives from the following departments was formulated: (1) Survey of Pakistan, (2) Punjab Forest Department, (3) Revenue Department, and (4) WWF – Pakistan.

The involvement of Survey of Pakistan (legal authority for mapping and surveying in the country) fulfilled legal requirements in the delineation survey. More than two hundred officials, technicians, and field staff were engaged in the field surveys from the mentioned four departments.

13.3.5 Boundary Delineation Survey

Officials from the Punjab Revenue Department (authority for land administration in the country) and Punjab Forest Department identified/demarcated the state forest boundary pillars using maps (*latha* maps, *mussavi* maps, SoI topo maps, and forest history file maps). Two types of survey data (GPS receiver and total station survey) were collected for identified pillar marks using cyclic mechanism in parallel as shown in Fig. 13.7. The project team decided to develop a set of geodetic network controls and at least two forest bench marks (especially to serve forest boundary lines in MFD) as a baseline to total station survey using a technique called fast (rapid) static DGPS surveying with post-processing solution. It was made sure that Positional Dilution of Precision (PDOP) is not more than two meter for base station of DGPS setup (Poole 2012). Positional accuracy criterions were defined and projections were standardized. According to the national mapping standards, SoP officials observe boundary coordinates in planner coordinate system ("Lambert Conformal Conic" projection with "Everest Definition 1962" datum/spheroid) (Mugnier 2009). Total station survey was completed maintaining a centimeter level accuracy on the basis of DGPS baseline. More than 4,300 GPS and 2,400 total station coordinates/points were collected covering the whole study area. These coordinates were processed and transformed to standard projection (geographic coordinate system with WGS 1984 datum/spheroid). Necessary attributes were also associated.

The team developed a consensus to resolve the state and community land conflicts wherever it was possible (Fig. 13.8). The ambiguities like missing *sehaddas*[6] were also resolved using *mussavi* maps of adjacent villages. The remaining areas (about 1 %) were left under ambiguity status due to security issues and missing revenue records. Meanwhile a number of sessions on basic level GIS and GPS training were also arranged for forest officials for capacity building.

13.3.6 Forest Boundary Line Delineation and Finalization

Processed survey coordinates were connected to form forest boundary lines using topological vector data editing techniques. Hand draft sketches from the field helped to connect the coordinates accurately. Figure 13.9 shows the delineated forest boundary lines.

To eliminate human errors and other possible misconnections in pillar marks/boundary lines, the delineated boundary was subjected to a number of screen overlay tests to finalize. The very first overlay test was to verify delineated boundary line against the *Patwari* maps (*latha*/*mussavi* maps). The lengths between two adjacent

[6] *Sehadda* (village tri-junction) is a revenue stone placed at the junction point of three adjacent village boundaries. Normally a village map (*latha*) is marked with the help of three *sehaddas*.

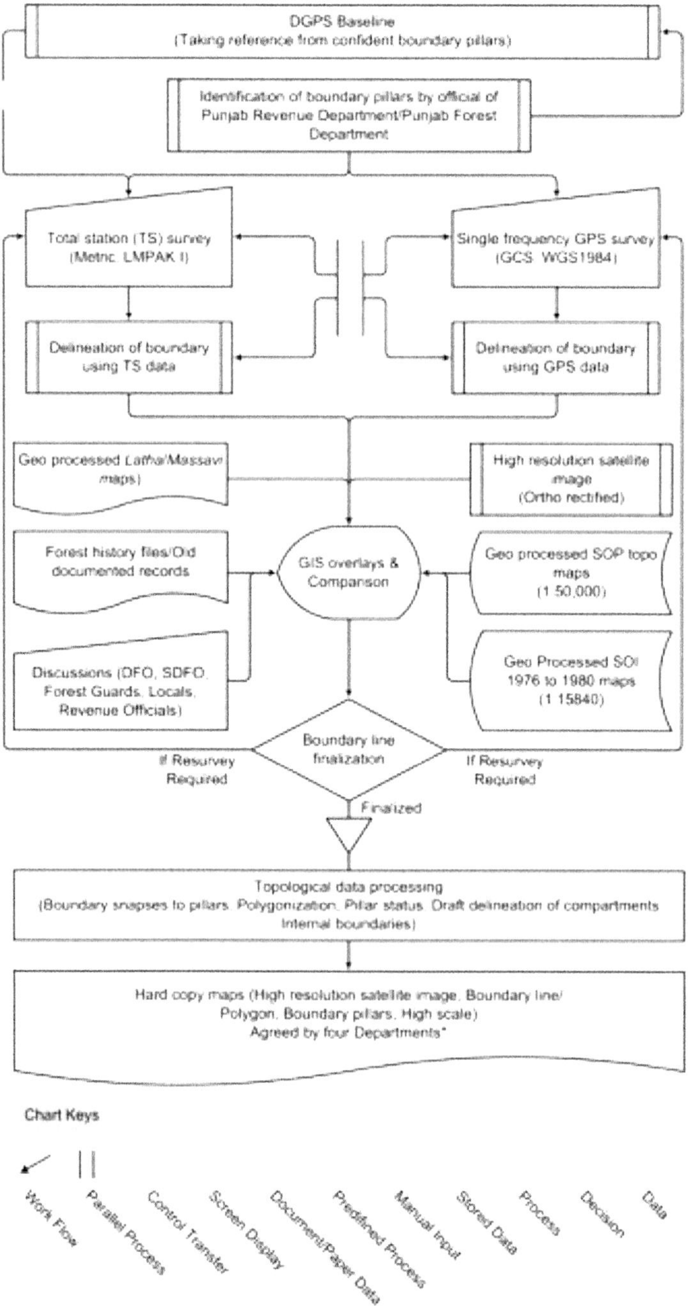

Fig. 13.7 Flow diagram of boundary delineation process

Fig. 13.8 (*Left*) Revenue map discussion and TS observation of the pillar. (*Right*) GPS coordinate of the pillar is being recorded (*Source*: WWF – Pakistan)

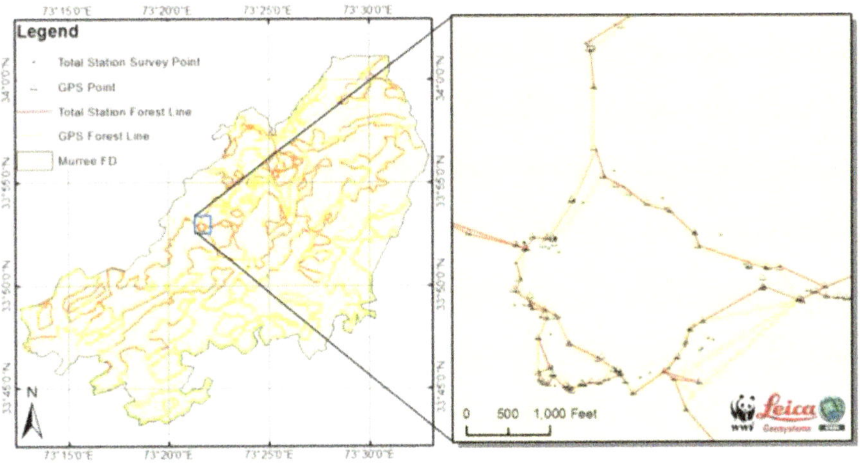

Fig. 13.9 Map showing the delineated forest lines

pillar marks were compared with the lengths mentioned on *mussavi* maps (Fig. 13.10). In a similar way, if a line qualified overlay with the *mussavi* map, then it was subjected to overlay with historic topographic maps of SoI and SoP topographic maps and finally to high-resolution ortho-rectified GeoEye-1 satellite imagery. If a line failed to comply any of the above overlays, then the area was subjected to another survey to acquire accurate coordinates of possible pillar marks. Ground knowledge was an important input to these tests. The lines were finalized on the basis of ground knowledge of MFD officials as well after a consensus between SoP, Punjab Revenue, and Punjab Forest department officials (Fig. 13.11). In some inaccessible areas, physical features like the Jhelum River in the north-eastern edge of the study area provided support as reference for boundary lines. The state forest land was then polygonized using the finalized boundary lines. Finally, these polygons were given attributes using existing tabular records such as forest name, forest

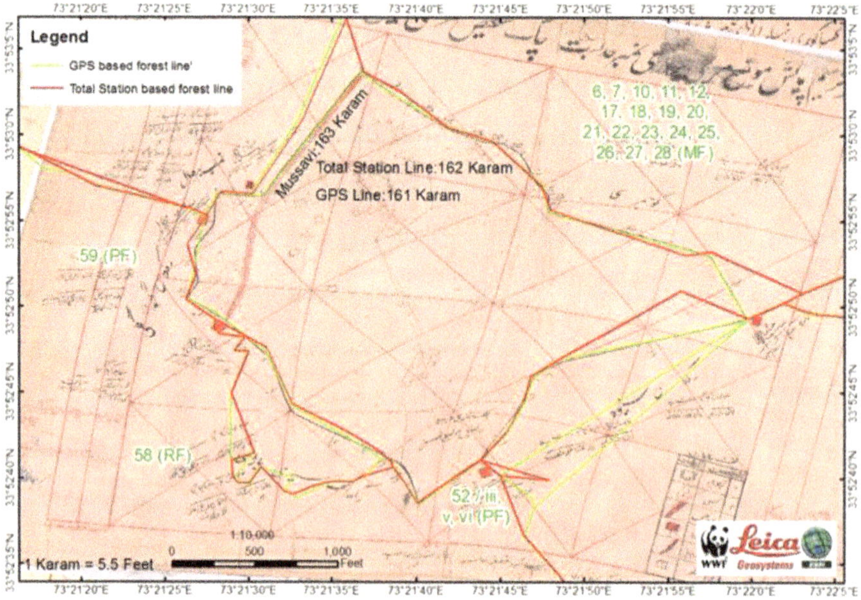

Fig. 13.10 Map showing the overlay of delineated lines on *mussavi* map

Fig. 13.11 (*Left*) Discussion on boundary using *latha* map. (*Right*) Boundary finalization on computer machine incorporating comments from officials of SMC (*Source*: WWF – Pakistan)

numbers/compartments, forest administrative block, subdivision information, area, and forest type. The status attribute, i.e., constructed, degraded or old, of the pillar marks was also associated to each pillar coordinate.

13.3.7 Identification of Encroachments

The team also identified encroachment areas in the MFD by using overlay of finalized boundary with satellite images of GeoEye-1. These lab-identified areas (constructed area, forest clearing, and agriculture fields) were also confirmed on ground. Figure 13.12 illustrates the identified encroachments in MFD.

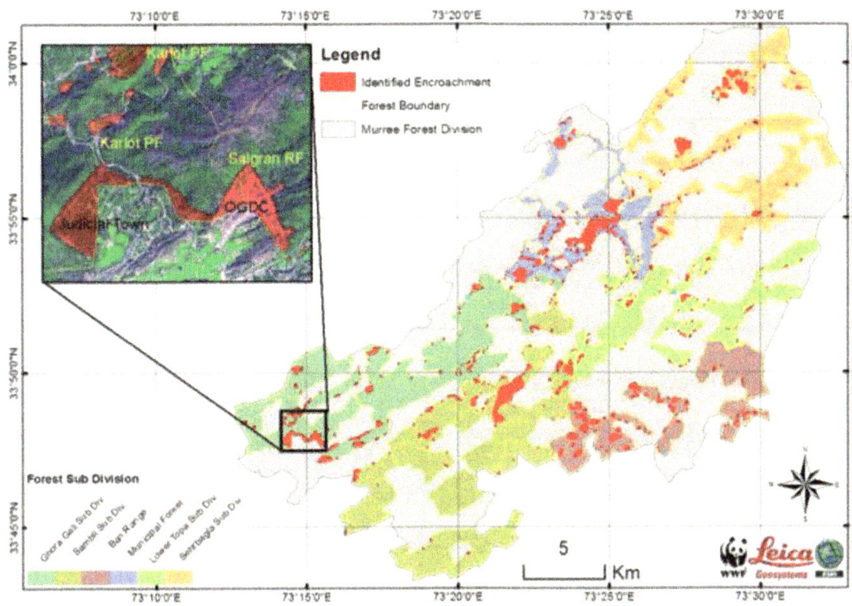

Fig. 13.12 Identified encroachments in MFD highlighted with *solid red color*

13.4 Results and Discussions

The joint team conducted field surveys delineating a total forest boundary length of 625 km from October 2011 to November 2012, and approximately 4,390 GPS/total station survey points were observed. So far, the Punjab Forest Department has erected 1,928 pillars over a 4-year demarcation period from August 2008 to November 2012. By January 2013, 140 FIRs have been registered in the police stations of Murree Tehsil. The Punjab Forest Department has so far retrieved 518 ha (1,279.21 acres) of encroached area out of a total of 1,158 ha (2,862 acres) identified. The delineation process identified 2,325 number of encroachers in and along the forest boundaries currently occupying 641 ha (1,584 acres) of State Forest land in MFD. Historic records of forest maps showed the marks of 2,149 boundary pillars, whereas only 23 pillars were verified on ground during survey, while the remaining pillars had possibly been deteriorated or removed. Table 13.2 gives the statistical information of forest subdivisions in MFD in detail after delineation.

About 99 % of the forest boundaries in MFD were delineated. Total station was used for priority areas (as referred in Fig. 13.6) and single frequency GPS receiver was used for the rest. Both of the data were corrected using DGPS baseline. To estimate the accuracy in areas with reference to revenue records, lines of three internal

Table 13.2 Encroachment and other stats of forest subdivisions/ranges in MFD

Forest subdivision/ range	Area (ha)	Boundary marks/pillars	Number of encroachers	Area under encroachment (ha)	FIRs	Retrieved land area (ha)
Ghora Gali	4,606.61	939	329	200.56	56	171.02
Sambli	5,369.98	786	99	183.62	43	225.92
Lower Topa	2,439.12	620	453	48.09	27	61.75
Sehr Bagla	3,182.12	666	207	21.92	2	11.52
Municipal range	2,160.99	1,054	1,027	146.52	12	43.17
Ban range	2,368.53	414	210	40.22	0	4.30
Total	20,127	4,479	2,325	641	140	518

Table 13.3 Comparison of areas

Locality/forest	Revenue record area Kanals	DGPS survey area Kanals	GPS survey-1 area Kanals	GPS survey-2 area Kanals	Total station survey area Kanals	Delineated area Kanals	Accuracy (%)
Abbasi colony	72.5	73.92	71.4	71	73.95	72.48	99.97
Ahata Noor Khan	803	–	810		801.5	808	99.37
Hildholo	456	–	526	603	–	455.4	99.87
	Area (ha)						
Entire MFD (RF/PF)	19,135	–	–	–	–	20,110	>94.9

1 *kanal* = 20 *marlas* = 5,445 sq. ft. = 505.86 sq. m
1 acre = 8 *kanals*
1 ha = 2.471 acre

chacks[7] were taken as example. Table 13.3 gives the area calculations of different internal *chacks* in comparison with the total forest area delineated after the surveys. Values indicate that all of the techniques have some inherent errors due to certain factors. For example, the obstacle in the way of reflector may be one reason for observing a location in offset against the original location. Foliage cover may be causing multipath effect for GPS and DGPS measurements (SESD 2011). Anyhow, correction of the lines was done in the GIS laboratory after overlaying the *Mussavi* Maps and mutual discussions. Overall an accuracy of greater than 99 % in the delineated areas of state forests was acquired after delineation process as evident from the areas of internal *chacks*.

The reason for the difference between areas of entire MFD (about 975 ha or 2,410 acres) is the area of unmeasured internal *chacks*. Gill (2013) quotes a figure of about 100 plus internal *chacks* within MFD which are pending for demarcation.

[7] Reserved and protected forests in MFD occupy many small enclosed localities usually called internal *chacks* whose rights were declared under the ownership of local villagers at the time of settlement.

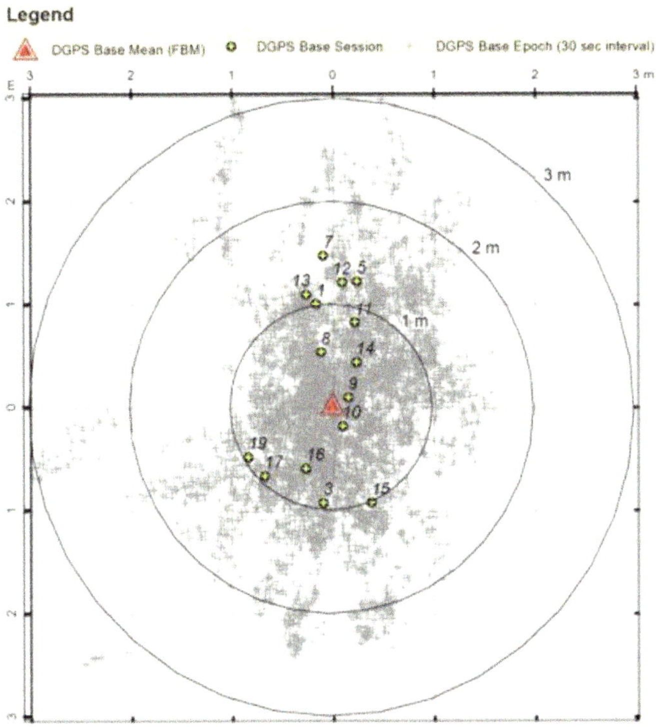

Fig. 13.13 Position calculated from epoch of interval 30 s from DGPS base

Two forest bench marks (one in Ghora Gali forest at an elevation of 6,400 ft and the other at the Patriata Forest Rest House at 7,800 ft elevation) were established using Leica SR20 Post-Processing Differential Solution (the manufacturers of the SR20 unit claims an accuracy with root mean square distance of 40 cm in measuring the position of any location in ideal conditions (Leica 2012)). Positioning precision and accuracy are highly affected by foliage in forests as compared to open area (Rodríguez-Pérez et al. 2006). Data acquisition of geodetic network controls was completed with minimal errors.

Considering DGPS Base for forest bench mark (FBM) processing, the receiver acquired its location in 14 sessions, and each session lasts for a few minutes to several hours. A spatial distribution of just three percent of the total data acquired for base is shown in Fig. 13.13 above along with the mean FBM.

For the geodetic network design, fast (rapid) static GPS technique was used which has an accuracy of about first order. For a maximum baseline length of 30 km, the limit of error in position may be calculated as follows (Mumma 2005):

$$\text{Accuracy standard for first order GPS Survey} = 0.035\,\text{m} + 1:100{,}000$$

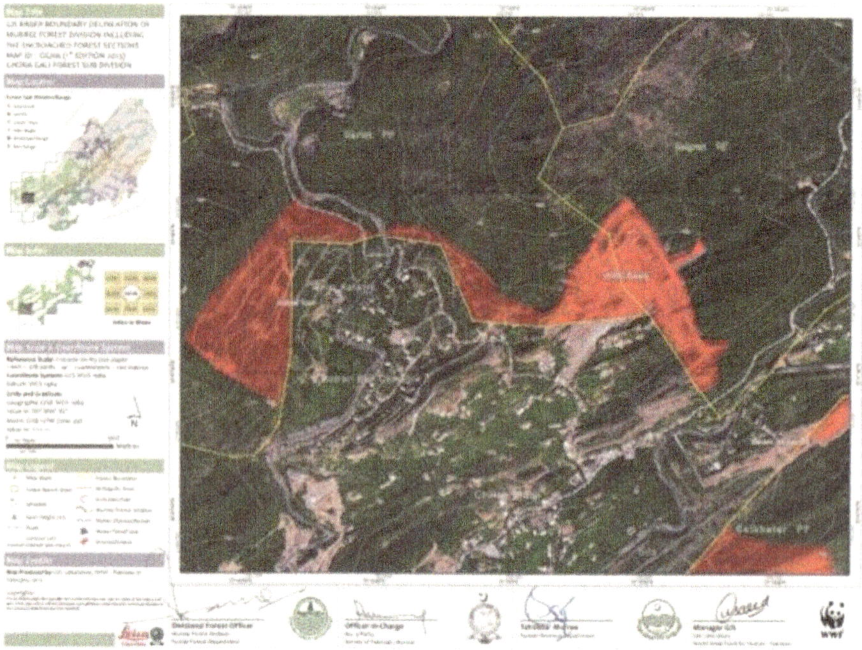

Fig. 13.14 Sample map from the atlas of legal boundaries of Murree

$$\text{At 95\% confidence level the deviation limit in baseline is} = \pm\sqrt{(0.035)^2 + \left(\frac{30{,}000}{100{,}000}\right)^2}$$

$$= \pm 30.20 \text{ cm}$$

Hence, a network of DGPS base and rover has an uncertainty of ±30 to ±50 cm with all the ambiguities included in the boundaries delineated using total station solution.

To accomplish the study aim of providing the court a document of evidence against encroachments, a map atlas containing the forest boundary, pillar marks, and encroachment information was designed and developed. Officials from the four departments (Punjab Forest Department, Punjab Revenue Department, Survey of Pakistan, and WWF – Pakistan) endorsed this document by signing each of more than 150 maps. The scale of these maps was high (1:10,000) to clearly depict the encroached and built-up land overlaid on fine resolution (0.46 m) GeoEye-1 satellite image. Figure 13.14 shows a sample map from the atlas for the reader's reference. To understand and interpret the maps in the atlas, the working bench of the Lahore High Court on the case was given a briefing on the basics and applications of GIS in forensic sciences. Recently the atlas is shared with the court and the decision on this case is expected soon. The team and stakeholders are hopeful and looking forward to the retrieval of encroached State Forest land.

The atlas is also shared with the four lined departments/stakeholders. Currently, the management of MFD is working to safeguard and for possible retrieval of the forest land using this atlas. Furthermore, a scanned copy of the signed atlas is also published online (see WWF – Pakistan website) for general public.

13.5 Conclusions and Recommendations

The present work on Murree State Forests is the first study of its kind in Pakistan. In this study, GIS is used as an effective tool for replacing the conventional methods of forest land record keeping; hence, it will lead to a better management of the natural resources of the area. A big hurdle in the demarcation/delineation process was misleading land records that were unintentionally or perhaps deliberately tempered or made unavailable. Hence, the study resulted from no/scattered data to detailed baseline information in a single atlas, which has legal implications too. Literature revealed that previously no such extensive spatial data records were built for any other forest in Pakistan. The same exercise should be carried out at provincial as well as national levels.

The boundary pillar data collection activity was carried out at large scale (with the involvement of more than 200 officials/technicians/field staff of the four departments). Without systematic storage and management, it was impossible to complete this process in time. It took a total duration of about 6 months to complete the boundary delineation survey. However, this study highlights that there is still a need to erect 2,462 boundary pillars at their delineated locations. The boundary lines delineated on the pillar marks will ultimately help in forest monitoring to highlight and discourage further encroachments. Earlier, in each court case regarding forest land encroachment in MFD, multiple maps with contradicting information by officials (from the forest and revenue departments) were provided to the Lahore High Court and the Pakistan Supreme Court. This was delaying or misleading in court decisions. These duly signed maps of the atlas have resolved this issue by providing single concrete forest boundary extents.

After the submission of initial alarming temporal maps, the Lahore High Court banned single tree cutting in MFD (court order dated April 08, 2010). The encroachment issue was well highlighted in the print and electronic media of Pakistan. Moreover, during the boundary delineation activities, the awareness among communities and stakeholders for the environmental laws and protection status of the forests was enhanced. A clear reduction in encroachment pressures is one of the positive outcomes. Developed FBMs were very helpful for the study and would be highly useful for a geographic reference in future activities. It is recommended that monitoring and maintenance of these boundary references should be carried out on regular basis.

The development of encroached societies is illegal as per "The Forest Act 1927." According to Section 4-A and 4-B of "West Pakistan Forest Rules under the Acts," no one in any state can partition or clear jointly on State Forest lands without the sanction of the provincial government. Hence, the land occupied by illegal societies in RF/PF and *Guzara* Forest should be recovered immediately and must be rehabilitated to the original state.

All forest field staff should be trained in using geospatial technology to make them resourceful for better forest management. A basic level GPS training is strongly recommended for the management staff. It will help them to translate forest boundaries on the ground with the help of GPS and maps. The study furthermore suggests that the Provincial Forest Departments in Pakistan should be equipped with the latest technologies, more specifically well-established GIS laboratories with dedicated and well-trained human resources. On policy level, it is suggested to have a national level neutral institution, namely, "Forest Survey of Pakistan," that keeps records and monitors the state forests on a regular basis just like the Forest Survey of India.

Acknowledgments A number of departments and individuals provided support and contributions to this study. The following text attempts to acknowledge the contributors for the successful completion of the project. However, apologies are extended in advance if an individual is missed unintentionally.

This study was made possible principally through the enthusiastic participation of the field staff of the Punjab Forest Department (three DFOs, five SDFOs, three RFOs, 14 block officers/foresters, 128 forest guards, 15 clerks/stenos, and 3 drivers), Survey of Pakistan (9 technical persons), officials of Punjab Revenue Department (27 in number), and WWF – Pakistan (19 technical persons).

Supporting Organizations:

Special thanks are due to the Punjab Forest Department (PFD), Punjab Revenue Department (PRD), and Survey of Pakistan (SoP) for the support in conducting joint field surveys, data processing, and validation. The authors are grateful to the National Council of Conservation of Wetlands (NCCW) Ministry of Climate Change for providing Differential Global Positioning System (DGPS) for the study. The team is thankful to the Punjab Forest Department and Partnership Fund, Indus for All Programme of WWF – Pakistan for the funding/logistics support.

Project Monitoring Committee:

Mr. Raja Khalid Hussain, Chief Conservator of Forests Northern Zone, Rawalpindi, Punjab Forest Department
Mr. Ghulam Sarwar, Director, Survey of Pakistan
Mr. Imdad Ullah Bosal, Commissioner, Rawalpindi
Dr. Omar Jahangir, Assistant Commissioner, Murree
Mr. Saqib Zafar, District Coordinator Officer, Rawalpindi
Mr. Shahid Rashid Aawan, Deputy Secretary (Planning), Forestry, Wildlife, and Fisheries Department, Lahore, Punjab

Legal Advisory Team:

Mr. Ahmer Bilal Soofi, LLM, Cambridge, UK (Advocate Supreme Court of Pakistan)
Mr. Khalid Mian (Advocate High Court)
Ms. Amna Warsi, LLM, Pb. (Advocate High Court)
Ms. Ayasha Warsi, LLM, Pb. (Advocate High Court)

Map Production Team and Cartography:

Irfan Ashraf, Urooj Saeed, Kaif Gill, Naeem Shehzad, Wasif Yousaf, Syed Muhammad Irteza, and Zain-ul-Abdin
Geographic Information System Laboratory, WWF – Pakistan

Map Validation:

Muhammad Javed Gill (DFO Murree), Raja Akram (Incharge SoP Murree), Rana Nayab Khan (Tehsildar Murree), and Urooj Saeed (Manager GIS, WWF – Pakistan)

Software and Imagery Support:

The authors would like to thank ESRI for the software grant. Google Earth images were significant for the initial assessments of the encroached lands. ESRI base map imagery was used for forest boundary overlay at outer peripheries of Murree Forest Division (where satellite images of GeoEye-1 were missing).

The authors are thankful to Maj. (R) Shah Nawaz Badar (Secretary Govt. of the Punjab, Forestry, Wildlife and Fisheries Department), Mr. Ali Hassan Habib (Director General, WWF – Pakistan), Major Muhammad Ilyas Ch. (Deputy Surveyor General, Survey of Pakistan), Mr. Ghulam Sarwar (Director, Survey of Pakistan), Mr. Imdad Ullah Bosal (Commissioner Rawalpindi), and Ms. Uzma Khan (Director Biodiversity, WWF – Pakistan) for their guidance and support throughout the study.

References

Abbasi A (2010) Prime Murree land worth billions of rupees grabbed (ed. Shakil-ur-Rahman M), The News, 27 January

Ahmed J, Mahmood F (1998) Changing perspectives on forest policy – policy that works for forests and people series no. 1. IUCN Pakistan and International Institute for Environment and Development, Islamabad/London

Ali J, Benjaminsen TA (2004) Fuelwood, timber and deforestation in the Himalayas. Mountain Res Dev 24:312–318

Bhatti ES (2011) An overview of forests in Pakistan. Lectures – World Environment Day 2011. Pakistan Engineering Congress, Lahore/Punjab, Pakistan. http://pecongress.org.pk

Bukhari A (1997) The role of NGOs in promoting fuel wood production in Pakistan, Wood Fuel Production and Marketing in Pakistan-RWEDP Report No. 49. The FAO Regional Wood Energy Development Programme in Asia, Faisalabad, pp 31–34

Christopherson RW (2009) Geosystems: an introduction to physical geography. Prentice Hall, Upper Saddle River

Down, T. (2010). Murree leaves 17 societies shivering, December 16, Murree, Punjab, Pakistan

FAO (2010) Global forest resources assessment 2010 main report. FAO, Rome

Gill MJ (2013) Divisional Forest Officer, Murree Forest Division, Rawalpindi, Issues of forests in Pakistan (Interviewer: Ashraf I), 13 July, dfo_murree@hotmail.com

GOP (2002) The Forest Act, 1927. Punjab Laws Online. http://punjablaws.gov.pk/laws/40.html. Accessed 23 June 2012

Hasan L (2001) Analysing institutional set-up of forest management in Pakistan. Pakistan Institute of Development Economics (PIDE), Islamabad

Hasan L (2008) An anatomy of state failures in forest management in Pakistan, 23 AGM papers. Pakistan Society of Development Economics, Islamabad, pp 1–25

Hussain T, Khan GS, Khan SA, Masood N, Ashfaq M, Sarwer N (2012) Farmers' agro forestry in Pakistan, farmers' role-trends and attitudes. Curr Res J Soc Sci 4(1):29–35

Irshad M, Khan A, Inoue M, Ashraf M, Sher H (2011) Identifying factors affecting agroforestry system in Swat, Pakistan. Afr J Agric Res 6(11):2586–2593

Khan M (2012) Punjab gets its first center for land record management information system. ProPakistani. http://propakistani.pk/2012/06/18/punjab-gets-its-first-center-for-land-record-management-information-system/. Accessed 19 Aug 2012

Leica (2012) Leica SR20 single frequency survey solution. Retrieved 14 March 2012, from Navigation Systems: https://rs6.eporia.com/company_5//SR20_Brochure_en.pdf. Accessed 19 Aug 2012

Mugnier CJ (2009) Grids and datums. http://www.asprs.org/a/resources/grids/. Accessed 19 Aug 2012

Mumma B (2005) Highway surveying manual – chapter 7. Washington State Department of Transportation, Environmental and Engineering Service Center, Design Office, Olympia

PADs (2011) Agricultural statistics of Pakistan 2010–11. Pakistan Bureau of Statistics: http://www.pbs.gov.pk/content/agricultural-statistics-pakistan-2010-11/. Accessed 16 Aug 2012
Poole I (2012) GPS accuracy, errors & precision. Radio-electronics: http://www.radio-electronics.com/info/satellite/gps/accuracy-errors-precision.php. Accessed 2 Sept 2012
Roberts TJ (1992) The birds of Pakistan. Oxford University Press, Michigan City
Rodríguez-Pérez JR, Alvarez MF, Sanz E, Gavela A (2006) Comparison of GPS receiver accuracy and precision in forest environments. Practical recommendations regarding methods and receiver selection. In: Proceedings of XXIII FIG Congress. FIG International, Munich, p 16
SESD (2011) SESD operating procedure Global Positioning System. U. S. Environmental Protection Agency – Science and Ecosystem Support Division, Athens, GA. http://www.epa.gov/region4/sesd/fbqstp/Global-Positioning-System.pdf. Accessed 26 Feb 2012
WWF-Pakistan (2011) NEWS – an agreement signed between Punjab Forest Department and WWF – Pakistan to map and monitor the forests of Murree forest division. http://www.wwfpak.org/newsroom/061011_mougs.php. Accessed 9 Mar 2012

Annotated Bibliography of Selected Court Cases Involving Geospatial Technologies

Provided below are selected court cases in which the use of geospatial technologies was either in dispute or provided evidence to resolve a dispute. The technologies are grouped into three categories: GPS, remote sensing, and GIS. Within each category, cases are sorted chronologically from the earliest case to the most recent. There is a mixture of supreme court cases, district court case, and appeals court cases, and many different US states are represented. Although there are certainly many other court cases and geospatial technologies that could be provided, it is hoped that this list will provide a starting point for most readers interested in discovering the diverse ways geospatial technologies have been represented in the US judicial system.

Global Positioning Systems (GPS) Cases

Johnson v. State (1986) Fla. App., 492 So.2d 693

After police obtained a warrant permitting a single GPS tracking device to be installed on an airplane, a second backup device was also installed as a precaution. The first GPS device did fail and data from the second GPS device was introduced during trial. The Florida Appeals Court determined the second GPS device constituted "an illegal entry beyond the scope of the warrant."

State v. Campbell (1988) Or., 306 Or. 157, 759 P.2d 1040

The opinion of the court was that the use of any tracking device without a warrant or in an emergency situation was a violation of the state constitution.

United States v. McIver (1999) 9th Cir., 186 F.3d 1119

McIver sought to exclude GPS tracking evidence on the basis of being an illegal search and seizure under the Fourth Amendment. Police attached a magnetized GPS to the undercarriage of the defendant's truck while parked in his driveway. The court rejected the contention since the defendant made no attempts to shield

G.A. Elmes et al. (eds.), *Forensic GIS: The Role of Geospatial Technologies for Investigating Crime and Providing Evidence*, Geotechnologies and the Environment 11, DOI 10.1007/978-94-017-8757-4, the undercarriage of his vehicle from police and since the device did not interfere with the operation of the vehicle. The court determines that subsequent monitoring was not a search.

State v. Jackson (2002) Wash. Ct. App., 46 P. 3d 257, 261
Appeals court decided GPS tracking did not violate Fourth Amendment rights or state constitutional provisions. Instead, GPS was viewed as an augmentation of visual tracking. The defendant's use of public road revealed information the police would have been able to document without any such tracking devices.

Whitehead v. State (2002) Ga. App., 258, 271, 574 S.E.2d 351
With consent, police attached GPS device to the vehicle of an informant. The informant was guided to the home of a drug dealer home by a friend. After purchasing 100 pounds of marijuana, police tracked the vehicle and eventually stopped it to recover the drugs. Using the seizure, police arrested the drug dealer. The appeals court affirmed that there was probable cause to arrest the drug dealer.

State v. Clifton (2003) 158 N.C. App., 580 S.E.2d 40
Defendant used counterfeit checks to purchase a vehicle equipped with a GPS system. Police located the vehicle through the manufactured GPS. The North Carolina Appeals Court affirmed the conviction of the defendant for obtaining property by false pretenses.

State v. Jackson (2003) Wash.,76 P.3d 217, 224
Court ruled that a warrant was required prior to using a GPS tracking device. Unlike visible observation of binoculars, GPS did not merely augment visual tracking but instead provided a technological substitute that was capable of extensive intrusions into private affairs.

State v. Jackson (2003) Wash., 150 2d 251, 76 P.3d 217
Defendant reported his daughter missing. Police obtained a search warrant and attached GPS devices to Jackson's vehicles while impounded for the search. After returning the vehicles, police informed Jackson that if the missing girl was hastily buried, animals would dig her up and she would be found and recovered quickly. Data from the GPS revealed Jackson traveled to two separate remote locations. Police recovered the missing body at the second location. Both the district and appeals court determined a warrant was not required under the Washington State constitution. The Washington Supreme Court made the final determination that GPS tracking without a warrant was a violation of the state constitution. However, since police did have a warrant to install the GPS, the use of GPS data as evidence was permissible.

People v. Lacey (2004) N.Y., No. 2463 N/02, WL 1040676
First New York case seeking constitutional protection against GPS devices. The New York court decided a warrant was required before law enforcement could attach a GPS to a suspect's vehicle and track movements. The court noted that a person is entitled to feel secure knowing their movement is not being tracked without a warrant based on probable cause.

United States v. Bennett (2004) U.S. App. 9th Cir., 363 F.3d 947
A joint task force targeting smuggling activity from Mexico into Southern California boarded Bennett's boat and, after an extensive search, found over 1,550 pounds of marijuana. Defendant was convicted and appealed the admission of GPS testimony indicating Bennett had been in Mexico prior to his arrest. During trial, the custom officer testified about the GPS data but did not produce the data. When asked, the officer stated he was not the custodian but had witnessed the data. The trial court determined the government is not excused from the best evidence rule's preference for the original data and therefore contended that the GPS-based testimony was inadmissible under the best evidence rule.

United States v. Berry (2004) D. Md., 300 F.Supp.2d 366
Police obtained a warrant to place a GPS device on the vehicle of Berry who was suspected of trafficking heroin. The court regarded GPS tracking as an extension of visual tracking and supported the probable cause to obtain the warrant. Although GPS data was recorded after the warrant expired, that data was not introduced at trial.

Brown v. Texas (2005) Tex. App., 163 S.W.3d 818
The defendant was a truck driver convicted of murder. At trial, GPS records from his truck were presented as evidence to prove he was at the location the victim was found around the time the victim would have been strangled. Two witnesses testified to the GPS records: an employee of the trucking company who stated the GPS records were business records, and a geographer. Defendant appealed the qualification of the geographer as an expert witness. The appeals court affirmed that the geographer, who used GPS on a daily basis, had a degree in geography, and demonstrated knowledge of how GPS worked and its reliability was a qualified expert witness.

Chism v. State (2005) Ind., 824 N.E.2d 334
As part of his sentence for conspiracy to deliver drugs, Chism was placed on home detention and ordered to wear a GPS anklet. Chism argued the state code did not permit GPS monitoring. The appeals court agreed that no authority existed for the trial court to assign GPS monitoring. A review by the Indiana Supreme Court affirmed the original trial court's order to wear a GPS anklet, citing the GPS device did not monitor movement but only broadcasted alerts if the defendant left his residence.

Elgin v. St. Louis Coca-Cola Bottling Co. (2005) U.S. Dist., Case No. 4:05CV970-DJS
Defendants installed GPS trackers on company vehicles after vending machines experienced cash shortages with no sign of forced entry. Despite the installation of trackers on Caucasian employees, the plaintiff alleged the act was racially motivated. The investigation by the defendant cleared the plaintiff of any wrongdoing and he received no adverse employment action. The court opined that the installation of tracking devices on their own vehicles did not intrude upon the employee's expectation of privacy.

People v. Gant (2005) N.Y. Co. Ct. [Westchester County] WL 1767655 NY Slip Op 25307; 9 Misc. 3d 611; 802 N.Y.S.2d 839

Grant was indicted for criminal possession of narcotics. Defendant motioned to suppress all GPS evidence. The court ruled that no warrant was required to place a GPS device on a vehicle when investigating a crime, the defendant had no expectation to privacy on public roads, the defendant was unable to demonstrate he owned the vehicle with the attached GPS, and the defendant was not entitled to suppression of evidence obtained from the use of the GPS device.

State v. Jackson (2005) Tenn. Crim. App., No. M2004-00562-CCA-R3-CD

Defendant was convicted of possession of cocaine within 1,000 ft of a school. On appeal, the testimony of the surveyor who measured the distance from the school to the offense location with a GPS was determined to have a scientific foundation. The expert witness had been a professional surveyor for 16 years, passed several tests, and used GPS devices for 7 years. The witness explained the GPS was accurate to 1½ in. The appeals court affirmed the trial court did not abuse its discretion in allowing the testimony of this witness to prove the offense occurred within 1,000 ft of a school.

Turner v. Am. Car Rental (2005) 92 Conn. App.

The plaintiff rented a car from the defendant. The car was equipped with a GPS device to track the position and speed of the vehicle. According to the contract, the defendant would impose a penalty for speeding. The plaintiff alleged the defendant never stated the penalty would be drawn directly from his bank account and he furthermore alleged the GPS unit in the car violated his privacy. The appeals court concluded the plaintiff failed to show that he had any expectation of privacy on a public highway. As for the fee, the original court sided with the plaintiff and determined the defendant was engaged in deceptive practices.

United States v. Moran (2005) N.Y. U.S. Dist., 349 F.Supp.2d

Moran protests a GPS device attached to his vehicle by law enforcement personnel without a warrant. The GPS device tracked Moran's vehicle for 2 days. Law enforcement personnel could have conducted a visual surveillance of the vehicle as it traveled on the public highways. The district court found that Moran had no expectation of privacy in his vehicle on a public roadway. Thus, there was no search or seizure and no Fourth Amendment implications in the use of the GPS device.

Medina v. State (2006) Fla. App., 920 So. 2d 136

The trial court rejected on objection by the plaintiff's counsel that GPS tracking evidence did not satisfy the Frye standard. Florida Court of Appeal affirmed the trial court decision on the grounds that GPS technology was neither new nor novel and had long been accepted within the scientific community and was therefore not subject to a Frye hearing.

People v. Randolph (2006) Cal. App., Unpub. 4th Appellate District, Division 2

Defendant pleaded guilty to abuse of his spouse and was required to submit to continuous monitoring by a GPS device. The defendant appealed, arguing the

GPS monitoring was invalid, unreasonable, and unconstitutional. The appeals court rejected the defendant's claims citing the GPS was necessary to help reform the defendant by discouraging him from concealing future criminality.

State v. Harte (2006) 395 N.J. Super. 162, 928 A.2d 157
Defendant was convicted of eluding police after a high-speed pursuit. The state introduced audio and video evidence. The audio evidence was a recording of the pursuing officer in which pursuit speed was announced. The video of the chase also included the officer's GPS unit displaying the speed of the pursuing officer's vehicle. Although this case was an appeal on the authenticity of the audio and video recordings, the pursuit speeds indicated by the officer were verified through the video capturing the GPS speeds.

Doe v. Schwarzenegger (2007) U.S. Dist., 476 F. Supp. 2d 1178
Defendants had been convicted of sexual offenses decades prior to the passage of Jessica's Law in California which required sexual offenders to wear GPS monitoring devices for life. After the passage of the Sexual Predator Punishment and Control Act (SPPCA), defendants were notified that they would be required to wearing a GPS anklet. After review of the SPPCA, the district court determined the SPPCA could not be retroactively enforced and the requirement for the defendants to wear GPS anklets was inapplicable.

Gerardi v. City of Bridgeport et al. (2007) Conn. Super. Ct., CV084023011S
The City of Bridgeport acquired vehicles equipped with GPS devices for use by city employees. Pending disciplinary hearings and termination resulting from GPS data indicating the defendant was not properly performing his job, the plaintiff alleges he was not notified of the GPS unit and therefore the city violated state statutes. The court found the statutes claimed to be violated by the plaintiff did not apply to this case, and the court was therefore unable to grant the injunctions sought by the plaintiff.

John Doe et al. v. Walsh in her capacity as Chairperson et al. (2007) Mass. Super. Ct., Opinion No.: 99601
Four convicted sex offenders on parole sought an injunction against the defendants requiring them to wear GPS monitoring devices after a state law was passed requiring paroled offenders to continuously wear GPS monitoring devices. The court determined GPS monitoring devices meant to keep sex offenders out of exclusion zones was not a violation of individual liberties nor a violation of privacy since data would only be transmitted if the offender entered an exclusion zone.

Morton v. Nassau County Police Department (2007) U.S. Dist., 05-CV-4000 (SJF) (AKY)
Suspected of being involved in a series of burglaries, police attached a GPS device to Morton's vehicle while parked on a public street. Morton argued GPS was a violation of the Fourth Amendment. The district court dismissed the plaintiff's claim, declaring the use of a GPS device was not an unreasonable search or seizure.

State v. Martin (2007) 182 Vt. 96, 377; 944 A.2d 867
Defendant was convicted of boating while intoxicated and causing death from an accident. The defendant appealed, stating the accident did not occur within Vermont's boundaries and therefore Vermont did not have jurisdiction. The prosecution presented GPS recordings from the rescuers and their testimony. All GPS recordings placed the boat accident in Vermont. The appeals court determined the testimony and GPS recordings established beyond a reasonable doubt that the boat accident did occur within Vermont's boundaries.

United States v. Garcia (2007) U.S. App., 474 F.3d 994
The defendant appealed his conviction of crimes relating to the manufacture of methamphetamine. The only issue on appeal was whether evidence obtained as a result of a GPS tracking device attached to the defendant's car should have been suppressed as an unconstitutional search. The appeals court affirmed the district court determination that the GPS tracking device did not constitute a seizure. The GPS device did not affect the car's driving qualities, did not draw power from the car, did not take up space that might have been occupied, did not alter the car's appearance, and did not "seize" the car in any intelligible sense of the word.

Vitka v. City of Bridgeport et al. (2007) Conn. Super., CV0804022961S
The City of Bridgeport acquired vehicles equipped with GPS devices for use by city employees. Pending disciplinary hearings resulting from GPS data indicating the defendant was not properly performing his job, the plaintiff alleges he was not notified of the GPS unit and therefore the city violated state statutes. The court found the statutes claimed to be violated by the plaintiff did not apply to this case and the court was therefore unable to grant the injunctions sought by the plaintiff.

Stone v. Maryland (2008) 178 Md. App. 428, 941 A.2d 1238
Wanted in connection to several burglaries, a Maryland Trooper obtained Stone's cell phone number. After "pinging" the cell phone, the trooper located Stone's vehicle and attached a GPS tracking device. Stone was arrested in his vehicle after being located by a GPS device. A subsequent search of the vehicle yielded additional drug charges which Stone seeks to dismiss. The Maryland Appeals Court ruled GPS tracking was not a violation of the Fourth Amendment because Stone was traveling on a public road. The court further determined the cell phone "ping" was irrelevant to the case because Stone's vehicle was in a motel parking lot in full public view. The "ping" only helped narrow down the search area.

United States v. Coleman (2008) U.S. Dist., Case No. 07–20357
DEA agents obtained a warrant to locate the defendants vehicle through the manufacturer installed OnStar GPS. Once the vehicle was located, local sheriff deputies arrested the defendant and found heroin during their search of his vehicle. The defendant contends that he had a reasonable expectation of privacy in utilizing the GPS and that the tracking of a third party's factory-installed GPS receiver was constitutionally vague. The appeals court concluded the original warrant was valid based on probable cause and that tracking with a warrant is not constitutionally vague.

People v. Weaver (2009) N.Y., 909 N.E.2d 1195, 1199–1200
Police tracked suspect's vehicle for a 2-month period without a warrant. The court concluded the extended tracking constituted a search, requiring a warrant. Although the court admitted the defendant's expectation to privacy was diminished while on a public road, that expectation was not reduced to zero.

Foltz v. Commonwealth (2010) Va. Ct. App., 698 S.E.2d 281
Appeals court determined the weeklong warrantless use of GPS tracking device placed on the bumper of Foltz vehicle to monitor the movement of Foltz was not an unconstitutional search or seizure. The police could have obtained the same information by following the vehicle.

United States v. Marquez (2010) 8th Cir. 605 F.3d 604
The use of a GPS device to track the movements of Marquez vehicle did not violate the Fourth Amendment. The defendant was suspected to be involved in the interstate transport of drugs. The court ruled that the extended tracking of the defendant's vehicle was not random or arbitrary, was noninvasive, and resulted from reasonable suspicion. Furthermore, the device was attached to the vehicle while parked in public.

United States v. Maynard (2010) D.C. Cir., Nos. 08-3030, 08-3034, WL 3063788
With an emphasis on privacy expectations, the court ruled that the warrantless use of a GPS tracking device was an unconstitutional search under the Fourth Amendment. It was concluded that the extended use of GPS provided more information than the police would have possibly been able to obtain through observation of movements.

United States v. Pineda-Moreno (2010) 9th Cir., 591 F.3d 1212, 1217
Defendant observed purchasing a large quantity of fertilizer. Recognized as plant food used for the growth of marijuana, DEA agents placed a GPS tracker on the defendant's vehicle. GPS devices were placed on the vehicle in both public areas and the defendant's driveway. Agents noticed there were no access controls (e.g., fence/gate) and no posting against trespassing. Agents eventually used the GPS records to locate a marijuana field the defendant traveled to. The appeals court affirmed a person does not have an expectation of privacy in their unsecured driveway. Furthermore, the GPS tracking did not constitute a violation of the Fourth Amendment.

United States v. Cuevas-Perez (2011) U.S. App., 640 F.3d 272
Federal agents and the police came to suspect the defendant of being involved in a drug distribution operation. Without a warrant, a detective attached a GPS tracking unit to defendant's vehicle while it was parked in a public area. While the defendant was on a trip, the police took up visual surveillance and discontinued the use of the GPS device. The device had been in use for a total of approximately 60 h. The appeals court determined that suppression of the GPS evidence was not warranted because the warrantless use of the GPS tracking device did not violate defendant's Fourth Amendment rights since (1) GPS tracking does not constitute a search, (2) the surveillance was not lengthy, and (3) real-time information is exactly the kind of information that drivers make available by traversing public roads.

United States v. Jones (2012) 132 S. Ct., 945; 181 L. Ed. 2d 911
Federal Bureau of Investigation (FBI) agents obtained a warrant for the installation of a GPS tracking device to the undercarriage of a vehicle registered to Jones's wife while parked in a public parking lot. Agents tracked the vehicles movement for 28 days despite a 10-day effectiveness period for the warrant. At the trial, GPS evidence was presented and Jones was convicted. Jones appealed the evidence and the DC Appeals Court overturned the conviction, finding the admission of evidence obtained by the warrantless use of a GPS device constituted a search and was a violation of the Fourth Amendment. The US Supreme Court reviewed the case and affirmed the opinion of the appeals court.

People v. Moorer (2013) NY Slip Op 23048, 39 Misc. 3d 603
Defendant accused of murder. Police "pinged" his cell phone to determine his location. Judges ruled the cell phone here was not surreptitiously attached to an unwitting individual. Moreover, public ignorance about cell phone technology can no longer be maintained in this day and age—cell phones are voluntarily carried by their users and may be turned on or off at will. By a person's voluntary utilization, through GPS technology, of a cell phone, a person necessarily has no reasonable expectation of privacy with respect to the phone's location—vis-à-vis the pinging—even though he maintains what may be a reasonable expectation of privacy in the content of his phone conversations. Therefore, the pinging of defendant's cell phone does not implicate or violate defendant's rights under the New York State Constitution.

United States v. Baker (2013) U.S. App., 713 F.3d 558
During the course of an investigation of armed robberies, police placed a GPS tracking device on the car of the defendant's girlfriend. The GPS data allowed police to link the car to an armed robbery. The defendant was tracked, pulled over, and found to be in possession of a firearm. The defendant appealed the admission of the GPS evidence, stating it violated his Fourth Amendment rights. Due to the fact the defendant did not raise the issue during the original trial, the appeals court determined the defendant waived his rights to raise the issue during an appeal.

United States v. Miller (2013) U.S. App., No. 12–50238
Upon release of imprisonment, the defendant was required to submit to GPS monitoring during a 3-year period of supervised release. The defendant appealed, asserting that the GPS monitoring constituted a greater deprivation of his liberty interest than was reasonably necessary. A review of the defendant's background found that any impairments of the defendant's privacy were outweighed by the benefits of GPS monitoring; accordingly the appeals court affirmed the decision of the district court.

Remote Sensing Cases

United States v. Reserve Mining Co. (1974) D.C. Cir., 380 F. Supp. 11
Satellite imagery presented to provide evidence of violations of the Clean Water Act by Reserve Mining Co.

United States v. Kilgus (1978) 9th Cir., 571 F.2d 508
Court refused to admit thermal imagery evidence due to officer's lack of training in interpreting the results and not being able to understand the underlying theories and principles.

Velsicol Chem. Corp. v. New Jersey D.E.P. (1982) N.J., 442 A.2d 1051
Although the court allowed aerial photographs to be presented during trial, an accompanying scientific report based on the interpretation of the photographs was excluding due to the lack of a witness responsible for the report being present to testify about the methods and analysis discussed within the report. The court expected an expert witness to authenticate and explain the remotely sensed data.

California v. Ciraolo (1986) 106 S. Ct., 476 U.S. 207
An airplane was used to fly over the suspect's property and take aerial photographs which conclusively showed the presence of marijuana. According to this supreme court opinion, anything capable of being viewed from a public space is not protected by privacy laws in the United States. Therefore, aerial observations and photographs are legally permitted to monitor and document areas considered private property and present any evidence during trial.

Dow Chem. Co. v. United States (1986) 106 S. Ct., 476 U.S. 227
Supreme court ruled the use of commercial aerial photography used to search private property did not constitute a search if the private property was observable with the naked eye and therefore the photography was admissible as evidence.

United States v. Sanchez (1987) U.S. Court of App. 9th Cir., 829 F.2d
Through the use of thermal imaging (FLIR), a US Customs plane observed an unidentified airplane make a landing in the Arizona desert. The Customs pilot observed (through the FLIR) a pickup pull-up next to the plane and the departure of the plane. Agents on the ground arrested the occupants of the truck and found marijuana. The only evidence linking McCall, the defendant, to the unidentified airplane was the FLIR. The defendant objected to the admission of any evidence obtained through the use of the FLIR, arguing the government had not laid a sufficient foundation for its admission (e.g., testimony of an expert witness). The Customs pilot testified FLIR could only identify generic objects, which satisfied both the trial court and appeals court since FLIR was accepted by the scientific community for the detection of generic objects. Furthermore, the pilot visually witnessed the events through the FLIR that he was operating and highly familiar with. The appeals court affirmed the decision of the trial court, stating the pilot's expertise was adequate to lay a proper foundation for admission of the FLIR evidence.

United States v. Ishmael (1995) Court of Appeals 5th Cir., 48 F.3d 850
Acting on an informant tip, DEA agents used a thermal imaging device to determine if the plaintiff might be operating an indoor marijuana operation. The results suggested he was and agents acquired a warrant to search the building. Defendants argue they had an expectation of privacy and demonstrated secretive activities to ensure privacy. The only probable cause the DEA had was the results of the

thermal imager. The district court suppressed all evidence seized from the property and the results of the thermal imager. DEA appealed the district court decision. The appeals court ruled that the use of thermal imaging did not violate the Fourth Amendment and subsequently found the DEA did have probable cause needed to obtain a search warrant.

West-Oviatt Lumber Co. v United States (1998) 40 Fed. Cl. 557, 566
US Forestry Service (USFS) introduced satellite imagery; however, they failed to ground-truth information derived from the imagery. The court decided not to rely on the satellite imagery and subsequently reached a verdict against the USFS.

St. Martin v. Mobil Exploration (2000) 5th Cir., 224 F.3d 402, 407
Aerial photographs introduced as evidence against allegations of land degradation caused by Mobile Exploration. The aerial photographs and accompanying expert testimony showed erosion was caused by defendant.

Nutra Sweet Co. v. X-L Engineering Co. (2000) 7th Cir., 227 F.3d 776
Nutra Sweet presented aerial photographs as evidence to confirm a history of dumping of hazardous waste on X-L Engineering Land which in turn migrated through groundwater onto Nutra Sweet's land.

Kyllo v. United States (2001) 121 S. Ct. 2038, 533 U.S. 27
The supreme court ruled the use of a thermal imaging device did require a warrant since the technology revealed details of the interior via technology that was not in the general public use.

United States v. McCall (2008) 5th Cir., No. 553 F.3d 821
After being convicted of distributing drugs within 1,000 ft of a school, McCall appealed and argued the government failed to prove the offense was within 1,000 ft of a school. The appeals court concurred. During trial, an aerial photograph was presented without any indication of scale and a detective testified he had driven the distance and although he never measured it, he opined it was within 1,000 ft. The appeals court ruled the opinion of the detective was a lay opinion and not sufficient for a jury to reasonable conclude the distance was less than 1,000 ft. Furthermore, without scale, the court stated the aerial photograph was useless.

United States v. Kattaria (2009) Court of Appeals 8th Cir., 553 F.3d 1171
An informant notified police that Kattaria had an indoor marijuana operation. Thermal imaging was used to confirm the allegations. Kattaria challenged the warrant; however, the appeals court found the police had conducted additional investigations to determine the reliability of the informant. The investigation revealed that Kattaria had a previous conviction for marijuana trafficking and also had an unusually high utility bill. Based on the combined information, the warrant was supported by probable cause.

Banks v. United States (2010) 94 Fed. Cl. 68
Plaintiffs motioned to excluded expert testimony pertaining to the use of Lidar data, contending the technology is inaccurate and unreliable. The court determined the plaintiff's reliability criticism was insufficient to support a motion to strike.

Nicholas v. State (2010) Ark. App. 696, CA CR10-324
Nicholas was pulled over for speeding based on the results of a Lidar speed gun. Nicholas was subsequently arrested for driving while intoxicated and motioned to suppress evidence based on the officer's alleged failure to calibrate the device. The officer initially testified he was trained and certified to operate the Lidar gun and calibrated it on a daily basis. The appeals court found the officer was qualified and dismissed Nicholas' motion to suppress the Lidar evidence.

People v. Mann (2010) 397 Ill. App., 3d 767; 922 N.E.2d 533
Mann cited for speeding. Appealed trial court's decision to admit Lidar speed gun measurements based on the argument Lidar guns had not passed the Frye test for reliability. Citing numerous cases involving Lidar speed measurements, the appeals court ruled Lidar was a generally accepted technology based on accepted scientific principles.

Commonwealth v. Danier (2011) 78 Mass. App. Ct. 1121, Unpub.
Defendant accused of speeding based on the results of a Lidar speed gun. Defendant admitted speeding but appealed the reliability of the Lidar gun and claimed the trial judge never instructed her that she could cross-examine the officer about reliability. Appeals court found the Lidar gun results sufficiently proved the defendant was speeding and concluded any doubts the defendant could appeal would only be speculative.

Pfender v. Secretary, Pa. Dep. of Corr. (2011) U.S. App., 443 Fed. Appx.749
A cell search of an inmate discovered an aerial photograph of the prison site in the inmate's possession. Pennsylvania Department of Corrections classifies aerial photographs of State Correctional Institutes as contraband that threatens security. Investigations revealed the photograph was supplied by the defendant and therefore visiting rights of the defendant were suspended for 2 years. The defendant offered alternative explanations that could explain how the inmate acquired the photographs; however, no substantive evidence was submitted to support the alternative explanations. The appeals court affirmed there was a rational basis for suspending visitation and the rights of the defendant were not violated.

State v. Branch (2011) 243 Ore. App., 309; 259 P.3d 103
Branch arrested for possession and delivery of cocaine within 1,000 ft of a school. Officers utilized Lidar to determine the distance was within 1,000 ft. Branch appealed, citing the state had not established the scientific validity of Lidar. Appeals court exercised judicial notice, citing the well-known use of Lidar qualifies it as admissible.

Geographic Information Systems (GIS) Cases

United States v. Asarco Inc. (1998) 9th Cir., 214 F.3d
The US and Coeur d'Alene Indian Tribe filed suit against Asarco mining for environmental damages. The defendants sought access to the Coeur d'Alene Indian

Tribe's GIS database as a public document. The tribe argued the database was tribal property since they created and paid for it. Defendants argued the US government paid for it. The court ruled the GIS database was a public administrative record, regardless of who created or funded it.

State v. White (2000) Mo. App., 28 S.W.3d 391; 2000 Mo. App.
Defendant was convicted of delivery of a controlled substance within 2,000 ft of a school. On appeal, the defendant argued the state failed to prove he knew he was within 2,000 ft of a school. According to Missouri law, the offender must knowingly commit an offense within 2,000 ft of a school to receive an enhanced penalty. While a GIS map was presented as evidence during trial to prove the distance, the appeals court found it insufficient for proving the defendant knew he was near a school. The GIS map did not indicate topography or visibility around the school. The court reasoned a jury could not reasonably infer whether the school was a large visible complex or a single classroom in the basement of a residential building. Because the map and testimony were insufficient to establish that the defendant knew about the school, the court reversed the ruling against the defendant.

State v. Wright (2000) Conn. App., 752 A.2d 1147, 1156
Wright was convicted of drug offenses. A computer-generated GIS map was used to determine the offense occurred within a school zone. Wright appealed the trial courts admission of the GIS map. The Appeals court determined the map was properly authenticated and admissible since the GIS technician went to the site to verify the maps fairly and accurately represented the correct distance. Furthermore, formulas used to create the map were verified correct by state and private engineering companies.

State v. Perry (2001) Tenn. Crim. App., 2000 WL 1246577
Defendant arrested for possession of cocaine and additionally charged with being within a school zone. GIS was used to create a map with a 1,000 ft buffer zone around a school. The city GIS manager testified about the GIS map and accuracy of the 1,000 ft buffer zones. Defendant was convicted and, on appeal, convictions and sentences were affirmed.

Swaggert v. County of Hubbard (2003) Minn.Tax., File No. C8-02-619
After an increase in real estate property tax resulting from an increased valuation of property, the plaintiff challenged the county's decision. GIS was utilized to determine usable lakefront property which was incorporated into the methodology for determining property value. Although the court did not find fault with the methodology, it did order the value of the property reduced and the recomputation of property taxes.

Microdecisions, Inc. v. Skinner (2004) Fla. App., 889 So. 2d 871
Microdecisions requested a copy of county GIS data, which they sell on their website. Skinner, the county property appraiser, requested Microdecisions enter into a licensing agreement before the data was delivered. The issue before the court was whether the county property appraiser could require prospective commercial users to enter into a licensing agreement for the resale of public records. Basing

their decision on Florida copyright laws, the appeals court decided that Skinner had no authority to assert copyright protection in the GIS maps and could not require a licensing agreement.

Town of Greenwich v. Freedom of Information Commission (2005) 274 Conn. 179, 874 A.2d 785

A citizen requested a copy of all GIS data for the Town of Greenwich. The director denied the request, so the citizen filed a complaint with the commission. The director claimed the data was exempt from disclosure under the Freedom of Information Act. The commission found the data was not exempt. The appellate court agreed with the commission. The Connecticut Supreme Court reviewed the case and affirmed the appellate court decision. The GIS data could have been obtained in separate portions from various town departments. The GIS database was simply a convenient compilation of information already available to the public and therefore not exempt from disclosure. The director further argued the release of the GIS data presented a potential public safety concern. With the director unable to provide any specific statistical data correlating criminal activity or terrorist activity with the disclosure of GIS, the Connecticut Supreme Court determined the director was unable to demonstrate how disclosure of GIS data would compromise security.

Forest Guardians v. United States Federal Emergency Management Agency (2005) U.S. App., 410 F.3d 1214

Forest Guardians (FG) requested GIS data pertaining to the National Flood Insurance Program (NFIP) from the Federal Emergency Management Agency (FEMA) in order to study development and the loss of flood plain areas. FEMA provided printed maps. FG made a second request in which FEMA denied in accordance with Freedom of Information Act exemptions. According to FEMA, the data would compromise the privacy of individuals by revealing specific geographic point locations for NFIP-insured structures. The appeals court agreed with FEMA, noting the nearly nonexistent public interest in the data was negligible since hard-copy maps were provided. The privacy interest of individuals was a concern. The GIS data would reveal names and addresses, ownership, flood risks, and the manner in which property was purchased. Providing such information would constitute an invasion of personal privacy. Furthermore, disclosing the GIS data could subject individuals to unwanted contacts or solicitations by private insurance companies. The appeals court affirmed the judgment of the district court and denied FG request for the GIS data.

Liston v. Town of Pomfret (2006) Conn. Super., CV054001612

The plaintiff learned the adoption of a new electronic GIS map changed zoning codes. The town advertised and conducted public hearings for the purpose of adopting the GIS maps as the official map. The plaintiff alleges the maps were actually a completely new zoning map and not a revised electronic map. The superior court agreed with the plaintiff that the new GIS map effectively changed zoning classifications and was not a revision of the previous zoning maps used by the city.

State v. Lindsey (2006) Tenn. Crim. App., 208 S.W.3d 432

The defendant was convicted of the sale cocaine within 1,000 ft of a school. The city GIS manager presented aerial photographs and testified that the location of the offense was 480 ft from a school.

Dallas v. Texas (2008) Tex. App., No. 05-07-00037-CR

Dallas was convicted of possession with intent to deliver cocaine within 1,000 ft of a school. Dallas argued the evidence was insufficient to prove he was within 1,000 ft of a school. A GIS map with aerial photographs, land surveys conducted by the city surveyor's office, and 1,000 ft buffers around schools was used to provide spatial context to the location of the offense. The location of the Dallas's offense clearly fell within the 1,000 ft buffer zone. The GIS map maker testified the map was accurate to within 3 ft. According to Texas state law, any map produced or reproduced by a municipal or county engineer for the purpose of showing the locations and boundaries of a drug-free zone is admissible into evidence. Upon adoption of a resolution or ordinance approving the map as an official finding and record, the map becomes prima facie evidence. The appeals court concluded a fact-finder could rationally conclude beyond a reasonable doubt that the offense occurred with 1,000 ft of a school and therefore rejected the appeal.

Multi Ag Media v. United States Department of Agriculture (2008) 380 U.S. App. D.C. 1, 515 F.3d 1224

The US Department of Agriculture (USDA) denied a request by the plaintiff to provide the Farm Service Agency's (FSA) GIS database of farm data. The USDA contends the data would compromise the privacy of the family-owned farms by disclosing information that could be used to assess the financial situation of the family farms (e.g., crops and field acreage). The plaintiff argued the GIS database was used to monitor program compliance and such information would allow the public to more easily determine if the USDA was catching cheaters and lawfully administering its program. The appeals court sided with the plaintiff, noting the benefits of providing the data would outweigh personal privacy interests.

Seago and Real Estate Information Service, Inc. v. Horry County (2008) 378 S.C. 414, 663 S.E.2d 38

The plaintiff requested GIS data from the county under the Freedom of Information Act in order to resell the data on the Internet. The county obtained copyrights for the data to protect its investment and the integrity of the data. The county informed the plaintiff the GIS data could be obtained for a fee, but the plaintiff would have to sign a licensing agreement acknowledging the county's copyright. The Supreme Court of South Carolina determined the county could copyright the data; however, the fees violated South Carolina Codes. The issue of fees was remanded back to circuit court for further proceedings.

County of Santa Clara et al. v. Superior Court of Santa Clara County (2009) Cal. App., No. H031658

The county received a request for GIS data under the public records acts. After a court order to provide the GIS files, the county filed this appeal. The county requested an end-user license agreement and placed a fee of $250,000 on the GIS data request to cover costs. The county also asserted the GIS data was intellectual property and copyrightable. The appeals court affirmed the superior court's judgment and ruled the California Public Records Act (CPRA) did not allow the county to request an end-user license agreement, nor was the data copyrightable as no such provision existed in the CPRA. The county could recover fees associated only with the direct cost of recovery. The appeals court remanded the case back to the superior court for determination of the recovery fees.

Sierra Club v. Superior Court (2011) 195 Cal. App. 4th 1537, 125 Cal. Rptr. 3d 913
Encouraging local governments to develop and maintain computer mapping systems, the California Legislature exempted computer mapping systems from public disclosure. This exemption meant local governments were not required to provide public records when requested for cost of duplication only. The local governments were allowed to recoup costs associated with building and maintaining computer mapping systems and implement a nondisclosure agreement. The Sierra Club was able to receive print out versions of GIS maps as public records subject to duplication costs only. However, the printed versions lacked analytical capabilities sought by the Sierra Club and therefore they petitioned for writ of mandate compelling the county to provide the GIS database for a fee consisting of only the direct costs of duplication. The trial court denied the petition; therefore, the appeals court was asked to decide the matter. At issue was the interpretation of the language used to define "computer mapping systems." Examining the intent of the legislative act (to encourage development), the appeals court affirmed the GIS database was a computer mapping system exempt from public disclosure. The Legislature implemented the act to encourage development because of the costs and time involved in developing and maintain computer mapping systems. Based on the intent of the act, GIS databases were deemed computer mapping systems exempt from public disclosure. The court opined the legislature, not the judiciary, should determine public policy in matters of disclosure and exclusions of databases.

Oregon Natural Desert Ass. v. McDaniel (2012) U.S. Dist., 282 F.R.D. 533
Oregon Natural Desert Association (ONDA) received GIS data from the Bureau of Land Management (BLM). Review of data discrepancies determined the GIS data was "interim" data produced years earlier and therefore inaccurate. BLM confirmed the data was "interim" and offered to provide the final data set for an additional fee. ODNA sought motions to compel the BLM to complete the record and include additional route coverage. The court disagreed with the BLM claim that the GIS data was adequate and approved the motion to compel the BLM to provide the complete record to the ONDA at no cost since the BLM had the data when requested by ONDA. However, the ONDA was not granted the supplemental data they asked for but could propose a more limited modification at a later date after they received and reviewed the complete record from the BLM.

Kane County v. United States (2013) U.S. Dist., Case No. 2:08-cv-00315

Kane County and the State of Utah sought title to highway rights-of-way located on federal land under the Quiet Title Act of 1866. Although the Quiet Title Act was repealed in 1976, valid rights-of-way existing prior to the repeal were grandfathered in and deemed to continue in effect. The United States provided quiet title to some of the 15 roads but denied quiet title to others. Kane County and the State of Utah filed action to obtain the remaining roads. Aerial photographs and USGS topographic maps compiled from aerial photographs prior to 1976 conclusively proved the existence of roads subject to be grandfathered under the Quiet Title Act. Recent GPS measurements of the road centerlines also proved the current roads fall within permissible range of errors that exist in map making. Based on the evidence provided by the aerial photographs, Kane County and the State of Utah were able to gain quiet title from the United States. Despite testimony, additional roads that were not proven through aerial photographs remained under the control of the United States and were not quiet titled to the plaintiffs due to the lack of evidence establishing rights of claim.

Index

G.A. Elmes et al. (eds.), *Forensic GIS: The Role of Geospatial Technologies for Investigating Crime and Providing Evidence*, Geotechnologies and the Environment 11, DOI 10.1007/978-94-017-8757-4, © Springer Science+Business Media Dordrecht 2014

D